Integrated Regional Risk Assessment, Vol. II

Environmental Science and Technology Library

VOLUME 4/2

The titles published in this series are listed at the end of this volume.

Integrated Regional Risk Assessment, Vol. II

Consequence Assessment of Accidental Releases

by

Adrian V. Gheorghe

ETHZ, Swiss Federal Insitute of Technology,
Zürich, Switzerland

and

Michel Nicolet-Monnier

PSI, Paul Scherrer Institute,
Villigen, Switzerland

KLUWER ACADEMIC PUBLISHERS

DORDRECHT / BOSTON / LONDON

A C.I.P. Catalogue record for this book is available from the Library of Congress

ISBN 0-7923-3718-2 (Vol. II)
ISBN 0-7923-3717-4 (Vol. I)
ISBN 0-7923-3719-0 (Set)

Published by Kluwer Academic Publishers,
P.O. Box 17, 3300 AA Dordrecht, The Netherlands.

Kluwer Academic Publishers incorporates
the publishing programmes of
D. Reidel, Martinus Nijhoff, Dr W. Junk and MTP Press.

Sold and distributed in the U.S.A. and Canada
by Kluwer Academic Publishers,
101 Philip Drive, Norwell, MA 02061, U.S.A.

In all other countries, sold and distributed
by Kluwer Academic Publishers Group,
P.O. Box 322, 3300 AH Dordrecht, The Netherlands.

Printed on acid-free paper

Printed in the Netherlands

CONTENTS

PREFACE

Industrial development is essential to improvement of the standard of living in all countries. In a given region, old and new plants, processes, and technologies have to coexist. Technological penetration and substitution processes are generally taking place; they are entirely dynamic and this trend is going to stay like this.

People's health and the environment can be affected, directly or indirectly by routine waste discharges or by accidents. A series of recent major industrial accidents and the effect of pollution highlighted, once again, the need for better management of routine and accidental risks. Moreover, the existence of natural hazards complicate even more the situation in any given region. In the past effort to cope with these risks, if made at all, have been largely on a plant by plant basis; some plants are well equipped to manage environmental hazards while others are not.

Managing the hazards of modern technological systems has become a key activity in highly industrialized countries. Decision makers are often confronted with complex issues concerning economic and social development, industrialization and associated infrastructure needs, population and land use planning. Such issues have to be addressed in such a way that ensures that public health will not be disrupted or substantially degraded.

While hazard managers and risk assessors have been successful at identifying hazards and reducing overall risk exposure, economic growth and technological development have led to a new risk situation characterized by:

- an increasing number and variety of hazards
- hazards giving rise to a broad range of partial and temporal risks
- public dissatisfaction with hazards managers and hazards' owners.

Due to the increasing complexity of technological systems and the higher geographical density of punctual hazard sources, new methodologies and a novel approach to these problems are challenging risk managers and regional planers. Risks from these new complex technological systems are inherenttly different from those addressed by the risk managers of the 1960's and 70's. Recent awareness of environmental problems by a large public has led worldwide to dissatisfaction and to the formation of all kind of pressure groups that do exert a strong political influence and are quite often not ready to accept any compromise. As a result of such approach, interesting and vital research project were stopped (Swiss nuclear moratorium), plans are buried and decisions delayed for many years (i.e., repository for nuclear wastes). Nowadays it becomes increasingly difficult to site new plant facilities which are perceived as risky or undesirable by the local population. Project of national interest cannot be launched or even achieved, once started. On the other hand distrust has become so problematic, that at times, risk managers are no more considered as impartial person, or as reliable source of information regarding risk protection aspects.

It was felt that existing hazard management techniques need to be supplemented with concepts and methods that are integrative at a regional level. Integrated regional risk assessment and safety/hazard management (IRRASM) represents a coordinated strategy for risk reduction and safety/hazard management in a spatially-defined region across a broad range of hazard sources (during normal operation and accidental situations) that includes synergistic effects.

> ***Integrated Regional Risk Assessment** and Safety/Hazard Management (**IRRASM**)*
> *IRRASM is a multi-disciplinary process: engineers, computer scientists and modelers play a central role in the risk assessment stage. Social scientists can contribute with practical advice to the embedding process concerning hazard sources and help communal organizations to deal with such problems, taking into account local economic conditions and political reality. They can make important contributions towards a better understanding of how the practice of regional risk management can meet the needs and expectations of society/community. In this way other kinds of regional planing tasks (for emergency preparedness or remedial actions) or the implementing of utilities as part of the infrastructure (such as canalization network, sewage plants, waste disposal units, power plants and drinking water distribution, etc.) can be integrated more easily after reaching the consensus of agreement of the population.*

In view of the above mentioned optic a joint project, *PPR&S (Polyproject on "Risk and Safety of Technical Systems)*, was launched together with the participation of the following institutions:

- ♦ Swiss Federal Institute of Technology, ETHZ (Zürich, Switzerland)
- ♦ Paul Scherrer Institute, PSI (Villigen / AG, Switzerland)
- ♦ EAWAG (Dübendorf / ZH, Switzerland)

There are a number of national and international efforts to deal with risk assessment at the regional level. The ETHZ (Swiss Federal Institute of Technology - Zürich) - "Poly-project on Risk and Safety of Technical Systems" took the initiative to research on various aspects related to regional risk assessment and safety management. A series of basic questions had been asked. Answers had to be given; they had to comprehend the following main issues:

- what is integrated area risk assessment and safety management
- how to define a region/area for study
- type of activities and targets at risk
- objectives and scope
- the need for risk impact indicators
- the need for a comprehensive methodology.

Regional risk assessment and safety management seems to be a medium which helps to integrate people, issues and decisions in area risk assessment.

PPR&S is the discursive description of knowledge in addition to the development of interdisciplinary and practical methods for the application of risk management for a large variety of technological systems (e.g., rail and road transportation of dangerous goods, chemical plants, nuclear power plants, biotechnology, landfill, etc.). The scope of these applications are to be presented as a book series and is intended to be an integrated regional risk assessment and safety management guideline manual complemented by specialized software, databases, literature reviews and a novel methodological framework with due regard to the existing conditions prevailing in Switzerland. Further goals aimed at in this project are namely:

i) encouraging and promoting multi-disciplinary work among the different departments and institutes at the Swiss Federal Institute of Technology (ETHZ), Zürich, and among other research institutions;

ii) establish and confirm the technical competence of the ETHZ in the field of risk and safety of technological systems including their relationship with natural disasters;

iii) encouraging scientific and academic contacts to other polytechnic institutes universities, industry, governmental agencies, and political institutions within Switzerland and internationally;

iv) supporting the teaching aims in the interdisciplinary field of risk analysis at ETHZ.

Plans are underway to develop together with the Swiss Federal Institute of Technology of Lausanne, EPFL, a postgraduate study program in the field of "Risk and Safety". For the most part, the PPR&S has now developed from a local to a National interest in disasters' prevention and emergency planing activities.

In the framework of the PPR&S it was decided to collect and review basic technical information and topics concerning the *Integrated Regional Risk Assessment Process* and to publish the results in book form, as being part of a books' series presented under the auspices of PPR&S. This series forms a whole and covers different aspects of risk assessment, management, risk acceptance, as well as legal and societal aspects thereof. The present book, entitled " INTEGRATED REGIONAL RISK ASSESSMENT " is subdivided into two volumes:

- Vol. I : "Continuous and Non-point Source Emissions: Air, Water, Soil", and
- Vol. II : " Consequence Assessment of Accidental releases "

Assessing the risks of a region implies the use of a complex methodology dealing with risks to health and to the environment, normal operation and accidental situations, a large variety of industries, impacts, regulations and actors involved in the decision making process.

As opposed to other existing approaches (e.g., UN- interagency project on risk assessment of large industrial complexes), the PPR&S project took the initiative to design procedural guidelines for IARASM implementation by means of identifying tasks and integrating them into a comprehensive and systematic approach. By contrast, other existing guidelines take a problem solving oriented approach which is too global and does not always assist systematically the analyst or the project manager.

This approach (i.e., task oriented approach) allows a systematic analysis of the problem of regional risk assessment, offers flexibility and efficiency in the implementation process, allows initiatives and ad hoc modeling and simulation.

Integration of risk can not be done through a single risk indicator. Integrated regional risk assessment should be considered as a process whereby decision aiding techniques (ranging from simple brainstorming, the Delphi method to Multicriteria Decision Analysis, and Decision and Knowledge Support Systems) should play an important role. Various techniques have to complement expert judgment, public participation and risk communication. It is a need to balance hard approaches (models, calculations) versus soft approaches (acceptability) in regional risk analysis. In the process of risk integration an important role should be played by the task of comparative risk assessment. PPR&S made successful experiments in working with such tools and approaches.

The advent of new information technology, e.g., artificial intelligence (expert systems, fuzzy logic, neural networks), multimedia, virtual reality, GIS, specialized relational databases, computer graphics, or ISDN technology, would play a significant role in the future of regional risk assessment and safety management practice. The experiments made within the PPR&S with some of these techniques are showing promising results.

The PPR&S project has had important inputs from existing Swiss practice and legislation. The need for a comprehensive regional risk assessment methodology has been highlighted in different occasions during the time of the Poly-project.

Some of the Poly-project lessons we have learned, when dealing with above issues are:

- When getting involved in a regional risk assessment do not take a simplistic approach
- On regional risk assessment, try not to exclude political or human interactions at all stages. Develop a risk triplex, namely: *"safety culture, environmental awareness, and emergency culture"*.
- When running a research or a case study on regional risk assessment do not entirely rely on a self-organizing effect within the project / case study. A strong interactive project management framework is needed from the beginning.
- Due to technical accidents or natural disasters, or their synergism, there is an emerging need for national and international organized research and activities in the above field. Risks from normal operation or from accidents may have some transboundary effects.
- Legal issues at the local level / national level have to be solved and harmonized before any implementation of a comprehensive regional risk assessment methodology.
- There is a need for specialized databases; their use might diminish the uncertainty in results.
- Recent advancements in information and telecommunication technologies (GIS, ISDN), multimedia, virtual reality, neural networks) could play an important contribution to the modeling of various risks.

- In regional risk assessment, all risks (local, regional, global) should be taken into consideration.
- Safety culture, public participation and risk communication are relevant issues in the overall landscape of the regional risk assessment process. Emergency culture, preparedness, and planing is an integral part of regional safety management.

Within the Poly-project we experimented various aspects as previously highlighted. As compared with similar projects in the world (e.g., the UN Inter-Agency on Risk Management), the present work brought new answers to this interdisciplinary subject. Work done within the PPR&S is complementary to the numerous activities developed in Switzerland.

Further information on the Poly-project and its publication series can be obtained from:

Poly-project *"Risk and Safety of Technical Systems"*
ETH-Center
CH-8092 Zürich
Switzerland
Phone: +41 1 632 2356
Fax: +41 1 632 1094

FOREWORD

In recent years, the community has become increasingly aware of the risks of locating hazardous industries near heavily populated environmentally sensitive areas. This new awareness means a novel approach to safety planing for hazardous industries, looking at the problem from the point of view of integrated regional risk assessment, which should include beside the risks arising from natural events (like earthquake, flood, forest fires, etc.) also the risks arising from processing plants, storage and transportation of dangerous goods.

The purpose of *Volume I* is to highlight the main procedures for risk assessment of health and environmental impacts from continuous emissions of pollutants into air, water, and soil referring to normal operation conditions.

Volume II is concerned with the assessment of the consequences of accidental releases. The matter treated should help to find an answer to questions, such as: - what can go wrong?; - what are the effects and consequences?; - how often will it happen?. The main procedural steps are supported by relevant methods of risk assessment recognized on an international level; this document gives also an overview of criteria and guidelines for implementation of risk assessment and management at different stages.

Information contained in Vol. I and II is based on a wide range of scientific publications and references, and particularly on contributions provided by the *Biomedical and Environmental Assessment Division of Brookhaven National Laboratory, USA; UN Inter-Agency Programme (UNEP / WHO / IAEA / UNIDO) on the Assessment and Management of Health and Environmental Risks from Energy and other Complex Industrial Systems.*

Both volumes shall be valuable to students, engineers, and scientists in charge of developing new methodologies for hazard analysis and risk assessment; practitioners active in the field of environmental protection; local or governmental Authorities in charge of implementing environmental risk impact assessment procedures and guidelines.

It should be noted that, although consideration of the continuous emissions from nuclear power stations and other nuclear facilities form an important part of the assessment of the integrated risks from large industrial areas which contain nuclear facilities, they are not considered here. This is because nuclear risk assessments are currently carried out at a higher level than that used for other facilities (e.g., in Switzerland by the HSK, *Hauptabteilung für die Sicherheit der Kernanlagen*, Villigen, CH) and would be available for use in integrated risk assessment at community level.

Complementary readings which are strongly suggested are: "Management and Control of the Environment, (WHO, 1989)" and "Rapid Risk Assessment of Sources of Air, Water and Land Pollution, (WHO, 1982, 1993)", Guidelines for "Integrated Risk

Assessment and Safety Management for Large Industrial Complexes and Energy Generating Systems" (IAEA, 1995).

ACKNOWLEDGEMENTS

The authors wish to express their sincere gratitude to Professor Wolfgang Kröger, Chairman of the Executive Committee of the "Polyproject, Risk and Safety of Technical Systems (PPR&S)", ETHZ - Swiss Federal Institute of Technology, Zürich, for his guidance and critical reviews during the different phases of this work.

We are also greatly indebted to all people, who through their support and many valuable suggestions for corrections and improvements of the manuscripts, helped us to finalize this work. We wish also to acknowledge more specifically the following individuals:

Prof. R. Hütter, Vice-President Research, ETH Zürich.
S. Chakraborty, HSK.
Dr. Hans-Jörg Seiler, Project Manager for the PPR&S.
Prof. J. Schneider, Institut für Baustatik/Konstruktion, ETH, Hönggerberg, Zürich.
Mr. H.A. Merz, EBP, Ernst Basler & Partner Ingenieurunternehmen.
Prof. Dr. B. Böhlen, former Director of BUWAL (Bundesamt für Umwelt, Wald und Landschaft), Bern.
H.R. Wasmer, Deputy Director, EAWAG, Dübendorf (ZH).
Prof. K. Hungerbuehler, Institut für Technische Chemie, ETHZ.
Dr. H. Künzi, Konzern Sicherheit und Umweltschutz, Hoffmann-La Roche AG., Basel.
Mr. K. Cassidy, Head Major Hazard Assessment Unit, Health and Safety Executive, London.

One of us (AG) would like to express special consideration and high appreciation to Mrs. Françoise Bordier for her exquisite support and distinguished encouragement in his professional activity while in Switzerland. Finally, in the preparation of this book, the authors are greatly indebted to Mrs. I. Kusar (PSI), who skillfully prepared the drawings and pictures for illustrating the manuscript.

LIST OF ABBREVIATIONS

BLEVE	: Boiling Liquid Expanding Vapor Explosion
CEC	: Commission of European Communities
DNIOH	: Danish National Institute of Occupational Health
EDMZ	: "Eidgenössische Drucksachen- und Materialzentrale", Bern, Switzerland
EIA	: Environmental Impact Assessment
FEMA	: Failure Mode and Effect Analysis
HAZOP	: Hazard and Operability Study
HIP	: Hazard Identification Phase
IRASM	: Integrated Area Risk Assessment and Safety Management
NFPA	: National Fire Protection Association
PHA	: Preliminary Hazard Analysis
PHI	: Potential Hazard Index
PSC	: Probabilistic Safety Criteria
QRA	: Quantitative Risk Analysis
RA	: Risk Assessment
SRD	: Safety and Reliability Directorate of the United Kingdom Atomic Energy Authority
UVCE	: Unconfined Vapor Cloud Explosion
WHO	: World Health Organization

CHAPTER 1

HAZARD IDENTIFICATION AND ANALYSIS

1.1. Introduction

Government, industry and the community now recognize the need to identify, assess and control the risks to both people and the environment which come from potentially hazardous industries. Appropriate plant location selection and comprehensive risk assessment and safety management are therefore essential in ensuring orderly development and at the same time the safety of people and the environment.

The next chapters provide guidance information on the methods and procedures for the identification and analysis of hazards, and the quantification and assessment of risks from major accidents in the process industry. The methods outlined here are based on a large number of sources included in the reference listing placed at the end of each chapter.

Further reading should particularly focus on relevant publications by UNEP, WHO, IAEA, UNIDO (see list of further reading) particularly in the recent UNEP publication: *"Hazard Identification and Evaluation in a Local Community"* and IAEA reports on *"Procedures for the conduct of Probabilistic Safety Assessment (PSA) of NPP' s"*; The *"Role of PSA and PSC in NPP Safety"*, which is to be published in the IAEA Safety Series.

Good industry safety practices, engineering safety codes and standards, design and operating procedures remain at the core of safety management. The increase awareness of hazards and of the accidents that may result in significant loss of life and property, have led to the development and application of systematic approaches, methods and tools for risk assessment. These methods termed *hazard analysis* or *quantified risk assessment* are hazard evaluation tools. *Figure 1.1* is an overall scheme of the risk assessment process, which involves: system description, the identification of hazards and the development of accident scenarios and events associated with a process operations or a storage facility; *the estimation of the effects or consequences* of such hazardous events on people, property and the environment; *the estimation of the probability or likelihood* of such hazardous events occurring in practice and of their effects - accounting for the different operational and organizational hazard controls and practices; *the quantification of ensuing risk levels, outside the plant boundaries, in terms of both consequences and probabilities;* and, the assessment of such risk levels by reference to quantified risk criteria.

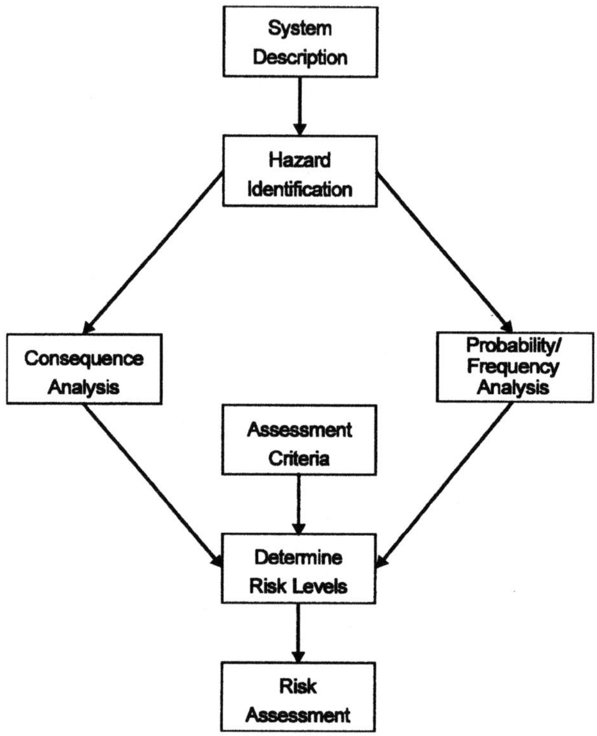

Figure 1.1. Methodology for the overall risk assessment process

The process of quantified risk assessment is *probabilistic* in nature. It recognizes that accidents are rare but possible events and that risk cannot be entirely eliminated. Because major accidents may or may not occur over the entire life of a plant or a process, it is not appropriate to base the assessment process on the consequences of accidents in isolation. The *likelihood or probability or such accidents* to actually occur should be taken into account. Such probabilities and resultant risk levels should reflect the level of design, operational and organizational controls available at the plant.

Problematic of Risk Assessment
There are a number of uncertainties associated with the quantification of risk. Amongst the most important sources of such uncertainties are the mathematical models in estimating the consequences of major accidents including dose-effect relationships and the setting of probabilities for different accident scenarios and for the probability effects of such accidents. Significant procedural and methodological advances have been developed in order to address and reduce the effect of such uncertainties. The risk assessment process should in all cases expose and recognize such uncertainties.

It is to be noted, that the main value of the quantified risk assessment process should not rest with the numerical value of the results (in isolation). *Rather, it is the assessment process itself which provides significant opportunities for the systematic identification of hazards and evaluation or risk.* The most significant advantages in this regard relate to the optimum allocation of priorities in risk reduction in that the assessment process provides for the clear identification and recognition of hazards and as such enable the allocation of relevant and appropriate resources to the hazards control process. The quantified risk assessment process also provide a useful tool for risk communication.

Risk and Environmental Impact Assessment

Risk Assessment (RA) and Environmental Impact Assessment (EIA) are two approaches which have as an aim to delineate policy in the field of environment, protection against pollution and safety against external hazards. The European Commission has converted these two approaches into Directives: "The major-accident hazards of certain industrial activities" (82/501/CEE; 87/216/ECC first amending; 88/610/CEE, second amending), and the "The assessment of the effects of certain public and private projects on the environment" (85/337/CEE).

Project types subjected to RA and EIA are not the same. RA includes industrial plants, while EIA includes some categories of projects that are not subject to RA. RA and EIA may have different temporal application.

- ***RA** is usually prepared for a project already built, or at least in presence of a final design, before its operational phase: the main purposes being to define constructional and operational characteristics in order to reduce the probability of occurrence of an accidental situation or to reduce and control its consequences, and to prepare an emergency plan to prevent likely harm to man or environment.*
- ***EIA** is instead carried out before or during a planing phase, and the main purpose of it is to verify a compatibility of the project with the environment (including the consideration of environmental acceptability of accidental aspects).*

Both EIA and RA are forms of predictive analysis. A systematic identification of possible environmental impacts as well as a rigorous analysis of their magnitude and probability is required for both processes.

Impacts identification is made on the basis of *check-lists. Matrices* are used to display activities along on axis with appropriate environmental factors listed along the other axis of the matrix. Many variations of the simple interaction matrix have been utilized in environmental impact studies. *Networks* are used to integrate impacts causes and consequences through identifying inter-relationships between causal actions and the impacted environmental factors, including those representing secondary and tertiary effects.

Prediction of impacts can be made on the basis of the following main types of methods:

- *Physical models* in which illustrative or working-scale models are constructed to represent the environment (visual representations of environment by pictures, photographs, films, or working models, using wind tunnels or waves chambers).

■ *Experimental methods* involving practical field or laboratory work (field experiments, laboratory experiments).

■ *Mathematical models* in which the relationship between causes and effects is represented in the form of one or more mathematical relationships (empirical models relationships are established from statistical analysis of observations on the environment, or internally descriptive relationships are based on explicit representation of the mechanism of the processes occurring within the environment).

■ *Evaluation methods* used to calculate the value of an environmental aspect that will be lost or disturbed as the result of an activity. The main categories are: Atmospheric effects (emissions, air quality, higher order effects); surface aquatic effects (discharge and run-off, effects on water flow and quality, higher order effects); sub-surface effects (leachate, effects on soil and groundwater); acoustic effects (emissions, effects on sound and noise in the environment); direct effects on plants and animals; direct effects on landscape; accidental effects

■ *Inventory approach*, i.e., a simple environmental effects' description in terms of the number of receptors in a given environmental area that will be affected.

■ *Other approaches* such as: expert evaluation of effects (asking an expert to provide an estimate of the likely size of an effect based on his knowledge; prediction by analogies (extrapolating from a similar activity); comparison with standards (an effect is predicted to be acceptable or unacceptable by direct evaluation against pre-determined standards or norms).

On identifying and assessing severe impacts it is important to remember that a risk is defined as a product between probability of occurrence and magnitude of harm, and that the probability of occurrence of a severe impact is not necessarily connected to an accidental condition at the facility, but to the particular features of the environment too (ecosystems in particular have a defined resilience and resistance). *Figure 1.2* shows as an example the environmental fate screening assessment decision network for *Food Chain*. Literature on this subject can be found in Belvisi and Boeri[1] (1993).

1.2. Hazard Identification

1.2.1. PREVIEW

Subsequent to the compilation of process plant information (system description) required for risk assessment, *the first and most essential step in any risk assessment is the identification of all relevant hazards applicable to a particular plant or operation, as basis for further analysis*. In all cases, it is necessary to establish:

- what dangerous situations exist within a plant or a process operation, and
- how these situations may come about.

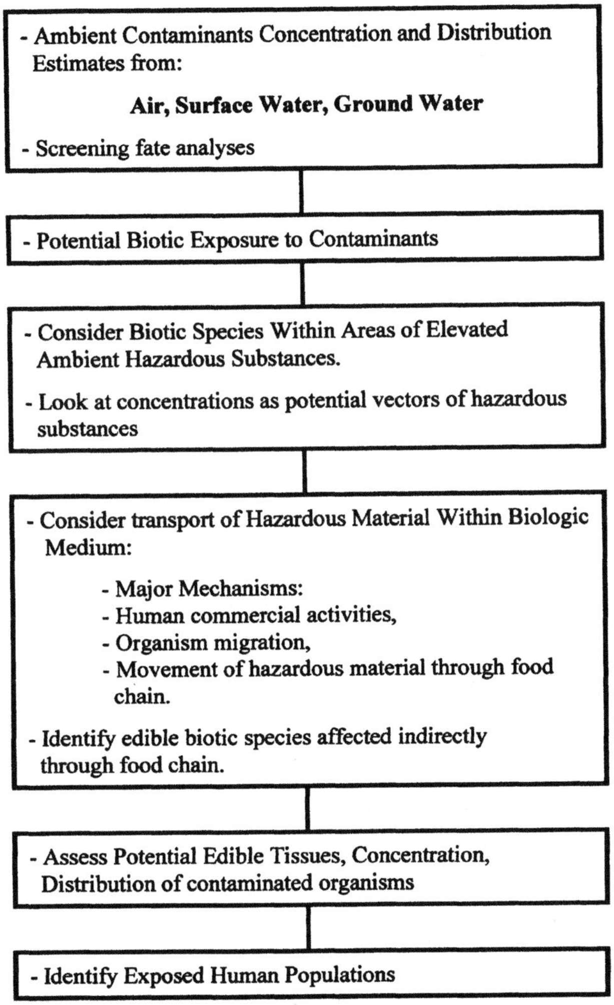

Figure 1.2. Environmental fate screening assessment decision network for Food Chain (U.S. Environmental Protection Agency, 1988)

This component of the analysis, termed *Hazard Identification,* involves consideration of all situations in which the potential for harm may exist in order to identify those which are hazardous, followed by a systematic analysis of the sequence of events which could transform this potential into an accident. Once an accident scenario has been established, the *likelihood* of such an accident occurring in practice (accounting for design operational and organizational safeguards) and its consequence (impact effect) should it occur, can be estimated. *Figure 1.1* indicates the context of hazard identification within the overall risk assessment process.

This section provides guidance on the role of the hazard identification process, the tools and techniques available to undertake hazard identification and the relevance and scope of application of these techniques. The review presented here is intended to provide a basic procedural framework to assist in undertaking hazard identification for both existing and new proposed plants. It does not intend to duplicate the extensive body of reference material available on the subject. A list of the most relevant references which should be consulted is included.

> *It must be particularly noted that there is not a fixed golden rule as to which particular technique should be adopted. There are, however, useful and important guidelines. It may be necessary to use a variety of approaches to improve the hazard identification process. Techniques may also be used in isolation or in complement to each other.*

1.2.2. OBJECTIVES OF HAZARD IDENTIFICATION

Hazard identification is the corner stone in the assessment of the safety of an installation. It is essential to have clear understanding of the type and nature of hazardous incidents associated with the operations of a plant and of the initiating and contributing events that can lead to such hazardous incidents. Without such an understanding the formulation and implementation of any risk management strategy is in many cases not possible and certainly inefficient. *The main objectives for identifying hazards at an early stage of the assessment process are basically*:

> ***Main Objectives for Identifying Hazards at an Early Stage of the Assessment Process***
> *a)* ***Provide the basis for the design and operation*** *of appropriate operational (hardware) and organizational (software) safety mechanisms. Safeguards must be appropriate and relevant to each type of hazards, and unless such hazards are identified and recognized, safeguards may be irrelevant or sub-optimal.*
> *b)* ***Risk quantification and evaluation.*** *Estimations of likelihood and consequences of hazardous incidents cannot be undertaken unless each hazard has been identified in the first instance.*
> *c)* ***Accidents prevention.*** *Accidents can be prevented by anticipating how they may occur. A systematic understanding of the major contributors to hazardous incidents and of the interaction of contributing events (concurrently or sequentially) enable the formulation of appropriate mitigating measures (e.g., shut-off systems) that may prevent such events escalating into major hazards.*
> *d)* ***Prioritization of hazards for further analysis and control.*** *Systematic identification of hazards enables the formulation of risk management strategies based on optimum resources allocation on a priority control/management basis.*
> *e)* ***Risk communication.*** *The process of identifying hazards may also be used for safety training purposes, as a tool for communicating safety information to the general public and as a basis for emergency procedures and emergency planning.*

1.2.3. METHODOLOGICAL ISSUES ON THE HAZARD IDENTIFICATION FOR REGIONAL RISK ASSESSMENT

Potential Hazard Sources in an Area Study

A main step in the area risk assessment and safety management process is the delineation of an appropriate area. The appropriate basis for area selection depends on the particular circumstances of each case. There are not absolute rules: any definition of study area will inevitably be arbitrary to an extent. Several factors which should be considered are suggested next:

- the area should be selected for its physical and industrial, economic, demographic, or environmental characteristics (administrative boundaries may be used if appropriate)
- hard boundaries should not be drawn before the initial hazard identification and hazard analysis of the area which may be affected will not have been identified
- transport systems of dangerous goods to and from the area should only be considered within the study area
- some risk sources will have potential for effects well beyond the immediate area
- community concern may also be a factor to be considered.

These procedures aim to provide guidance in the setting up a methodological and practical framework for hazard identification in an area study (e.g., fixed installations, storage, transportation of dangerous goods).

Hazard analysis in the content of these guidelines is dealing with the identification of undesired events that lead to the materialization of the hazard (a physical situation with a potential for human injury, damage to property or environment). It also consider the appropriate mechanisms by which those undesired events could occur and the estimation of consequences and relative likelihood of any harmful effects.

Major hazard installations in a study area have the potential to cause a major accident in one of the following categories:

a) Large releases of toxic substances with lethal or harmful effects even at considerable distances from the point of release
b) Release of extremely toxic substances in small quantities which are lethal or harmful even at considerable distances from the point of release
c) Large releases of flammable liquids which may either turn to produce high levels of thermal radiation or form an explosive vapor cloud
d) Explosion of unstable or reactive materials.
e) Competent authorities in the area study should pay special attention to major hazard installations. A major hazard control system in the area must be established. Local/national financial and technical resources could play sometimes a restrictive role.

Procedure for Hazard Identification

■ ***Task 1**: Define Objectives of Hazard Identification.* Determination of the objectives of the *Hazard Identification Phase (HIP)* in a IARAMS (Integrated Area Risk Assessment and Safety Management) program together with its intended and potential uses is the main important step in the process of performing a hazard identification study.

The objectives and uses of the HIP will determine in turn the scope of the analysis, the appropriate procedures and methods, and the personnel, expertise, funding and time required for the analysis, as well the associated documentation necessary.

General Objectives of HIP. Hazard identification phase is an efficient and necessary procedure to assist risk analysts and decision making for risk assessment and safety management in a study area. A number of major objectives could be identified:

- to establish what dangerous situations exists within a plant or a process operation
- to establish how these dangerous situations may come about
- to assist in the assessment of the safety of a hazardous installation.

The *first general objective* aims at extending the general understanding of the important issues and situations that might affect the risk analysis process for individual plants and processes; the synergy of individual hazards to the area study level has its special significance. Design and operational problems can be identified and a hazard classification scheme can be considered.

The *second objective* contains elements of risk assessment and deals with accident scenario development and interpretation of results. Consequence evaluation of various accidents and their impact propagation in time and space has special significance in the hazard identification phase.

The *third objective* aims at providing information that can later assist further steps in risk assessment and plant operations safety management. For example, this may be in the form of improving the scenario specifications for risk analysis or identifying appropriate safety measures to comply with given risk criteria (e.g., individual or societal), or advice for emergency preparedness and accident management.

The Plant Life-Cycle. A HIP can be performed at various stages of the plant life-cycle, namely:

- the plant at conceptual / early design stage
- the plant at the final design stage
- the operating plant (the integration of the plant into a complex industrial site).

Hazard identification phase, involves considerations of all situations (scenarios) in which the potential for harm exist in order to identify those which are hazardous, followed by a systematic analysis of the sequence of events which could transform this potential into an accident.

Once an accident scenario has been established, the likelihood of such an accident occurring in practice (accounting for design, operational and organizational safeguards) and its consequence (impact effect) should it occur, can be estimated. It is generally considered desirable to start a HIP process as early in the plant life-cycle as possible. Design, procedural weaknesses, potential human errors that are recognized as early as possible in the system's life-cycle can be corrected or improved less expensively than those that remain until the plant is in operation. During the operational phase of the plant, identified procedures which may lead to accidental situations can be carefully managed and impacts can be avoided.

While a HIP can be started during any of a given life-cycle stages, is it recommended that the HIP models and documentation be maintained and updated throughout the operating life of the plant to provide continued benefit.

Specific Objectives and Uses of HIP. Specific objectives and corresponding uses of HIP related to the first general objective of assessing the hazards and establishing dangerous situations for a plant / process / technology are as follows:

- *Identification of Specialized Methods and Techniques.* There is not a fixed golden rule as which particular technique should be used. There are, however, useful and important guidelines. It may be necessary to use a variety of approaches to improve and refine the hazard identification process. Techniques may also be used in isolation or in complement to each other.
- *Identification of the Potential Hazards.* Hazard identification is the corner stone in the assessment of the risk of an installation/process. It is essential to have clear understanding of the type and nature of hazardous incidents associated with the operations of a plant and of the initiating and contributing events that can lead to such hazardous incidents.
- *Assessment of Important Dependencies (system, process, man-machine, external events).* Important dependencies between components, systems, chemical and physical processes, and between humans and technical systems, that may affect the safety of the plant are assessed. Other important elements for assessment in view of hazard identification relate to the quality of substances, and their nature (e.g., toxic, flammable, explosive), population density around the given plant, safety practice, loading / unloading of substances, etc.
- *Analysis of Severe Accidents.* The results of HIP can help in identifying the consequences of accidents which could be man made or due to the interaction with external factors.
- *Design Modifications.* HIP for plants or processes at the design stage can be used to evaluate the potential hazards induced by various design modifications; this activity has to be considered as an iterative process.
- *Prioritization of Hazards at the Plant or Area Level.* After hazards were identified, a prioritization scheme would allow to focus attention and allocate resources to the most important hazards (e.g., high consequences, low probabilities etc.,) associated with the plant/process. A systematic identification of hazards enables the formulation of risk management strategies based on optimum allocation on a priority control/management basis.

These sub-objectives are applicable to all three stages of the plant life cycle.

Specific objectives and corresponding uses of HIP related to the second general objective are:

• *Providing the Basis for the Design and Operation of Appropriate Operational and Organizational Safety Mechanisms.* Accident propagation and description should provide insights to safeguard procedures which are appropriate and relevant to each type of hazards, and unless such hazards are identified and recognized, safeguards may be irrelevant or of a second importance.
• *Quantification and Evaluation of Risk.* Estimations of likelihood and consequences of hazardous incidents cannot be undertaken unless each hazard has been identified in the first instance.
• *Prevention and Mitigation of Accidents.* Accidents can be prevented by anticipating how they may occur. A systematic understanding of the major contributors to hazardous incidents and of the interaction of contributing events (concurrently or sequentially) enable the formulation of appropriate mitigating measures (e.g., shut-off system) that may prevent such events escalating into major hazards.

The foregoing uses are applicable to all three life-cycle stages; different level of confidence should be expected. Comparing the level of consequences for individual identified hazards is meaningful if the assumptions, techniques, models and primary data and information used in the different HIPs are compatible.

Specific objectives and the corresponding uses of HIP related to the third general objective of assessing hazards at plant or process levels are:

• *Adopting an Integrated Approach to the Control of Hazardous Industry.* Incorporating environmental and health risk impact assessment, requires that i) all hazards associated with the operations of a potentially hazardous installation are identified, ii) hazards are analyzed in terms of their consequences to people, property and the biophysical environment and their likelihood of occurrence, and iii) risks from the operations are quantified and assessed in terms of location and land use planning implications.
• *Emergency Preparedness and Accident Management.* Results and associated insights from HIP provide an effective framework for risk analysis training, developing operational procedures and a rational basis for emergency planning and accident mitigation. HIP results can be used for training purposes.

This is not a comprehensive list of potential objectives and corresponding uses of HIP.

■ ***Task 2**: Define the Scope of Hazard Identification.* After defining objectives, the definition of the scope of the HIP study is the second most relevant element in the management / organization and implementation of the HIP. The scope of the Hazard Identification in a complex risk analysis study can be described mainly in terms of the following parameters:

i) Potential sources of hazards (e.g., Radioactive releases, toxic substances, fire, explosions)
ii) Plant / process damage states,

iii) Initiating events
iv) Potential consequences
v) Prioritization of hazards.

The relevant factors that determine the extent to which these parameters are included in the HIP are:

i) Objectives and intended uses of the HIP
ii) Availability of appropriate information and data
iii) Available resources and expertise.

Issues Regarding the Scope of HIP. Hazard identification requires the consideration of all relevant information regarding the facility (e.g., plant / process). This might typically include:

- Site and plant layout
- Detailed process information in the form of engineering diagrams and operating and maintenance conditions
- Nature and quantities of materials being handled
- Operational, organizational and physical safeguards
- Design standards.

The identification process should not be limited to the activities at the facility, but should also consider:

- Natural events (e.g., floods, avalanches, earthquakes, landslides, lighting strikes)
- Technological events such as vehicle impact on a support structure or impact of aircraft
- Malicious acts
- Hazardous events on neighboring sites (e.g., loss of outside power for a nuclear power plant or loss of outside heat for a chemical installation).

The process of hazard identification and its analysis is based on a number of recognized principles. It should:
- *be comprehensive, holistic and systematic*
- *be qualitative, quantitative and site-specific*
- *be complementary to other safety studies*
- *use consistent and well - documented data collection methods*
- *review adequacy of safeguards*
- *utilize all opportunities for risk reduction.*

Hazards associated with waste and transportation should be included in the analysis. The identification of possible sources of accidental emissions which may be hazardous to the environment requires systematic analysis.

Factors Influencing the Scope of a HIP. The following factors should be considered in determining the scope of the HIP:

i) The objectives and the intended uses of the HIP generally set its scope. Important benefits from performing a HIP can be obtained even if not all parameters identified as characterizing its scope are investigated
ii) The availability of the information required for a particular study scope (e.g., Design stage of the plant / process; operational procedures, test and maintenance procedures; modeling of the man-machine interface; internal fires and floods, etc.).
iii) The availability of expertise and resources constitutes an important factor influencing the scope of a HIP study. The harmonization of various types of resources including models, methodology, computational procedures is of relevant importance in hazard identification at plant, process or area level
iv) Various stages of the life-cycle require the use of specific techniques in the HIP. For proposed developments, the assessment process has the following elements of analysis:

- a Preliminary Hazard Analysis (PHA
- a Hazard and Operability study (HAZOP)
- a fire safety study
- emergency plans and procedures
- a final hazard analysis
- a construction safety analysis
- hazard audits.

At the design stage the techniques required to be carried out are HAZOP fire safety study, emergency plans and final hazard analysis. The PHA is required with the development application. In the usual risk assessment practice the construction safety is required before construction starts and the hazard audits are done throughout the life of the plant. In summary, it is essential that, at the outset of the planning of a hazard identification study, the scope of the HIP is precisely defined in accordance with the integrated area risk assessment and safety management procedural guidelines. The hazard identification is followed by risk prioritization of installations and in depth risk analysis of a selected number of installations / processes.

***Task 3**: Identify Key Factors to be Considered.* The key factors to be considered in a HIP are:

- type of activity (e.g. process, storage, transportation of dangerous goods)
- substances involved (e.g. toxic, flammable, explosive)
- quantities involved
- distance from the populated area
- meteorological conditions
- safety records for individual activity/plant, etc.

■ *Task 4: Evaluate Information Needed in the Study Area.* In order to be able to identify possible hazards of specific activities in a study area (see *Task 5*), one must obtain information of a general nature for each activity:

i) *Fixed Facilities.* A general description of the nature of activities at the site includes:

- nature, type and quantity of substances used (main input stream and auxiliary materials), processed, stored (including transportation vessels) and produced
- type of materials produced as waste, air emission and water emissions (average and maximum quantities)
- main methods of waste treatment and disposal
- transport of materials in and out (including pipelines)
- number and type of transportation vessels with hazardous materials that can be present
- surrounding land use(activities, main roads and dwelling areas).

ii) *Transport of Hazardous Materials.* Use as the basis information to identify transported hazardous materials the UN list of hazardous materials. Identify the main control factors in the specification of corridor data:

- physical route features
- traffic composition
- material properties in transit
- tanker design features by mode of transport
- prevailing weather conditions
- population and land development patterns.

and, if possible, the main origins and main destinations. Road, rail, barge, ship, pipelines and conveyors as well as main transfer facilities should be considered.

Special attention should be given to chlorine, ammonia, LPG and other liquefied flammable gases, toxic gases, flammable liquids and gases. Major hazard installations are commonly associated with the following activities:

- petrochemical works and refineries
- chemical works and chemical production plants
- LPG storage and terminals
- stores and distribution centers for chemicals
- large fertilizer stores
- explosives factories
- works in which chlorine is used in bulk quantities
- nuclear power generation.

■ *Task 5: Identify Activities to be Considered for a Hazard Identification Scheme.* For the specific plant identification processes some background information is required. The desirable set of information includes:

- general environmental quality (e.g., air, water, land)
- geographical information
- demography (population density and distribution by age and most sensitive exposed population)
- transportation routes
- topography
- river systems (and other waterways)
- climatic and meteorological data
- actual and intended land use and zoning
- general location of industrial facilities.

The above list is not exhaustive and should be considered only as indicative examples.

Activities to be Considered for a Hazard Identification Scheme. The following list gives an indication of types of activities which should be considered for inclusion in a HIP for a given study area:

- agricultural activities
- biochemical and biotechnology and / or pharmaceutical
- defense
- explosives and fireworks
- food and drink
- gas works
- manufacturing
- mining and primary processing
- petrochemical and chemicals
- pipelines
- power generation and distribution
- research facilities
- storage
- waste treatment and disposal
- water treatment
- transportation of dangerous goods
- nuclear power plants.

■ ***Task 6**: Collect Basic Information on Activities and Their Associated Risk in the Study Area.* A list should be made of all hazardous installations and processes in the study area. Initiating events and accident scenario development should be considered as basic information for a HIP in a study area. The analyst will need to have a good appreciation of the likely magnitude of the risks of each event, prior to undertaking detailed analysis.

Representative Initiating Events. Having identified the types of hazardous incidents which might occur, the next step is to define discrete initiating events to be used to represent the range of possible incidents. An incident can be identified in a large number of variations. The process of finding representative initiating events will involve a certain degree of screening; it is important that no significant events are omitted.

Remark: The risk and safety analyst will need to have a good appreciation of the likely magnitude of the risks of each event, prior to undertaking detailed analysis.

Scenario Development

Each identified initiating events has to be considered systematically, in order to describe how the incident will develop. The analyst should consider the propagation or the domino effects where one incident may initiate others in nearby plant and equipment.

The following types of fatalities or effects should be further investigated in dealing with accidental situations:

- acute fatalities
- health and long term effects
- property damage and economic loss
- biophysical damage (air/water/land).

The next general scheme will allow identifying basic information for the hazard identification process in relation to various types of fatalities.

Acute Fatalities. The total quantity of each hazardous material at a facility under investigation (or one transport unit) is an important indicator in a hazard identification scheme. Nuclear facilities will not need to be considered in an initial hazard identification process.

Step 1: If the quantity Q_i ($i = 1, 2, \ldots, n$) corresponding to potential hazard i is equal or greater than the quantity prescribed in the *CEC Directory*, use label "yes"; otherwise use label "no" for hazard identification description. Declare the potential hazard i, corresponding to quantity Q_i as a hazard.

Step 2: Use a simplified classification based on the threshold quantity values for different substances:

- flammable substances > 10,000 kg
- explosive substances > 1,000 kg
- toxic substances; based on LC_{50}.

If the quantity of substances is equal or greater than the threshold quantity from above, label it as "may be", otherwise label it as "no".

Health and Long Term Effects. If specific categories of materials such as carcinogenics, mutagens, teratogens, asbestos, combustion products are present use label "yes", otherwise, "no".

Property Damage and Economic Loss. If the following type of losses might occur, then the label "yes" is accepted; otherwise apply the label "no":

- structural damage/loss including corrosive and other effects
- contamination
- infrastructure loss or external costs
- factors of strategic significance (crucial plant loss)
- crops and stock losses
- social dislocation.

Biological Damage. If the following type of damage could occur fill in one on the followings labels "yes", otherwise "no", in doubt "maybe".

- possible destruction of large quantities of animals, plants or destruction of whole species
- possible serious disruption or destruction of ecosystems
- presence of materials such as biocides, PCBs†, heavy metals
- possibility of crude oil spills, etc.

Accidents and their Prioritization Importance
Acute fatalities is an indicator for hazards prioritization importance and potential accidents prioritization. The following steps should be considered when applying a hazard prioritization scheme:

Step 1: If the activity falls within the distance corresponding to the different specific quantities mentioned in the CEC Directory, label it "yes". Otherwise label it as "may be" and continue with *Step 2*.

Step 2: For activities labeled "may be" (see Step 1), calculate the Potential Hazard Index (PHI) as a function of distance to the nearest population area.

If $PHI(d) < 1$, label "no"
If $PHI(d) >= 1$, label "yes".

All activities labeled "yes" should be further analyzed by way of quantified risk assessment.

- *Long-term health effects* should be estimated by making the worst case accident scenario for the maximum number of people that can be affected, due to an accident.
- *Property damage and economic loss* are estimated by making the attempt to quantify the possible damage by the worst case accident scenario.
- *Biophysical damages*: these are calculated for the area affected by the worst case accident scenario.

A number of models, indicators, techniques have been developed for various hazard identification schemes which can be applied at the area level. A prioritization risk

† PCB: Polychlorinated biphenyl.

assessment procedure is often needed, in order to establish which activities should be the focus on detailed risk assessment.

■ <u>*Task 7*</u>: *Risk Prioritization from Major Accidents in an Area Study.* Large and complex industrial areas include various risk sources and activities (e.g., operating process plants, storage terminals, transport activities). The process goes also to the level of an individual plant. A cumulative assessment of such risks should include a detailed hazard analysis and quantified risk assessment for all industrial facilities and associated activities.

■ <u>*Task 7.1*</u>: *Types of Activities and Inventories in an Area Study.* Adopting a methodology to deal with risk prioritization in a region, has to be able to include numerous types of activities and installations. A list of specific activities, which should not be considered exhaustive, is given next:

a) Fuel storage (delivery station, car station, intermediate depot, main storage etc.)
b) Processing and storage of fuel (refinery, alkylation process, naphta cracker)
c) Transport of fuel by pipeline, water, rail/road
d) Food and stimulants (sugar industry, flour industry, extraction of oil/fats etc.)
e) Specific basic products (leather, wood, paper, rubber industries)
f) Metallurgical industries
g) Electronic industries
h) Specific chemicals (e.g., Fertilizers, sulfuric acid, ammonia, hydrogen chloride etc.)
i) Pesticides
j) Explosives
k) Harbor facilities (containers, tanks)
l) Transport (pipelines, road and rail, water).

When considering (A) large industrial areas, stationary installations dealing with flammable substances, and/or explosives, and (B) toxic substances, or transport installations handling such products (LPG, petrol, oil), some criteria for the selection of industrial activities should be adopted, such as: distance from populated areas, or traffic density.

In a systematic methodology known as *"van den Brand" methodology*, a detailed checklist for activities and most important substances has been proposed. This list is associated with a scheme for the classification of substances by effect categories (e.g., category I indicates that all the populated areas within the circle whose radius is the maximum distance of effect must be included in the estimation, category II, III- if the affected area is a sector of circle the user must choose the sector which maximizes the calculated consequences.

■ <u>*Task 7.2*</u>: *Consequences of Major Accidents to People.* In dealing with the external consequences of an accident a number of such consequences may result (e.g., number of fatalities, number of people being hospitalized, various types of damages to the ecosystem, financial losses, etc.). In the *"van den Brand" methodology*, the external consequences of an (potential) accident means the number of fatalities amongst people that are living or working in the area around the facility where the hazardous activity takes place.

The external consequences from an accident caused by the substance i for an identified activity j, can be calculated with the relation:

$$C_{i,j} = A d f_1 f_2 \tag{1.1}$$

where

- $C_{i,j}$ = number of fatalities/accident caused by the substance i for an identified activity j
- A = affected area (ha)
- d = population density in populated areas within the affected zone (persons per ha)
- f_1 = correction factor for the distribution in the affected zone
- f_2 = correction factor for mitigating effects.

The following procedural steps should be considered:

- select one activity taken into consideration the number of substances which can cause damage in the same activity. Special attention should be given to the case when a group of substances may act together (consider an equivalent substance)
- adopt a classification scheme for the substances by effect categories
- estimate the distribution of population in the circular area whose radius is the maximum distance of effect
- calculate the external consequences $C_{i,j}$ and repeat calculations for all substances and activities in the analyzed area.

■ ***Task 7.3***: *Consequences to Environment of Major Accidents.* The consequences of major accidents to the environment are more difficult to be estimated due to the variety of substances which can be involved, as well as the environmental impact indicators relevant in a given accidental situation. Environmental risk indicators which could be considered are: loss of biotopes, loss of groundwater quality, loss of animals etc. Usually, an utility scale is associated with various environmental consequences; the relevant utility scale could include events related to incidents, accident or catastrophic outcomes.

■ ***Task 7.4***: *Monetary Consequences.* Evaluating monetary consequences of (potential) accidents requires a detailed estimation of possible consequences and their associated costs. Not always a monetary value for special classes of consequences (e.g., loss of life, special biotopes) is accepted a priori. The monetary evaluation of consequences should include also external costs, which very often are very difficult to be assessed. The type of consequences are not considered in the *"van den Brand" methodology.*

■ ***Task 7.5***: *Estimation of Probabilities of Major Accidents for Fixed Installations in the Area Study.* Probability *(P)* or frequency of major (potential) accident for fixed installations given as the number of accidents per year can be calculated by using the related so - called probability number N^* for a hazardous substance i and the fixed installation j which causes the given consequences (e.g., fatalities). A "probability number" has always attached an equivalent frequency value P. The relationship between N^* and P is :

$$N^* = / \log_{10} P / \tag{1.2}$$

The value of N* can be calculated by using the equation:

$$N^* = N + \sum_{i=1}^{4} n_i \tag{1.3}$$

where:

N = average probability number for the installation and the k substance
n_i = probability number correction parameters, with $i = 1, 2, 3, 4$ (e.g., loading/-unloading, safety systems associated with flammable substances, organizational and management safety, wind direction towards the populated area).

Calculating the probability value requires a number of procedural steps namely:

- Select one activity taken into consideration the number of substances which can cause damage in the same activity: If a group of substances may act together, consider an equivalent substance
- Select an average probability number N for each hazardous substance (or group of substances) identified for each activity
- Estimate the correction factors n_i ($i = 1, 2, 3, 4$)
- Calculate the probability number N^* and convert it into probability value P
- Repeat the tasks for all stationary activities.

■ ***Task 7.6**: Estimation of Probabilities of Major Accidents for Transportation of Hazardous Materials.* Probability (P) or frequency of accidents during transport (number of accidents/year) of a hazardous substance i which results in a set of consequences, is related to the probability number N. The value of N can be calculated with the relationship:

$$N^* = N + \sum_{i=1}^{3} n_i \tag{1.4}$$

where

N = average probability number for the transport of the substance
n_i = probability number correction parameters, with $i = 1, 2, 3$ (e.g., safety conditions of the transport system, the traffic density, wind direction towards the populated area).

Calculating the value of probability for the transportation of hazardous materials requires a number of procedural steps, namely:

- Select one route and 1 km portion of that route; each hazardous substance transported on this route is analyzed separately
- Estimate the correction factors n, for $i = 1, 2, 3$

- Calculate the probability number *N* and convert it into probability *P*
- If a portion of the transport system is exposed to the risk of accident due to different substances, the frequencies calculated for each substance have to be grouped under classes of injuries.
- Repeat the calculation for all identified portion of the routes.

■ ***Task 8**: Select Individual and Societal Risk Criteria.* All activities have an associated risk. Risk can be assessed and managed, but never eliminated. Indeed, zero risk cannot be achieved even if the activity itself is eliminated.

Probabilistic Safety Criteria (PSC) is associated with a rational decision making process which requires the establishment of a consistent framework with standards to express the desired level of safety. Societal or group risks should be considered when assessing the acceptability of any hazardous industrial facility. A number of factors should be borne in mind when developing PSC based on societal risk, including public aversion to accidents with high consequences (i.e., the risk level chosen should decrease as the consequence increases). PSC for individual risk are proposed under the consideration that risks from accidents in hazardous installations should present only a small increment to the risk to which individuals are already exposed.

Whilst individual fatality risk levels include all components of risk (i.e., fires, explosions and toxicity), there may be uncertainties in correlating toxic concentrations to fatality risk levels. The interpretation of "fatal" should not rely on any one dose-effect relationship, but involve a review of available data.

A criteria for the acceptability of societal risk must be defined before the task of prioritization is performed. When dealing with a risk matrix, the priority assessment risk categories correspond to the upper right hand side of the matrix of probability versus consequence i.e., activities with relatively *high probability and high consequences.*

Observation 1. The concept of societal risk implies that risk of higher consequences, with smaller frequency, are perceived as more important than those of smaller consequences with higher probabilities.

One can choose among various criteria of acceptability:

i) by setting a threshold for the probability class only
ii) by setting a threshold for the class of consequence only
iii) by considering a combination of both classes.

In prioritization of risks the following procedural steps are involved:

i) identify on the matrix of frequency vs. consequence all the activities which do not meet the selected criteria (i.e., all the activities whose calculated risk is beyond the acceptability)

ii) the list of all these activities is the final product of this task.

Observation 2. In the process of risk prioritization, activities with risks of relatively high consequences versus high probability and those of relatively high consequences versus low probability may be considered for detailed assessment, in relation to those of low consequences versus high probability.

Observation 3. By applying a risk prioritization scheme, the following situations could be encountered:

i) activities (e.g., hazardous installations) are scattered throughout the risk matrix, enabling classification and prioritization
ii) all activities fall within one region of the PSC risk matrix. In this case an additional criteria should be considered for further prioritization. This could be based on either a line of consequences only, or probability only, or by shifting, the consequences versus probability criteria line to lower values, or by using a multiple-consequences scale.

■ *Task 11: Use of Ranking Method for Risk Prioritization of Units / Plants Elements.* For risk classification and prioritization of separate installations (elements) of a plant within an industrial complex other specialized models should be considered.

A simplified model (based on the Dow Chemical Index) for the risk ranking of units/plants elements should take into consideration the following steps:

Step 1: Subdivide the installation into logical, independent elements or units.

Step 2: Determine the fire and explosion index F and the toxicity index T.

Step 3: Determine the material factor (MF) index using flammability and reactivity properties (instability and water reactivity of a chemical).

Step 4: Determine the general process hazard index (GPH) for specific situations (e.g., exothermic, endothermic reactions, etc.).

Step 5: Determine the special process hazards (SPH) index for specific situations (e.g., process temperature, low pressure, etc.).

Step 6: Determine the toxicity index, T based on NFPA (National Fire Protection Association) hazard figure.

Step 7: By using F and T factors, perform classification in hazard categories I, II, III (category I is associated to plant elements with the lowest hazard potential and category III has the highest hazard potential).

■ *Task 12: Evaluate the Necessary Data for Analysis.* Data needed for evaluating the individual hazard at plant / unit level is dependent on the model or prioritization technique in use.

Evaluate Individual Hazard at Plant/Unit Level

The main value of the quantified risk assessment process should not rest with the numerical value of the results (in isolation). It is the assessment process itself which provides significant opportunities for the systematic identification of hazards and evaluation of risk. The most significant advantages relate to the optimum allocation of priorities in risk reduction in that the assessment process provides for the clear identification and recognition of hazards and as such enable the allocation of relevant and appropriate resources to the hazards control process.

The procedures for identifying hazardous situations which may arise in process plants and equipment are considered to be the most developed and well established element in the assessment process of hazardous installations:

- the procedures and techniques vary in terms of comprehensiveness and level of detail from comparative checklists through to detailed structured logic diagrams
- the procedures may apply, at various stages in the plant' s life cycle.

Techniques for hazard identification essentially fall into three categories:

i) Comparative methods
  - process/system checklist
  - safety audit/review
  - relative ranking
  - preliminary hazard analysis

ii) Fundamental methods
  - Hazard Operability Studies (HAZOP)
  - "What If" Analysis
  - Failure Mode and Effect Analysis (FMEA)

iii) Logic diagrams methods
  - fault tree analysis
  - event tree analysis
  - cause consequence analysis
  - human reliability analysis.

■ ***Task 13***: *Use Dedicated QRA*[‡] *Techniques to Evaluate Risk Level at the Plant/unit Level.* The reference list given above summarize the dedicated probabilistic safety analysis techniques in use for evaluating risk level at a plant or unit level.

■ ***Task 14***: *Use Expert and Engineering Judgment to Further Decide on Risk Analysis for Installations.* When further technical details are available one can combine them in the overall process for risk assessment of various hazards. Expert and engineering judgments can often be employed for further evaluation of risk for installations.

Documentation and Specifications for Hazard Identification and Prioritization

The fourth major procedural step includes aspects of documentation of HIP. This section is mainly directed towards giving specific guidance concerning the suitable forms for external documentation: for example, presentation of results to different partners involved in an integrated regional risk assessment and safety management study.

[‡] QRA: Quantitative Risk Assessment

■ ***Task 15***: *Objectives and Principles of Documentation.* This task identifies HIP users, applications and basic principles to be followed in the documentation effort. The primary objective of the HIP documentation should be to fulfill the requirements of a regional risk assessment and safety planning process and be suitable for the applications in question.

The potential users are:

- Various companies located in the region in question (management, operating personnel)
- Designers / vendors
- Regulatory authorities (in the field of industrial safety, environmental management, health policy), including other potential reviewers
- Various local / central governmental bodies
- The public in the region in question.

The documentation of HIP should be:

- Well structured (by various types of activities, processes, installations, units, etc.)
- Clearly and easy to follow to review and update
- Compatible with existing management information systems
- Integrating, when accessible, possible and necessary into advanced multimedia information technology, GIS or other computer aiding retrieval technologies.

In addition, means should be provided for possible extensions of the analysis, including integration of improved models, methodologies and data broadening the scope of the HIP in question, and use of alternate application in full agreement with the integrated regional risk assessment and safety management.

In the documentation process, some principles should be further considered:

- Conclusions from a HIP study should be distinct, reflect the complexity of the analysis and the relevance of such information for the further implementation of distinct steps in the regional safety planning process
- Emphasis be given to the analysis of uncertainties in the data and to sensitivity and prioritization analysis, where the effect of assumptions, set of initial potential hazards considered and conservatism in risk scenario design, methods and modeling involved are clearly demonstrated
- A distinction should be always made regarding the level of analysis in the HIP (e.g., Regional, plant or equipment level) and the relative estimation of the risk level in the prioritization process.

■ ***Task 16***: *Organization of Documentation.* In this task the specific and detailed (e.g., for process or transportation activities) organization of the documentation is established. The organization of the HIP documentation should be governed by the following principles:

- *Tractability:* for review, updating and implementation of the analysis it should be possible to trace any information with minimum effort
- *Sequentiality:* the order of appearance of the analysis in the final documentation should follow, as far as possible, that of its actual performance; that is:
  - Activities and inventories in the area study
  - Consequences of major accidents to people and environment
  - Monetary consequences
  - Estimation of probabilities of major accidents (fixed installation and transportation of dangerous goods)
  - Prioritization of risk at plant / unit level
  - Risk assessment for individual plants / installations.

It is recommended that the documentation of the HIP study be divided into the following:

- summary report
- main report (including geographical representation of the relative risk in the region)
- annexes with (detailed) calculations and assumptions regarded for activities / plant risk prioritization in the area study
- disks with appropriate information (text, maps, graphics, databases, images, models used for the prioritization, etc.).

Summary Report. It should provide an overview of the HIP study, assumptions, objectives, scope, results and conclusions to be integrated in further tasks of the risk assessment and safety management methodology with reference to the regional level.

The summary report is designed to:

- support review of the HIP at the area level
- communicate key aspects of the prioritization process and its findings.

The summary report is the last part of the documentation to be prepared.

Main Report. This should give an organized (stepwise) presentation of the HIP study, including area study description, study objectives, methods of risk prioritization and assessment used, types of consequences used, probability evaluation, probabilistic risk criteria, area study modeling results and conclusions.

The main report together with its annexes is designed:

- to support further risk analysis and safety management in the study area
- to communicate information on the overall risk prioritization and assist in further detailed risk assessment work
- to represent the relative importance of various installations in the area study and their associated risks to health and environment, due to accidental situations

– to facilitate choose the appropriate models and techniques for estimation of risk of individual plants / installations.

A good rules of thumb is to put information in the annexes, because most users will not need it or will not need to consult it regularly.

■ ***Task 17***: *Preparation of Document.* In *Task 17* the study documentation is generated on the basis of the objectives and principles defined in *Task 15* and in accordance with the organizational procedures of the documentation established in *Task 16.*

1.2.4. HAZARD IDENTIFICATION TECHNIQUES

The procedures for identifying hazardous situations which may arise in process plants and equipment are generally considered to be the most developed and well established element in the assessment process of hazardous installations. The techniques have been reviewed in a number of documents, notably: Lees[2] (1980), CONCAWE[3] (1982), AICHE[4] (1985), IAEA (1991), EFCE (1985) and SRD (1986), IAEA-TECDOC-727 (1993).[5]

It must be recognized that:

- The procedures and techniques vary in terms of comprehensiveness and level of detail from comparative checklists through to detailed structured logic diagrams.
- The procedures may apply at various stages of project formulation and implementation. From the early decision making process to determine the location of a plant, through to its design, construction and operation.

Techniques for hazard identification essentially fall into three categories. *Figure 1.3* indicates the most commonly used techniques within each category. Safety Audit/Review and Event Tree Analysis and Hazard Operability are discussed in more details as they represent prevailing trends in applications.

Hazard Identification Techniques

Category 1:	***Comparative Methods***
	- Process/System Checklist
	- Safety Audit/Review
	- Relative Ranking: Dow and Mond Hazard Indices
	- Preliminary Hazard Analysis
Category 2:	***Fundamental Methods***
	- Hazard Operability Studies (HAZOP)
	- 'What if' Analysis
	- Failure Mode and Effects Analysis (FMEA)
Category 3:	***Logic Diagrams Methods***
	- Fault Tree Analysis
	- Even Tree Analysis
	- Cause Consequence Analysis
	- Human Reliability Analysis

Figure 1.3. Techniques for hazard identification

■ *Category 1: Comparative Methods*

Process/Safety Checklists

Checklists are used to identify hazards and examine compliance or otherwise with standard procedures. Checklists are limited to the experience base of the checklist author(s). *Qualitative results* from this hazard evaluation procedure vary with the specific situation, including the knowledge of system or plant; they lead to a "yes-or-no" decision about compliance with standard procedures.

Safety Audit/Review

The safety/audit review includes systematic on-site examination of process plants, equipment and safety systems as well as interviews with different people associated with plant operations, including: operators, maintenance staff, engineers, management, safety and environmental staff and personnel. An examination of accident records, maintenance procedures, emergency plans, etc. is also undertaken. A walk-through on-site inspection can vary from an informal routine function that is mainly visual, with emphasis on housekeeping, to a formal comprehensive examination by a team with appropriate background and responsibilities. When a comprehensive safety review is undertaken, *it is referred to as safety audit/review, process review, or loss prevention review.* In addition to providing an overall assessment of the safety of the plant both operationally and organizationally such reviews intend to identify plant conditions or operating procedures that could lead to an accident and significant loss of life or property.

Various hazard evaluation techniques are used for safety auditing, including checklists, and "what-if" questions.

Relative Ranking Using Dow and Mond Indices
The Relative Ranking method assigns (i) penalties to process materials and conditions that can contribute to an accident and (ii) credits based on plant features that can mitigate the effects on an accident. An index for a relative ranking of the plant risk is derived from the combined penalties and credits. The method gives also *qualitative information* on equipment exposed to possible damage through accident propagation.

Preliminary Hazard Analysis
The preliminary hazard analysis method (PHA) is designed to recognize early hazards and it focuses on the hazardous materials and major plant elements, since few details on the plant design are available, and there is likely no information available on procedures.

It is an inductive technique performed anywhere from the system concept to the early design phase of a system. Its purpose is to search for and identify very early hazards that are present. It is not meant to effect the control of the hazard but to recognize all the implications of the hazard. The method consists of formulating a list of the hazards related to available design details, with recommendations to reduce or eliminate hazards in the subsequent plant design phase.

The input required for a PHA includes any available system design objectives, design specifications, equipment specifications, system energy sources, interface specifications, environmental data, operational concept documents, and governmental regulations concerning possible hazards. Any relevant historical data of like systems and their failures would also be of value. *The output of PHA is qualitative,* with no numerical estimation or prioritization, and consists of a list of hazards with recommendations for their possible containment or elimination, which may be used to develop system safety requirements, as well as general performance and design specification. The hazards may also be categorized according to their importance.

Fault Hazard Analysis (FHA)
FHA is also referred to as Subsystem Hazard Analysis (SSHA). It is meant to examine certain high priority hazards identified in the PHA, as well as ones which the certainty of the hazard is not well understood. The FHA tries to answer in greater details how an individual system component or module can fail, and what possibly can happen if it does fail.

The input to the FHA is the component description and function, the failure modes of the components, and any environmental data of relevance including critical human inputs which might be subject to error. *The output is a qualitative assessment* of what may occur when the component fails, and recommendations for containing or eliminating the failure and their implications.

The FHA concentrates more on individual component failures, whereas the System Hazard Analysis (SHA), described below, examines how system operations and failures can affect the safety of the system and its subsystems.

System Hazard Analysis (SHA)
A SHA is usually performed after FHA and continues until the design is complete and accepted. The SHA is a qualitative analysis which looks for common-cause problems, and correlated failure analysis. The aim is to tray to expose combinations of events or failures (dependent or independent), including the failure of safety devices, as well as degradation due to normal wear and tear, that may cause problems later.

■ *Category 2: Fundamental Methods*

Hazard and Operability Studies Techniques
Hazard and Operability Studies (HAZOP)is sometimes also called Operability Hazard Analysis (OHA), or Operability and Support Analysis (OSHA). ***It focuses on hazards that result from tasks, activities or operational system functions that can occur. HAZOP's purpose is to identify hazard and operability problems which could compromise the system's ability to carry on correctly and in a safe manner.*** It is a generally qualitative set of analyses which occur during the late design phase. A systematic search for hazards is generated, examining the system's instrument diagrams and flow charts at critical junctions, for effects which may be deviated from the normal operating parameters.

> ***HAZOP Study***
> *It is a systematic technique for identifying potential hazards and operability problems. It involves essentially a multi-disciplinary team which methodically "brainstorms" the plant design focusing on deviations from the design intention. The effectiveness of the hazard identification process relates strongly to the interaction of the team and the individual diverse backgrounds of the personnel involved. The method aims to stimulate reactivity and generate ideas. The ultimate objectives are to facilitate smooth, safe and prompt plant start up to minimize extensive last minute modifications, and ultimately to ensure trouble-free long term operation.*

The input to the HAZOP are the complete set of detailed engineering documentation (plans, design drawings, procedures, etc.).

The output of the OHA are possible deviations from the normal operating parameters, causes of these deviations, the consequences of the deviations, and possible containment strategies. If the design documentation is not complete, the analysis may be incomplete.

The study can be readily extended to quantify the possible magnitude of the release but the frequency will have to be obtained by further analysis. ***A full quantitative examination involving both magnitude and frequency is referred to as "hazard analysis" (HAZAN).*** The frequency for more complicated cases will have also to be developed applying a full fault tree analysis. The frequencies are usually quoted as annual failure rates. For relief valves it would be more correct to use a figure corresponding to the chance of failure on demand.

HAZOP studies are systematic techniques that were developed using a multi-disciplinary team for the evaluation of hazards and plant operability. *The HAZOP technique is based on the assumptions that:*

- the plant shall perform as designed in the absence of unintended events which might affect the plant behavior
- be managed in a competent manner
- be operated and maintained in accordance with good practice and in line with the design intent
- the protective systems will be tested regularly and kept in good working conditions.

> ***Remarks***
> *The standard practice and degree of completeness of a HAZOP study is very difficult to demonstrate conclusively to a non-participant because the results depend more on the experience and attitudes of the participants and on the leadership style adopted than on the procedure itself.*
>
> *For an effective HAZOP study, the participants should be selected to provide the necessary experience, knowledge, skills and authority in the following areas:*
>
> - *Process design*
> - *Instrument and control design*
> - *National and corporate engineering standards*
> - *Plant operation*
> - *Plant maintenance*
> - *Design and construction management*
> - *Project management*

The HAZOP procedure is more completely described in the Chemical Industry Safety and Health Council's: "A Guide to Hazard and Operability Studies (1977)"[6, 7] In simple terms, the HAZOP procedure takes a full description of the process and systematically questions every part of it to discover how deviations from the design intent can occur. The consequences of such deviations are then determined and if significant are reviewed and remedial action either recommended or flagged for further study.

All modes of plant operation must be considered:

- Normal operation
- Reduced throughput operation
- Routine start-up
- Routine shutdown
- Emergency shutdown
- Commissioning.

A comparison of the benefits and potential pitfalls of HAZOP studies are indicated in *Figure 1.4.*

Requirements to the HAZOP Team

- *A properly experienced and balanced study team with an experienced leader*
- *A positive, open and questioning attitude during the meetings and when deciding upon action to take on the points arising*
- Adoption of a systematic, detailed approach which concentrates on abnormalities
- The conscientious undertaking of pre HAZOP study preparation, including the suitability of guide words and methodology

Benefits

- A full description of the process under study with access to the design basis and intent
- A set of mechanical, piping and instrumentation drawings to allow an item by item review
- Safety and environmental hazards data sheets and specifications

Potential

- Ensure that the majority of design flaws are identified early in the project when design changes are still capable of being implemented
- Reduce the possibility of undesirable capital expenditure on major modifications at the startup phase

Pitfalls

- Potential excessive use of resource time if the HAZOP team is led by an inexperienced leader or if the participants do not have the necessary knowledge and experience levels
- The validity of the HAZOP is directly determined by the accuracy of the information used as input and from which problems can be inferred
- Management shortcomings often prohibit the availability of the most knowledgeable and experienced personnel
- The application of a cursory HAZOP of insufficient depth and understanding can lead to complacency and problems being overlooked
- Shortage of technical information or key design personnel can create frustration and expediency and reduce HAZOP effectiveness

Figure 1.4. HAZOP study - Advantages and drawbacks

"What If" Analysis
The main purpose of the method is to consider carefully the result of unexpected events that would produce an adverse consequence, by a detailed examination of possible deviations from the design, construction, modification, or operating intent. It identifies the hazards, consequences, and perhaps potential methods for risk reduction.

Failure Mode and Effect Analysis (FMEA)
FMEA also called FMEA/CIL (critical items list) or FMEA/CA (critical analysis), is aimed at studying the potential failures, both hazardous and non-hazardous, that occur

in order that their source can be eliminated. Used during design, implementation, and system operations, it identifies all of the ways a component of the system can fail, and each failure mode's effect. By applying a criticality analysis, the potential seriousness of the failure can be ranked.

FMEA is a tabulation of the system/plant equipment, their failure modes as a description of how equipment fails (open, closed, on, off, leaks etc.), the effect of failure mode (e.g., system response of accident resulting from the equipment failure). FMEA requires knowledge of system/plant function; it does not apply to a combination of equipment failures that lead to accidents.

The FMEA/CA procedure proceeds basically as follows: the failure mode is identified, the effect of failure determined, the cause of the failure resolved, the probability of the occurrence of failure established, the severity of failure rated, the possibility of detecting the error before it becomes a problem rated, the assignment of a risk probability number, and finally deciding what corrective action is required.

The input to FMEA is the system design, equipment list, function description, and operation concepts documents. The process is bottom-up. The result of using the method is *qualitative* and consists in a systematic reference listing of system/plant equipment, failure modes, and their effects. *The method is especially useful for the analysis of very critical processes.* The weakness of the FMEA is that it is very time consuming when applied on too broad a scale, it is not suitable for identifying combinations of errors, or for identifying operational input errors.

■ *Category 3: Logic Diagrams Method*

Fault Tree Analysis (FTA)

Fault trees are the best known and most widely used technique for developing failure logic. FTA is really a top-down technique applied during design and operational phases that is used in conducting some of the more general analyses above.

The technique uses a deductive approach which emphasizes cause or sequence of events causing the failure of a system, in contrast to PHA, which emphasizes the hazard itself and its effect. The basic process adopted is to select an undesired *"top-event"* and trace it back to the possible causes which can be component failures, human errors or any other pertinent events that can lead to the top event. This procedure should be followed methodically, identifying the immediate precursors to the top event, then the immediate precursors to these sub-events and so on.

It uses as input the complete knowledge of the system's functions, its failure modes and their effects (this can be obtained from FMEA). A fault tree is comprised of a complex of entities known as *"gates"* which, when the logical combination of the input conditions is satisfied, *produce a specified output which is propagated.* The construction of fault tree follows a defined methodology, but there is, as yet, no universally adopted set of symbols, which is unfortunate as fault trees provide a vivid method of communication.

The FTA approach is graphical in nature, with tree being used to diagram the logical connections, using Boolean logic, between failure modes of a system. The top of the tree, or the top event represents an undesirable event, such as equipment failure, human error, or function malfunction. The construction of the tree follows a systematic procedure which first identifies the general events; i.e., failure states, which, when logically combined, will result in the undesired top event. Then these secondary failure states will be investigated, and the failure states which are required to be combined to cause them will be identified and modeled. If there exists probabilistic information concerning events, system components, and/or subsystems failure's available, the a quantitative analysis can be performed. By analyzing the tree in some detail, one can determine the set of events which cause the top event to be explicitly identified. A fault tree can become very complex, thus practically the tree must be limited to one top event and one specific aspect of the system. The correct selection of the probability method related to a particular Boolean logic gate (conditional, joint or mutually exclusive) is very important.

The output of FTA is a list of equipment and/or operator failures that can result in a specific malfunction, ranked qualitatively by importance. Additionally, hidden failure modes that result from sub-system interactions and combinations of malfunctions may be uncovered.

The fault tree is not a model of all possible system failures or all possible causes for system failure as it only includes events which contribute to its top event. *A fault tree does not show sequences of events*, although some complex gates can be used to illustrate conditions on the sequence of input events. This can be confusing and often leads to difficulties in evaluating the tree.

A fault tree essentially shows system states and, because of this, other techniques are often more useful in cases where sequence is likely to be important, such as would generally be the case with batch processes. Fault trees may still be used for analysis of sub-systems on batch plants, however, such as protective instrumentation or services.

When compiled rigorously in "top-down" procedure, certain sub-events may appear more than once in the fault tree. This is not anomalous and does not prevent the tree being quantified. Attempts to draw the tree without repetition of events in the tree usually lead to the "top-down" approach not being rigorously followed, increasing the chance of omissions.

An important feature that the analyst must consider is *the possibility of failure of independent items by a single cause.*

Fault tree analysis provides an extremely powerful tool which is capable of handling most forms of combinatory events. It provides a good basis for quantification and is particularly useful where a small number of major outcomes are of concern, as is usually the case in hazard analysis. Very large trees can result, with a separate tree required for each top-event -relationships between different trees then need to be considered carefully. Only the outcome under consideration is shown - other outcomes from the causes in the tree will not be shown. Transition routes between states are not represented and the technique generally deals with binary states: partial failures and multiple failure modes can cause difficulties.

Computer codes for fault tree generation are still under development. These programs tend to be viewed with suspicion: the results are, of course, only as good a the logic input by the analyst. Their application would be expected to be limited to very complex systems, where development of the fault tree can be difficult. In such cases correct understanding of the failure logic is usually all the more important.

Event Tree Analysis (ETA)

ETA or incident sequence analysis identifies potential accidents by means of forward analysis from an initial event, such as equipment failure, human error, or function malfunction. Event trees make use of inductive logic (normally binary) and have been widely used in risk analysis of chemical and nuclear industries.

The event trees define the specific sequence of failures that is necessary for an accident to occur or, from an initiating event, the sequence of corrective responses to that event. The ETA differs from FTA in that the FTA uses deduction, or reverse analysis, to trace from an undesired event back to its basic causes. ETA by use of a decision tree, can depict their initiating events, and system safety functions, to determine the successes and failures of the safety functions as a malfunction occurs or is propagated through the system. The accident sequences and the chronological relationship between initiating and subsequent events can be described using ETA.

The method is *especially well suited for analyzing events that can have varied outcomes*. Accidents are ranked to determine the most important risks encountered. If the event probabilities are known, the expected probability of sequences can be defined. The *weakness of this mostly qualitative approach* is that it is poor at handling partial failures or time delays, which can occur between an event initiation and the event actually occurring.

Accident Scenario Modeling Using Event Tree Analysis

In many cases a single incident can lead to many distinct outcomes. *The process of developing possible accident scenarios is an essential element in the risk assessment process.* The event tree technique provides a logic framework for the determination and quantification of a sequence of events which can result in the occurrence of potential accidents. Two distinct applications can be identified which lead to the development of pre and post-accident event-trees. The basic steps of event tree analysis include:

Basic Steps of Event Tree Analysis

- *Identification of initiating events (hazard identification)*
- *Identification of functions or factors which can influence the sequence propagation*
- *Development of all possible outcomes*
- *Classification of outcomes in categories of similar consequences for further experience estimation*
- *Quantification of probabilities of each branch (using fault tree models, expert judgment, operational records on other means)*
- *Quantification of sequences (combining frequency or initiating event and sequence branch probabilities).*

Pre-accident event trees can be used to evaluate effectiveness of plant protective systems and operator actions against the occurrence of an accident initiator. Post-accident event trees can be used to evaluate types of accident outcomes that might arise from a release of hazardous materials.

Post-accident event trees can be appended to those branches of pre-accident event trees which led to unsafe plant states.

Cause-Consequence Analysis

It is a technique which combines the ability of fault trees to show the way various factors may combine to cause a hazardous event with the ability of event trees to show the various possible outcomes. Sequences and therefore time delays can be illustrated in the consequence part of the diagram. *A symbolism similar to that used in fault trees is used to show logical combinations.* The technique has considerable potential for illustrating the relationships from initiating events through to end outcomes. It can be used fairly directly for quantification, but the diagrams can become extremely unwieldy. Because of this, cause-consequence analysis is not as widely used as the first two techniques described, possibly because fault and event trees are easier to follow and so tend to be preferred for presentation of the separate parts of the analysis.

Human Reliability Analysis

The method consists in a systematic evaluation of the factors that influence the performance of human operators, maintenance staff, and other personnel in the plant and identifies error-likely situations that can cause of lead to an accident. It includes identification of system interface affected by particular errors and relative ranking of errors based on probability of occurrence or severity of consequences. Results are qualitative and quantitative and include a systematic listing of the types of errors likely to be encountered during normal or emergency operation.

1.3. HAZOP Study of Chemical Processes

1.3.1. HAZOP STUDY OF CONTINUOUS PROCESSES

Studies of continuous chemical processes are carried out in a series of meetings where mechanical and piping diagrams are examined line by line, vessel by vessel, using a list of guidewords to stimulate the hazard study teams' considerations of all conceivable deviations from design intent.

The list of guidewords depicted in TABLE 1.1 is worked through systematically by the team of mixed disciplines, led by a trained hazard study leader. Should potential problems be identified, then a review of the preventative or corrective measures designed to minimize the likelihood and consequences should be specified. Any further action should be noted and progressed outside the meeting.

The main information recorded on the *protocol form for the HAZOP minutes* is as follows:

HAZOP Minutes Record Sheet Information

- Deviation Guide Word
  - Possible Causes
  - Consequences
  - Existing Safeguards
- Action Required
- Responsible Person

Additional information is presented showing the persons present at the meeting and all relevant details concerning the line diagram under review.

1.3.2. HAZOP STUDY OF BATCH PROCESSES

The general characteristics of batch plants as compared with continuous plants are as follows:

■ The status of the various parts of the plants are changing cyclically with respect to time and therefore an engineering line diagram gives a very incomplete picture of the process operation

■ The processes are usually multi-stage and the individual units are often multi-purpose

■ Batch plants are often multi-product and reaction units usually have to be cleared out and modified when changing from one product to another.

From the above aspects it is clear that these can be several modes of operation for batch plants. At the very least, two fundamental states should be considered. These are:

– An "active" state when the item is in use, and
– An "inactive" state when the items is not in use.

This is in contrast to a continuous plant where, in steady state operations, a fixed mode in terms of flow, pressure, temperature etc. can be defined for each part of the plant.

The HAZOP methodology has been applied successfully to a diverse range of process operations including computer applications as well as plant procedures. The HAZOP technique identifies potential hazards and the possible mechanisms by which these hazards can occur.

The approach therefore adopted in a hazard and operability study (HAZOP) of a batch process is to apply the guide words initially to each step of the process. The additional guide words of "sooner than" and "later than" must be considered at each step. In addition the interactions between each of the steps need to be considered.

TABLE 1.1. HAZOP guide diagram for batch processes

Guide Word	Meaning	Example of Deviation
NO (NOT OR NONE)	The activity is not carried out or ceases	NO flow in pipe; NO reactant charged to process: Batch NOT cooled; Check omitted
MORE OF	A quantitative increase in an activity	MORE (higher, longer) quantity, flow, temp, pressure, batch, concentration, time
LESS OF	A quantitative decrease in an activity	LESS (lower, shorter) of above
MORE THAN OR AS WELL AS	A further activity occurs in addition to the original activity	Impurities present, extra phase (solid or gas in liquid phase) extra (unplanned) process operation
PART OF	The incomplete performance of an activity	Reduced strength, missing component, operation only PART-completed
REVERSE	Inversion of the activity	Back-flow or back-pressure; Heat rather than cool
SOONER / LATER THAN	An activity occurring at the wrong time relative to other activities	The activity occurs at the wrong time
OTHER (THAN)		Wrong material charged; Non-routine conditions, start-up, shutdown, maintenance; cleaning, etc.; Failure of services

This means that the HAZOP process for a full batch study is significantly more complex than for a steady-state continuous process. Considerably more detailed information is required in terms of batch operating procedures and valve status indications at each step of the process in order to meaningfully judge the potential process deviations (see TABLE 1.1).

1.4. Further Development in Process Safety Techniques

A further technique which is used to enhance hazard assessments and which focuses on key concerns in a process operation is the *fault tree analysis*.

This technique allows both a qualitative appreciation of the potential ways in which an incident may develop (as a logic tree) as well as a quantitative assessment where suitable failure rate and demand frequency data are available.

A further development of this technique has been to modify and interpret the fault tree in a positive sense as a *"hazard warning tree"*. A general outline of this technique is given below.

1.4.1. ADVANCES IN HAZOP TECHNIQUES

HAZOP Study of Procedures
The same basic HAZOP approach, as used for continuous process plant studies, can be applied to the study of "procedures". The aim is still to systematically and critically examine the intentions of the procedure and assess the hazard potential of possible deviations from these intentions.

However, the HAZOP study of procedures is sufficiently different to warrant special attention. In examining a continuous process, there is a steady state of conditions (e.g., temperature, pressure, flow, composition) and the HAZOP study seeks to identify deviations in these. A *"procedure"* is distinctly different and can be considered as having the following basic attributes:

- Starts from a defined state
- Follows a defined series of operations or activities
- Ends with another defined state

In any general procedure the defined starting point or state will include:

- the condition, provision, location and state of equipment, plant and materials
- the training and skill of the operator/technician
- the available support staff; lines of communication; etc.

The procedure will define the activities necessary to move from this defined starting state to the end state (which will include most of the factors considered previously).

> ***HAZOP Study of Procedures***
> *Thus the totality of the procedure must cover the flow of materials and information and the detailed activities. The HAZOP principle is then to thoroughly investigate all aspects of the procedure to discover possible deviations and to assess the consequences and likelihood of them. The first aspect will be the assessment of the initial state of the system. This implies some form of inspection by the operator/technician, presumably against a checklist. That is, the required state is defined and the procedure should ensure that it is met before proceeding. In applying the HAZOP study method, the question needs to be asked as to what the actions will be to remedy anything that is not in the required state and what the consequences would be if the state was other than as required.*

A *suitable guide words* to explore the initial state of the system may be:

- MISSING — Equipment, information, or material missing
- INSUFFICIENT — Insufficient supply/condition of materials, equipment, or information
- WRONG — Incorrect material, person, information, etc.

- TIME Insufficient time allowed, or available
- OTHER Deviation of some other variable.

Having been satisfied that the initial state of the system is appropriately set up, the *procedural aspects* must be studied. *HAZOP Guide Words* for procedures should be used effectively for preparing and examining operating procedures for plants (TABLE 1.2). Once the *final state* has been reached, the same basic approach used in assessing the *starting state* can be applied.

Responding to Deviations
As with all HAZOP studies, once a deviation has been discovered, its significance must be assessed.

The questions to ask at this stage are:

- If the deviation does occur, will it matter?
- If it does, how often is it likely to occur?

TABLE 1.2. HAZOP Guide words for procedures

PARAMETER	PURPOSE	DEVIATION
WHAT HAS TO BE DONE?	WHY; FOR WHAT PURPOSE?	WHAT IF OMITTED? WHAT ELSE? AS WELL / INSTEAD?
WHEN?	WHY THEN?	EARLIER / LATER?
WHERE?	WHY THERE?	ELSEWHERE?
HOW?	WHY THAT WAY?	SOME OTHER WAY?
HOW MUCH?	WHY THAT MUCH?	MORE / LESS?
HOW FAST?	WHY THAT FAST?	FASTER / SLOWER?
HOW OFTEN?	WHY THAT OFTEN?	MORE / LESS OFTEN?
WHO?	WHY THEM?	SOMEONE ELSE?
		WHAT ELSE CAN GO WRONG?
IF DEVIATION: DOES IT MATTER?	HOW OFTEN?	WHAT THEN NEEDS TO BE DONE?

Note: The statement on the initial state of the plant implies an inspection (against a check list) by the operator. It seems prudent to inquire what may happen if the operator finds any part of the plant in other than the required state and takes steps to correct the state, for example, opens a closed valve which should have been open before starting the procedure detailed.

Based on the answers to these questions, the need to introduce some form of check or balance is assessed. Exactly what can be done to either avoid the deviation, lessen its consequences or reduce its frequency is up to the study team to decide. Likewise, the appropriateness of any such action is up to the team.

Useful questions to probe any suggested corrective action are:

- Is it practical?
- Is it sensible?
- Is it cost effective?

Human Error
At all stages of the HAZOP study, the possibilities of human error must be considered. This does not imply that the people performing the task are either incompetent or inadequately trained. In fact, psychological studies have indicated that simple errors in well known routines can become more likely as our skill in the routine increases.

HAZOP Study of Selected Construction Activities
On some projects, construction work, or maintenance and repairs are necessary on existing processing plant or at close proximity to hazardous piping routes or processes. *It may then be necessary to examine selected construction/repair activities systematically at appropriate stages in the construction program.* Prior to bringing equipment and personnel on site (i.e., contract teams), each contractor should submit written proposals of how the activities are to be achieved and supervised. A series of meetings can then be held at appropriate stages in the construction program, to examine systematically the adequacy of the proposed detailed construction activities. Each activity should ensure that :

■ All reasonable provisions are being made to ensure the job is carried out safely, i.e.,:

- The contractor understands the implications of deviating from his defined method of working.
- The contractor and supervisors understand that safety and safe working practices have a higher priority than achieving target completion dates. The basic guidance is: "If in doubt or concerned, stop the job and seek operations management advice".

■ Sufficient thought is given to access/egress to the construction site, which on occasions involves checking the suitability/standards of the vehicles, linking with the plant control room, providing escorts for abnormally large vehicles and providing permits to work, etc.

■ Appropriate Site and Plant induction training is given in advance to all contract personnel employed on site.

A guide to the HAZOP style examination of the proposed construction activities and a list of factors to consider is available in literature.

HAZOP Study of Computer Based or PLC Systems
Studies of *computer systems* can be conducted in several ways depending on the nature of the system. The procedure can be assisted by the *use of block diagram representations of the equipment within defined cut points.* The interfaces between each item of

equipment can be systematically examined using an approach similar to that for batch processes where the basic guide words (LESS OF, MORE OF, etc.) trigger detailed consideration of the transfer of information/data, and the performance of critical items of equipment (e.g., power supplies, alarm systems, printers etc.).

The HAZOP Study guide words can be modified and used to prompt detailed consideration of the failure modes of modern computer based or PLC type control systems and this approach encourages a structured examination of each key unit in the control loop (e.g., DP cell, P/I, controller/ computer, I/P, control valve). Many new instruments contain PLCs (DP cells, density meters, controllers etc.) and their failure modes can be very different from conventional instruments (e.g., loss of input can default, such that automatic control reverts to manual without any audible alarm). Such novel failure mechanisms can only be revealed by lateral consideration of cause/effect deviations in input/output circuitry and software programs. In particular the wider implications of common mode failure should be addressed.

For micro-processor based systems, the effect of a hardware component or software failure on the output of the device is generally the most important consideration. Where a multi-input / multi-output device is being considered then each output (analogue and digital) should be considered separately.

Overall system safety integrity relies on:

a) Configuration (ergonomics, loop design)
b) Reliability and capability (performance, confidence)
c) Quality (information displayed, log)

Two key aspects of HAZOP studies of computer systems are to:

- Focus on any novel features of the device and examine the effects of their performance
- Systematically examine potential causes and effects of foreseeable fault modes which could result in potentially adverse output

A "novel feature" is an operation of the device which a user would not consider part of the standard functionality. It has normally been added by a manufacturer to give them an edge on their competitors. In many instances such features can add to the integrity of the device rather than detract from it. Examples of "novel features" are: set-point tracking, forced default to manual, memory sum-check failure, and specific action on initiation of "watchdog" (a software checking routine).

Procedure for Computer Based System

The following guide words can be considered in reviewing the safety and operability of a computer based system (TABLE 1.3):

These deviations can be applied in either (or both) of two general approaches :

a) A *"loop-by-loop" analysis* using the electrical/instrument loop diagrams as the major review item

b) A *"block-by-block" analysis* focusing on the potential for adverse interactions between sub-systems.

TABLE 1.3. HAZOP Guide words for computer based systems

Parameter	Deviation	Remark
1.	MORE OF	Blocks of data / transfer frequency
2.	LESS OF	Incomplete transfer / system crashes during transfer
3.	NONE OF	No transfer of data
4.	OTHER THAN	Mismatch due to re-format / software change / process variable change
5.	SOONER / LATER THAN	Questions how measurements are processed / time out / out of sequence / averaging assumptions
6.	CORRUPTION OF	Noise, magnetic fields, radio interference, welding ,lightning
7.	WHAT ELSE	Maintenance, simulation, earthing, high voltage due to fault condition
8.	REVERSE OF	Repeat steps 1 to 7 looking at data transfers in the opposite direction
REPEAT STEPS 1 TO 8 FOR ALL LINKS ACROSS COMPUTER CONTROL INTERFACES		

In either approach, allowance for human error (involving control room VDU layout and ergonomic factors) should always be considered.

Team Composition

The team composition will be biased towards participants with a strong background on computer, instrumentation, and electrical know-how. A senior process/operations adviser must be present. It is advisable to have an independent HAZOP Leader for significant computer based projects. Such a person should be conversant with computer based systems and ideally should have had previous experience and participation in similar reviews.

1.4.2. HAZARD WARNING APPROACH

Hazard Warning Concept

The fact that there are a likely to be a lot of lower level incidents prior to a major accident is a well known statistical axiom.

The possibility that a series of "intermediate level" or precursor incidents are likely to occur before any of the initiating events escalates to the major event is an essential concept in quantitative measurement of process safety known as *"Hazard Warning"*. The basic principle of the Hazard Warning Structure is that if the lesser events are not occurring then there is good assurance that the major incident will not occur either.

Hazard Structure
The statistical basis of the hazard structure has been explored by Heinrich[8] (1951) and many others in the occupational health and safety fields. The interpretation of the so called "pyramid of accidents" type structure reveals that a major hazard is in all probability going to be preceded by a series of preliminary warnings. These "warnings" are events that may occur more frequently than the top event (the major hazard) and usually terminate at various degrees of "near miss" or "minor damage" type levels (below the top event). This, of course, assumes that there are various levels of containment that need to be breached before the major hazardous event can occur.

It is recommended in the literature that a practical technique for monitoring operational and maintenance incidents be developed to provide warning as to potentially more serious hazardous events (UK Health and Safety Commission, 1984). Methods based on the fault tree approach, appeared at that stage to be the most promising. The goal is that industry develop the concept of hazard warning as an auditing tool in such a way that is convenient to use and operable by a wide selection of personnel.

In hazard warning, *a quantitative safety monitoring system based on the combination of the Fault Tree Method.* With the Poisson probability distribution has been proposed to predict the number of occurrences of minor events before some undesirable higher event would occur. This method, needed further development of its quantitative aspects to allow convenient or practical application.

The hazard warning system is based on the observation that most incident sequences in a fault tree terminate well before reaching the top event. These subsidiary events which do not propagate should be treated as hazard warnings, whereby the top event could have occurred, but in fact at that time it did not took place. By applying statistical analysis to the fault tree frequencies, the likelihood of the top event occurring, given the occurrence of some subsidiary event, can be computed. This is by definition a *failure of hazard warning* - as the top event is now assumed to have occurred. *The likelihood of failure of hazard warning should be kept as low as possible.*

Hazard Identification Based on Fault Tree

The fault tree methodology has a number of attractive features with regard to hazard identification, and these include:

- A mechanistic understanding of the fault sequence leading up to a major incident or equipment failure is identified
- An estimate of the likely frequency of occurrence of the major incident and of subsidiary incidents leading to the incident is obtained
- The true redundancy of backup safety systems is obtained (provided that common mode failures are identified)

- A written record of the analysis is produced, understandable by those who were not involved in the original study.

The fault tree method employs successive sequences and AND or OR gates to discover the combination of faults necessary for a particular event to occur. Normally, the top event or major incident is identified first (e.g., explosion, loss of key equipment item, etc.) and the immediate faults leading to this are then identified. The procedure continues down one level at a time with itemized faults listed explicitly, extending the detail of the previous level until base faults or initiating events are identified. These are events for which no further breakdown is warranted and for which estimates of frequency or probability can be made. Boolean algebra is used to compute all frequencies higher up the tree, eventually predicting the overall failure rate of the top event and of all preceding faults.

The General Hazard Attenuation Factor

Lees (1984, 1985), later published case studies which concentrated to a large extent on the most conspicuous quantitative feature of hazard warning. This is the clear attenuation that is apparent in a fault tree between the major outcome (top event) and any preceding events right down to the initiating events at the base of the fault tree. He argued the case that this type of information can be used quite effectively in the analysis of post accident outcomes and cited several examples from the Canvey Island study and other public reports to show its application.

Simply stated, the "attenuation factor approach" is to compare the frequencies of occurrence of two events at different levels in the fault tree; the ratio of the more frequent lower event to the mitigated higher event is the "warning factor".

How likely or unlikely a hazard will occur must depend on the structure of the hazard, i.e., on the incident pathway(s) to the undesirable occurrence of the major event. *This implies that some hazards which are mitigated against by a range of hardware or software measures may require many successive "levels" of mitigation failure for the major hazard to occur.* It would therefore be regarded as having a *"high warning structure"*. Other potential hazards may have little in the way of controls or protective systems and therefore constitute a *"low warning structure"*.

Lees therefore attributed to each accident scenario either a "high" or a "low warning structure" depending upon the number of incident escalation levels, each with mitigation features, and the overall incident attenuation factor.

The Statistical or Probabilistic Approach

While there are a number of possible statistical distributions to describe failures, *the Poisson time related distribution applies to the random occurrence of isolated events in a continuum of time* and is, by its similarity to the exponential distribution, also suitable for the examination of low probability events such as failures of protection systems (i.e., probabilities less than 0.1). Strictly speaking, this is not a Poisson distribution, but it is known as the *Erlangian distribution and it is simply the time-dependent form of the Poisson discrete distribution.*

Using the Poisson-related statistical theory, it is possible to calculate the probability that the upper level (top event or 1st level) will occur given that the preceding event (at the 2nd level) occurs no more than n times:

$$P_T(t,k<n) = \exp(-f_2 t) \sum_{k=1}^{n} p(k) \frac{(f_2 t)^k}{k!} \tag{1.5}$$

where

$$p(k) = \sum_{j=1}^{k} \frac{k!}{(k-j)!\,j!} p^j (1-p)^{k-j} \tag{1.6}$$

with

f_2 = failure frequency of 2nd level event
t = time duration of interest (for example, the plant life time)
k = number of "2nd level" event occurrences (k = 1 to n)

The probability mathematics appear rather awesome at first sight but can be easily handled by modern personal computer systems. Further explanation and examples may be found in the literature (Pitblado & Lake, 1987).

Benefits of the Hazard Warning Approach
The use of fault trees in industry is likely to expand greatly given the need to reduce major hazards and the prompting of governmental authorities and the international chemical engineering institutions and standards associations. Extensions to the fault tree method via the *"hazard warning approach"* appear to be eminently suitable for auditing of plant safety performance.

The major practical uses of developing hazard warning systems are the following:

■ It can assist in the communication of risk information with external authorities or public interest groups as a way of explaining the causes and likelihood of potential incidents. This technique is suitable for appealing the public's perception based on the fear that a "chemical disaster could happen tomorrow, without warning".

■ It can be used as an additional analytical tool for the risk assessment or evaluation of post-release scenarios or in the evaluation of documented case studies.

■ It can be used as an aid in chemical factory operations where it can be used to identify and assess the priority pre-release incidents (warning events) and in the setting of targets for the precursor events for specific incident scenarios. In a similar way it would be used in conjunction with conventional hazard auditing techniques.

■ It can be used as an aid to the factory development sections in determining quantified risk reduction benefits that can be used to assess the cost-effectiveness of additional safety related expenditure.

In conjunction with the fault tree analysis, the hazard warning approach offers a convenient means for examining the ongoing technical safety level of an operating plant.

1.5. Guidance on Implementation of Hazard Identification Techniques

The *appropriateness and relevancy* of any one particular technique of hazard identification largely depend on the purpose for which the risk assessment is being undertaken. The primary principle is to first examine the plant or operations from the broadest viewpoint possible and systematically identify possible hazards. Elaborate techniques as a primary tool may cause problems and result in missing some obvious hazards.

The *objectives of the analysis* must be clearly established at an early stage. It may also be necessary to adopt more than one technique depending on the level of detail required and whether the facility is a new proposed installation or an existing operation.

For example, a preliminary hazard analysis or a generalized `What if' analysis may be appropriate for a proposed new facility to assist in establishing a suitable location and when only preliminary design information is available. This could be followed by a detailed HAZOP at the design stage and then periodic safety audits and reviews at the operational stage. For an existing plant, HAZOP may be limited to when modifications are contemplated with safety audits and fault and event tree analysis undertaken as part of evaluating safety measures.

The following table provides a guidance framework to assist the most appropriate techniques for various situations (TABLE 1.4).

> *In stressing the need to identify hazards as early as possible in the development of a process, and especially at the process design stage, the impression may have been given that hazard identification ends when the design specification has been approved. In fact approval of a design means only that "at the time of the study the study team believe that, provided the plant is constructed and operated in accordance with their recommendations, the plant will be acceptable safe". The first uncontrolled change during construction or the first unapproved modification during operation negates this approval. Consequently hazard identification is a continuing ingredient of safe operations and should be applied, sometimes in a very simple form, to control any changes from the original intentions of the designers.*

The application of hazard identification methods (TABLE 1.5) can be both *time consuming and expensive* so that some sections of a plant will receive a more detailed study than others, the depth of the study depending upon an appraisal of the inherent hazards in the various sections of the plant. In the case of a highly sensitive reactor system the hazard identification study may be very detailed and often supplemented by a reliability analysis of the control system using a method such as fault tree analysis. On the other hand a water treatment plant might only be reviewed for operability and personnel protection. Therefore the *depth and scope of a study* is determined by an organization's perception of the hazards in a process and their appraisal of the need to control them.

TABLE 1.4. Guidance table on implementation of hazard identification techniques

Hazard Identification Technique	Site selection/ early design stage	Design stage of new plants	Operational stage of new and existing plants	Modifications to existing plants
Process system checklist	B	B	A	B
Safety Audit/Review	C	C	A	C
Dow and Mond Hazard Indices	C	B	A	C
Preliminary Hazard Analysis	A	C	C	A
Hazard Operability Studies	C	A	B	A
"What if" analysis	A	C	B	A
Failure Mode and Effect Analysis	C	A	A	B
Fault Tree Analysis	C	A	A	B
Event Tree Analysis	C	A	A	B
Cause/consequence Analysis	C	B	A	B
Human Reliability Analysis	C	A	A	B

Note: A: Best suited; B: Could be used; C: Last suited (not advised).

In applying the above mentioned methods one can calculate probabilities or frequencies related to some accidental release scenarios. These probability figures are to be combined later on with respective consequence assessment scenarios for a risk matrix representation (see also Chapter 4).

In making the risk assessment of any system, there are four main stages:

- To establish the probability of the conditions occurring that could lead to a fault developing in the unit of interest.
- To determine what is the probability of a fault developing and what will be the characteristics of the fault, given the conditions arise that are required for the development of a fault.
- If the fault occurs, to determine what will be the hazard potential of the fault in terms of energy or hazardous material released.

- To determine the consequences in terms of loss of life, injury and damage to property of the fault developing.

TABLE 1.5. Techniques of hazard identification

Techniques of Hazard Identification	**General Description**	**Data and Requirements**	**Output**
Hazard Operability	A systematic review of the plant design, section by section, using a series of guide words to identify possible deviations and establish necessary action to cope with such deviations.	Piping and Instrumentation and process flow sheets and diagrams. HAZOP relies on brainstorming amongst team of design/operational personnel	A comprehensive identification of possible deviations, their consequences, causes and suggested actions.
"What If" Analysis	Systematic examination of a process of operation, using "what if" prompt to suggest an initiating event, a failure from which an undesirable event sequence could occur	- Process flow sheets, pilot plans, PIDs; - two qualified analysts.	An identification of deviations with their consequences and recommended actions.

An allowance for errors will have to be incorporated for each stage of the analysis and the errors that have to be allowed for are likely to be larger when the plant being assessed is of a new untried design for which there is no relevant experience to call on. The risk implication for the population can be for the same plant design quite different depending on the location of the plant (population density, climate affecting the rate of release of hazardous material).

For the analysis of relatively large fault trees pertaining to a *nuclear reactor*, where the technology is well established, either analytical methods that compute failure rates directly or simulation methods such as *Monte-Carlo techniques* can be used. The simulation methods are more flexible but require considerably more computing time. Methods and data for quantitative risk assessment are available but present knowledge is insufficient for best estimate analysis.

Having identified some of the problems associated with the quantification or risk, two questions arise about what are the benefits of quantification of risk and what role can quantification of risk play.

Several techniques used for risk assessment of hazardous installations have been examined by J.C. Consultancy Ltd.[9] (1986) on behalf of the *Commission of the European Communities*. In their report they compare the techniques that can be used to asses the significance of risk in quantitative terms, and how these techniques are used in practice in the nuclear industry with the way they are used in those parts of the process industry that can be categorized as having a major hazard potential. The practices applied in France, Germany, and Britain have been reviewed too.

The examination of the techniques for quantitative risk assessment showed that there are three main problems related to the significance of risk:

- the data on which they have to be based may not be entirely relevant to the case being studied,
- the complete assessment of complex plant may require some simplification of the systems to keep the assessment costs within reasonable limits, and
- the techniques by their very nature do not present a comprehensive assessment of all the relevant technical, economic and socio-political factors.

From a *study on failure data related to various types of electronic equipment and small mechanical components,* Aitken[10] (1977) showed that typically observed failure rates (R) relating to a particular predicted failure rate spanned a 10 to 1 range, the variation being different for different types of equipment. The assessment of the probability of failure of even a relatively simple component involves the assessment of several factors of probability (failure due to design fault, material fault, fault in construction, operational fault, and probability of failure of inspection techniques). There is a ten to one range in the value of each factor, there can be a hundred to one range between the highest and lowest value of the probability of the component failing. In reality it is unlikely that all the low or all the high probability values will be combined. The most likely value will be somewhere between. The ratio of observed to predicted failure rate appears to follow a log-normal distribution. The median value for R (i.e., with probability $p = 50\%$) is 0.76 which indicates, on average, the predictions are pessimistic by about 30%. The chance of the ratio being within a factor of 2 of 0.76 is 70%. The chance of the ratio being within a factor of 4 of 0.76 is 96%. *An important fact concerning these data is that it is all related to instrument type equipment and should not be used outside that field.*

The *conclusions* that seems justified about the capability of quantitative techniques is that provided the techniques are based on accurate and relevant data, they can give useful assessment of the significance of faults and weakness in design and operation procedures. Fault tree and event tree methods provide a logical structure for assessing the significance of the relationship between the various parts of complex systems and what happens if there is some deviation in performance of a part of the system or its operating conditions. *It is mainly in terms of reduction in risk or improvements in safety that the benefits of applying quantified techniques of risk assessment can be judged.* The reduction in individual risk can be considered as a benefit of the quantified risk assessment process. Without a risk assessment the magnitude of the risk reduction would not be known and there would be no logical way of judging the benefit of the action taken.

Most of the *weakness of the techniques for risk quantification* stems from the fact that unless the meaning of the results obtained is clearly explained, they can be misleading. The techniques are only as accurate as the data they are based on. As an example, the failure probability of a pressure vessel is determined by a combination of nine factors, but there is only limited information on the probabilities associated with these factors (Nichols[11], 1975). The values used have to be based on engineering judgment. Van de Putte[12], (1981) suggest that probabilities smaller than 10^{-5}-10^{-6} / year should be treated with extreme caution as often some sub-probabilities or common mode failure has been overlooked. The weakness of this mostly qualitative approach is that it is poor at handling partial failures or time delays, which can occur between an event initiation and the event actually occurring.

1.6. Risks from Technical Systems: Integrating Fuzzy Logic into the Zurich Hazard Analysis Method

1.6.1. "ZÜRICH" HAZARD ANALYSIS AND FUZZY LOGIC[13]

Risk analysis[14] uses systems of classifications to characterize frequencies and consequences as elements of risk calculations. "Zurich Hazard Analysis" (ZHA)[15, 16] uses a specialized classification schemes to characterize risks of (undesired) events or incidents, e.g., among other things frequency is classified as "frequent" and consequence as "marginal". Such classifications are introduced to avoid using quantitative risk estimations; in the end the analyst deals with numerical suggestions what "frequent" or "marginal" means, depending on the surrounding field of application, type of industry, the use of expert judgment, etc. Additionally to the problem of numerating verbal expressions, it is often difficult to clearly separate classes (e.g., frequent, marginal) from each other. This is a question of boundaries and definitions.

In the classical theory of sets, a measured or observed value belongs unequivocal to a well-defined set but it is obvious that fixed definitions of boundaries are in practice approximately arbitrary. Because of this, the classical approach is generalized by the so called "fuzzy sets". The limits of sets within "fuzzy-logic" are fluid and an observed value belongs more or less to a defined class. This approach to define a classification scheme is a tool to deal with classes in risk analysis. Fuzzy sets allow making risk assessment and management more flexible and, because of naming membership grades, more transparent.

The "Zurich" Hazard Analysis (ZHA)

The ZHA is an inductive method developed by Zurich Insurance Company (Switzerland) in order to identify and assess the risks of an industrial plant and is based on MIL-STD-882B standards.[17] The method starts with the identification of hazards and their (possible) effects. *Figure 1.5* shows the procedural steps of the ZHA.

Plant Specific Risk Profile Matrix. The specific hazards[18] of an existing plant are considered *a priori*; the protection level, defined by the company's safety philosophy or set up as a goals, is also chosen in the same way (STEP 1 to 4 in *Figure 1.5*). The results of STEP 5 form the basis for an on-going analysis of hazard identification and risk analysis by means of fuzzy-logic.

The Zurich Hazard Analysis (ZHA)

Step 1: Scope Definitions

- *General overview vs. In depth approach for the plant risk assessment*
- *Extern scope depending on available experts, time, information and financial resources, etc.*

Step 2: Forming an Interdisciplinary Expert Team

- *Team leader*
- *Research, development, design*
- *Procurement, manufacturing, marketing, sales*
- *Quality control, safety, maintenance*
- *Others*

Step 3: Hazard Identification

Hazards are identified and entered into the hazard catalogue.

- *Contents are:*
- *Hazardous characteristics*
- *Malfunctions*
- *Environmental influences*
- *Use and operation*
- *Life cycle, etc.*

Step 4: Design of a Specific Protection Level Within the Given ZHA Risk Matrix

Step 5: Design of a Plant Specific Risk Profile Matrix

- *Risks are placed on the risk profile:*
- *Cataloguing of potential causes and effects*
- *Categorization related to frequencies and consequences*

Step 6: Design of a Risk Reduction Catalogue

- *Prioritize measurements.*

Figure 1.5. Procedural steps in the ZHA.

TABLE 1.6 shows an extract from a hazard catalogue with identified hazards and related effects (descriptions of cause-effects relationships are excluded). Such a list of mixed hazards is usually not enough in order to represent an overall evaluation of the plant risk.

To estimate risks of various hazards, additional information is required on the hazards frequencies and associated consequences. TABLE 1.7 indicates a possible relationship between various types of consequences (e.g., injury, damage, loss of image) and their classification by using an appropriate scale of magnitude (e.g., insignificant, catastrophic).[19]

Frequencies can be similarly classified. TABLE 1.8 shows part of a classification scheme.

TABLE 1.6. Extract from the hazard catalogue.

No.	Hazard	Effects
8	No emergency power supply	- loss of computer-data ≤1 day - computer standstill - work-delay - delivery-delay
14	Fire-extinguishers hard to find	- fire-fighting delayed - fast spreading of smoke/fire - production reduced - delivery-delay

TABLE 1.7. Examples for categories of consequences.

Types of Losses	Insignificant Category IV	Catastrophic Category I
1. Injury	none	death, total disability
2. Damage (SFr)	1000	> 1 million
3. Production or delivery standstill	restricted production	> 6 weeks
4. Computer-breakdown	none	> 4 days
5. Loss of image	none/minor	total, national/international

TABLE 1.8. Categories of frequency.

Categories	Frequent Category A	Unlikely Category E
frequency	once a week	once a decade

Having all relevant information about hazards i, ($i = 1, 2, ...,$ n) as given in TABLE 1.6 and TABLE 1.7, it is easy to enter them in a ZHA risk matrix. *Figure 1.6* shows the risk profile of the given plant by considering a number of n = 33 hazards.

The *protection level*, usually determined by the company's management, separates the acceptable (those below the protection level) from the unacceptable risks (those above the protection level) and reflects the company's safety philosophy. It takes into account the risk sensitivity of the society, plant owners, insurance programs, etc. Usually risks above this level are regarded as relevant to ZHA. The expert team is confronted with the following problems:

a) Because individual risks can only be placed in one field of the risk profile matrix, it is impossible to distinguish, from each other, those risks, which are in the same matrix field.

⇑ Frequency

A frequent				
B moderate			{21, 22, 25}	{20}
C occasional		{32}	{5, 19, 26, 30}	{12, 15, 17, 31}
D remote		protection level	{3, 4, 6, 8, 23, 24, 33}	{1, 2, 11, 13, 14, 16, 18, 27, 29}
E unlikely			{7}	{9, 10, 28}
F impossible				
	IV negligible	III marginal	II critical	I catastrophic

Consequence ⇒

Figure 1.6. Risk profile matrix, where {21, 22, 25}, etc., are the specific hazard in accordance with the hazard catalogue).

b) Using ZHA as a tool for risk management causes problems, because it is not possible to calculate the *overall risk* of a given plant. Therefore, the risks of one plant cannot be compared with those of an other plant.
c) Although ZHA tries to avoid numbers, an analyst has to deal with numerical suggestions about frequencies and consequences.
d) Non-generally accepted rules do exist. All results depend on the process of categorizing frequency and consequences of various potential hazards.

1.6.2. EXTENSIONS OF THE "ZHA"- METHOD BY USING FUZZY LOGIC

The designation of categories of consequences and frequencies used in the ZHA is done by verbal expressions, which can also be interpreted as *linguistic statements* (the field of fuzzy-logic).[20] For this, the integration of a classical method of risk analysis with this extended approach (i.e., fuzzy sets) helps building up an improved tool for assisting the decision making process.

In terms of fuzzy sets and systems, in order to avoid using a numerical value to characterize consequences and frequencies, a verbal description defined by a linguistic variable A is introduced. For example the linguistic variable "marginal damage" cannot be specified exactly, but it suggests, that its consequences lie within approximate known boundaries.

The membership-function μ_A describes mathematically the connection between a linguistic statement A (e.g., "catastrophic") and the conventions it stands for. By definition the degree of membership lies between [0; 1]. In classical set-theory a defined set belongs unequivocally to a specified category. A classification between two categories is not possible and therefore membership is always one or zero. The fuzzy-set-theory allows partial membership expressed by degrees of membership. As such, it can explore uncertainties in the categorization process.

Here we make use of the logical combinations introduced by Zadeh (1965)[21] for the fuzzy-logic operators:

- intersection: $C = A \cap B$

$$\mu_C(v) = \min\left|\mu_A(v);\mu_A(v)\right| \tag{1.7}$$

- union: $C = A \cup B$

$$\mu_C(v) = \max\left|\mu_A(v);\mu_A(v)\right| \tag{1.8}$$

where V indicates the universe of discourse.

For a finite fuzzy-set, the so called cardinality $|A|$ is defined as

$$|A| = \sum_{v \in V} \mu_A(v) \tag{1.9}$$

Additionally the relative cardinality is defined as

$$\|A\| = \frac{|A|}{|V|} \tag{1.10}$$

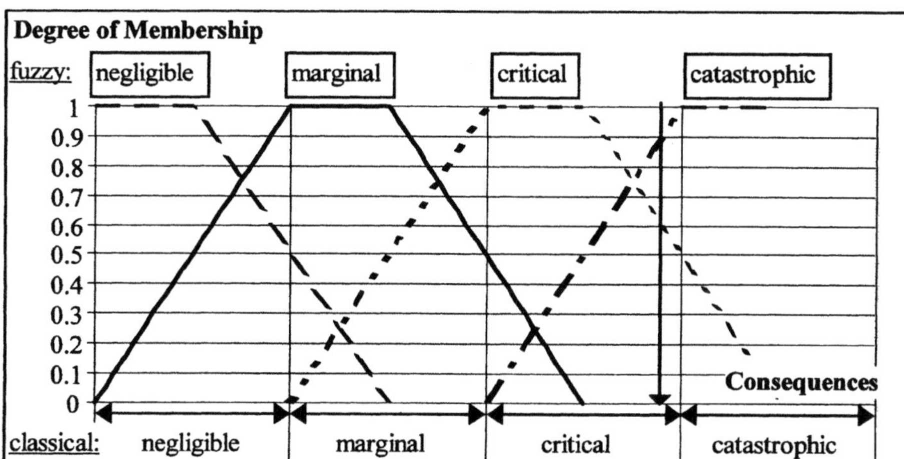

Figure 1.7. Degree of membership in classical, respectively fuzzy-set-theory.

where $|V|$ describes the number of elements v within V, because $\mu_A(v) = 1.0$.

Figure 1.7 shows the different approaches of classical and fuzzy-set theory related to the classification of consequences within ZHA. In the classical approach there are four identical columns all with the same degree of membership of one. Only "either/or" statements are possible here. For example: the consequence of an event would either be catastrophic or not catastrophic. Fuzzy-logic allows a more sophisticated description of the same event: the consequence of an event would be very close by catastrophic, but also near critical, but far removed from marginal and negligible. An equivalent fuzzy-logic set of this statement could be (see vertical arrow in *Figure 1.7*):

With *consequence* = {0.0/negligible; 0.0/marginal; 0.6/critical; 0.9/catastrophic}.

1.6.3. RISK ASSESSMENT AND HAZARD PRIORITIZATION WITH FUZZY-LOGIC

To transform individual hazards (e.g., *hazard number 8*) listed in the hazard catalogue of TABLE 1.6, into single risk values by means of fuzzy-logic, seven steps are required (PART I). Further three steps are needed to assess the overall plant risk[22] (PART II). The procedure is shown in *Figure 1.8*.

STEP 0:		Identification of hazards and associated risks, e.g., by means of ZHA
PART I:		**Transformation procedure from individual hazard to single risk**
Step 1	:	Construction of single risk membership functions for:
Step 1.1	:	- Consequences
Step 1.2	:	- Frequencies
Step 2	:	Construction of a acceptable risks' diagram
Step 3	:	Definition and agreement of a risk decision criteria set
Step 4	:	Calculation of individual decision function
Step 5	:	Calculation of an overall decision function
Step 6	:	Calculation of the linguistic assessment of a single risk
Step 7	:	Repeat STEPS 3 to 6 until each individual hazard of the hazard catalogue is transformed by using fuzzy representations.
PART II:		**Assessment procedure of the overall plant risk**
Step 8	:	Calculation of individual decision functions
Step 9	:	Calculation of the overall decision function of all single risks
Step 10	:	Linguistic assessment of the overall plant risk.

Figure 1.8. Procedural steps from individual hazards to overall plant risk (i.e., catalogue number of an individual hazard).

Assessment Procedure for Single Risks

As mentioned, frequency and consequences are elements of risk. In this context, these elements are looked at as fuzzy-sets and fuzzy-subsets, respectively (see TABLE 1.9).

TABLE 1.9. Classification of risk-elements.

Risk element	**G_i**	**Sets S_{ik}**
Consequence	G_1	{negligible, marginal, critical, catastrophic}
Frequency	G_2	{frequent, moderate, occasional, remote, unlikely, impossible}

Index i represents the risk-element, with $i = 1, 2$;
Index k represents the set of consequences, with $k = 1,..., 4$, *or the set of frequency set:* $k = 1,..., 6$.

Consequences. To develop membership-functions for a set of consequences, the categories need to be defined first, according to some injury, financial loss or environmental damage etc. Secondly, if using the classification scheme of the ZHA, it has to be determined where full membership of the categories "negligible" and "catastrophic" ends and starts, respectively. The next step is to define areas of full membership of the categories "marginal" and "critical", followed by the definition of overlapping areas of the different fuzzy-sets. The latter is purely based on common sense, e.g., it makes no sense to have the sets "catastrophic" and "negligible" overlap. Thus, consequence membership-functions have been defined as shown in *Figure 1.9*.

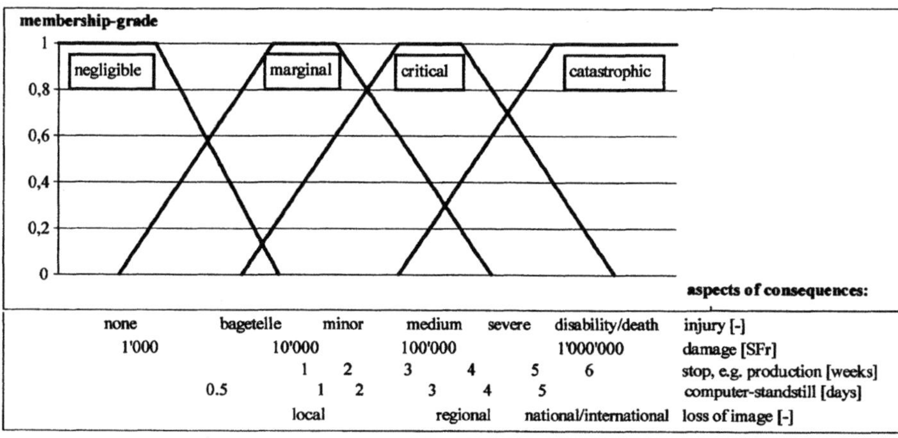

Figure 1.9. Defined consequence membership-functions

Frequencies. The inconsistency about judgment and meaning of verbally described frequencies makes it hard to define an appropriate scaling for these functions as required by STEP 1.2 (see *Figure 1.8*). When such classifications are made, there is a striking difference - up to two orders of magnitude - in judgment between the chemical industry and other industrial sectors. Synthesis based on the qualitative statements of Zogg[15] (1987) and Kröger[14] (1993) have been suggested and shown in *Figure 1.10*.

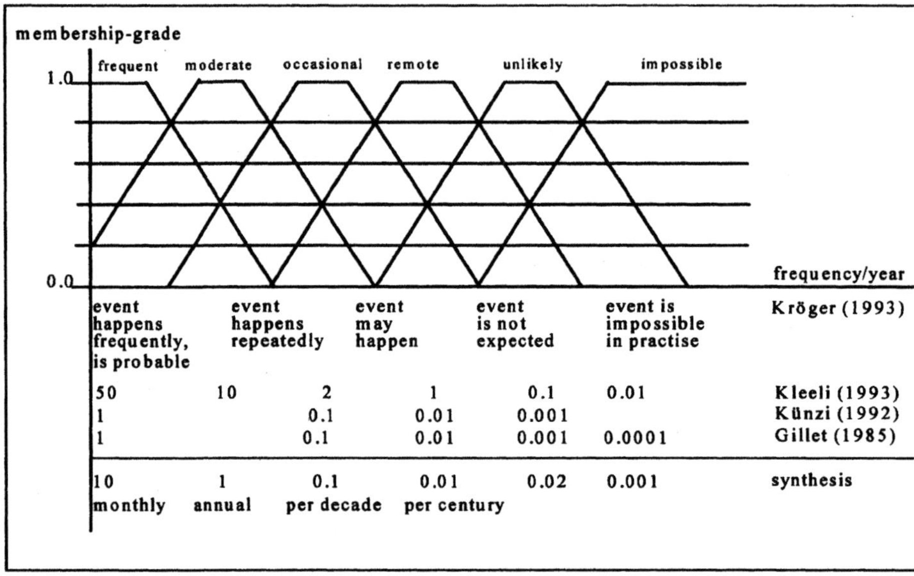

Figure 1.10. Defined frequency membership-functions.

A conservative approach in risk analysis makes it common that only input-data which represent highest risk-potentials are used to determine the category of a given hazard. TABLE 1.10 shows a detailed example of input data, which has been handled according to fuzzy rules (hazard number 1 in the risk profile matrix; see *Figure 1.6*). This hazard will be used in the subsequent steps in order to demonstrate the method.

TABLE 1.10. Example for input-data (hazard number 1) and the fuzzy output sets.

Risk Element	G_i	**Input Data**
Consequence	G_1	- loss of spare-part stock, value > 1 million SFr. - delivery interrupted for 3 to 4 weeks
Frequency	G_2	- once a century
Transformation		
Consequence:	G_1	{0.2/critical; 1.0/catastrophic}
Frequency:	G_2	{0.8/remote; 0.8/unlikely}

Risk Preference Diagram. In order to apply fuzzy-logic to the ZHA, risk has been defined as a linguistic set, i.e., *Risk = {very small, small, decreased, medium, increased, high}*. The above risk attributes are given preference functions (see TABLE 1.11) and are measured according to degrees of preference *v*. It has to be noted that the abscissa of the diagram in *Figure 1.11* merely indicates the relative position of risk and has no direct bearing on the overall plant risks.

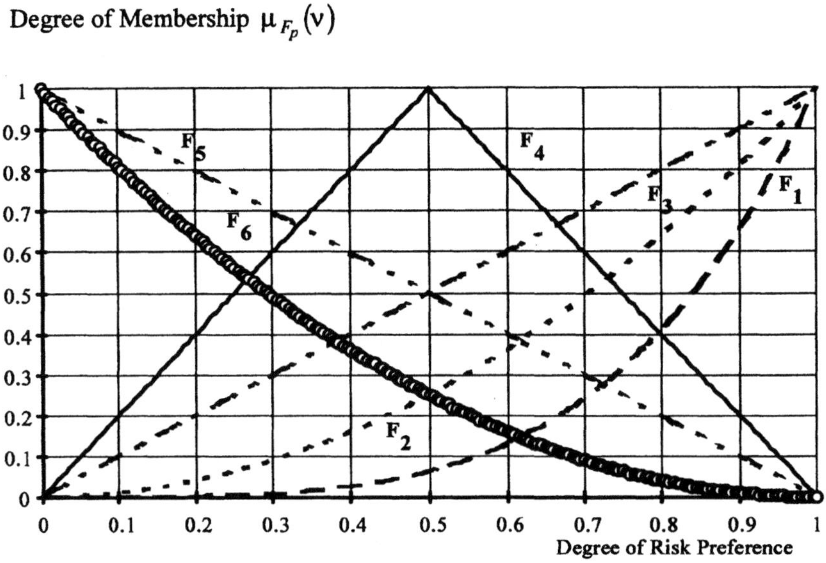

Figure 1.11. Risk preference diagram with risk-preference functions F_p.

The lower end of the abscissa (v = 0) represents "total risk" and the upper end represents "zero risk". "Total risk" statement (v = 0) does not correspondent to the risk classification F_1: "very small" and therefore the related degree of membership is 0.0. Shifting v from v = 0 to V = 1, the degree of membership increases. The diagram in *Figure 1.11* also outlines that F_4 ("medium") is well fitted if the relative position of risk is in the middle of "total risk" and "zero risk" (v = 0.5). Similar logic is be applied to all F_p (p = 1, 2, ..., 6).

Risk-preference functions (Figure 1.11) were suggested by Zadeh (1975)[23] for representing linguistic expressions as equations. The linguistic variables of the symmetrical functions F_3 and F_5, represent decreased and increased risk statements respectively. All verbal descriptions indicating an increase in risk are amplifications of F_3. This can be calculated by squaring, and the function F_4 "medium" can be understood as the average of F_3 and F_5. TABLE 1.11 gives the mathematical description of these risk-preference functions.

Risk Decision Criteria Set. Up to this point the elements of risk (consequences and frequency) have been determined, but they still need to be related to the attributes of risk. Usually this is done by applying *"if - then"* rules, such as:

[Rule d_{12}: If consequence of a hazard is "critical" and frequency is "remote", then the resulting risk is called "medium"].

TABLE 1.11. Mathematical description of F_p (v: degree of risk-preference).

Degree of Risk Preference	F_p	Mathematical Expression
very small	F_1	v^4
small	F_2	v^2
decreased	F_3	v
medium	F_4	$2v$ if $v<0.5$; $2(1-v)$ if $v \geq 0.5$
increased	F_5	$1-v$
high	F_6	$(1 - v)^2$

"If-then" rules can also be created by using pre-defined protection level. Thus, it is easy to create the following type of the statements:

["If a square box in the risk matrix touches the boundary of the protection level with one of its sides at the bottom, then it represents a 'medium' risk!"]

⇑ Frequency

A frequent	18 increased	21 high	23 high	24 high
B moderate	14 medium	17 increased	20 high	22 high
C occasional	10 decreased	13 medium	16 increased	19 high
D remote	6 small	9 decreased	12 medium	15 increased
E unlikely	3 very small	5 small	8 decreased	11 medium
F impossible	1 very small	2 very small	4 small	7 decreased
	IV negligible	III marginal	II critical	I catastrophic

Consequence ⇒

Figure 1.12. Classified risks in a matrix separating acceptable form unacceptable risks.

In the decision criteria set the classes S_{ik} of the risk parameter G_i (consequence: 4 classes; frequency: 6 classes) are combined leading to a number of 24 "If-then" rules. The degree of membership A(m) of an analyzed risk is determined by the union of fuzzy risk elements. This is expressed as:

$$A(m) = \min\left[\mu_{S_{1k}}, \mu_{S_{2k}}\right]$$

where

A(m) : Degree of membership of the risk-category R_m based on decision criteria d_m
$\mu_{S_{ik}}$: Degree of membership of the category *k* according to risk-element G_i.

As before, the decision-criterion d_{12} serves only as an example. Square box number 12 in the risk profile matrix is defined by the risk elements "critical" and "remote" (see *Figure 1.12*). Using the input data of hazard Number 1 (see TABLE 1.10) and their fuzzy transformation, the degree of membership A(12) can be expressed as:

$$\begin{aligned} A(12) &= \min\left[\mu_{S_{13}}(\text{critical}), \mu_{S_{24}}(\text{remote})\right] \\ &= \min\left[0.2, 0.8\right] = 0.2 \end{aligned} \tag{1.11}$$

and the associated risk is R_{12} = "medium".

The value 0.2 is the degree of membership of the analyzed individual hazard number 1, which corresponds to the fuzzy subset number 12 with the linguistic description "medium" risk. *Figure 1.13* shows the fuzzy risk profile matrix for hazard *i*, and the associated degree of memberships A(m). The scale on the axis relates directly to the fuzzy- modified input data in TABLE 1.10. The matrix in *Figure 1.13* shows the degree of spreading risk for hazard number 1: A(8) = A(12) = 0.2, A(11) = A(15) = 0.8, A(else) = 0.0.

⇑ frequency

0,0/A	A(18) min [0.0; 0.0]	A(21) min [0.0; 0.0]	A(23) min [0.2; 0.0]	A(24) min[1.0; 0.0]
0,0/B	A(14) min [0.0; 0,0]	A(17) min [0.0; 0,0]	A(20) min [0.2; 0.0]	A(22) min[1.0; 0.0]
0,0/C	A(10) min [0.0; 0,0]	A(13) min [0.0; 0,0]	A(16) min [0.2; 0.0]	A(19) min[1.0; 0.0]
0,8/D	A(6) min [0.0; 0,8]	A(9) min [0.0; 0,8]	**A(12) min [0.2; 0.8]**	**A(15) min[1.0; 0.8]**
0,8/E	A(3) min [0.0; 0,8]	A(5) min [0.0; 0,8]	**A(8) min [0.2; 0.8]**	**A(11) min[1.0; 0.8]**
0,0/F	A(1) min [0.0; 0,0]	A(2) min [0.0; 0,0]	A(4) min [0.2; 0.0]	A(7) min [1.0; 0.0]
7	**0,0/IV**	**0,0/III**	**0,2/II**	**1,0/I**

consequence ⇒

Figure 1.13. Fuzzy risk profile matrix for hazard number 1, with its degree of membership (Box: Degree of membership > 0).

The procedure has to be repeated with all individual hazards *i* (*i* =, 1, 2, ... , n) and, as a result, each hazard is described by a risk matrix of its own.

Individual risk decision function. The next task is to transform the degree of memberships of the risks into one single integrated linguistic statement. This can be achieved by applying the following transformation:

$$H(m,\nu) = \min \{1, [1-(A(m)-F_p(\nu))]\} \qquad (1.12)$$

where

$H(m,\nu)$	=	degree of membership (individual decision-function)
m	=	field number, decision-criterion d_m, respectively
$F_p(\nu)$	=	risk-preference function (see TABLE 1.11)
p	=	index of the risk-preference function, determined by R_m (or risk field)
ν	=	degree of risk-preference *(ν = [0; 1])*.

An individual decision-function $H(m,\nu)$ exists for every decision-criterion d_m or risk-field. As an example, the individual decision-function $H(11,\nu)$ is calculated using the input-data of the case introduced earlier (hazard number 1). For calculation purposes the interval ν = [0, 1] is subdivided into eleven "discrete" parts; V:= {ν} = {0.0; 0.1; 0.2; ... ; 0.9; 1.0}. The risk-category R_m and the index p of the risk-preference function F_p are given automatically by the index m of the decision-criterion:

m = 11 $\Rightarrow R_{11}$ = "medium" $\Rightarrow$ p = 4, which implies that:

$$H(11,\nu) = \min \{1, [1-(A(11)-F_4(\nu))]\}$$

$$H(11,0.0) = \min \{1, [1-0.8+0.0]\} = 0.2$$

$$H(11,0.1) = \min \{1, [1-0.8+0.2]\} = 0.4$$

...

$$H(11,0.9) = \min \{1, [1-0.8+0.2]\} = 0.4$$

$$H(11,1.0) = \min \{1, [1-0.8+0.0]\} = 0.2$$

The decision-functions $H(m=11,\nu)$ for all degrees of membership

$$A(m) = \min [\mu_{S_{1k}}, \mu_{S_{2k}}] \neq 0$$

are shown in *Figure 1.14*.

The next step (STEP 5 in *Figure 1.8*) summarizes all individual decision-functions $H(m,\nu)$ to a fuzzy-subset or decision-function *D(ν)*, respectively. This is done by the intersection *D* of all individual decision functions:

$$D = H_1 \cap H_2 \cap H_3 \cap H_4 \cap \ldots \cap H_m \qquad (1.13)$$

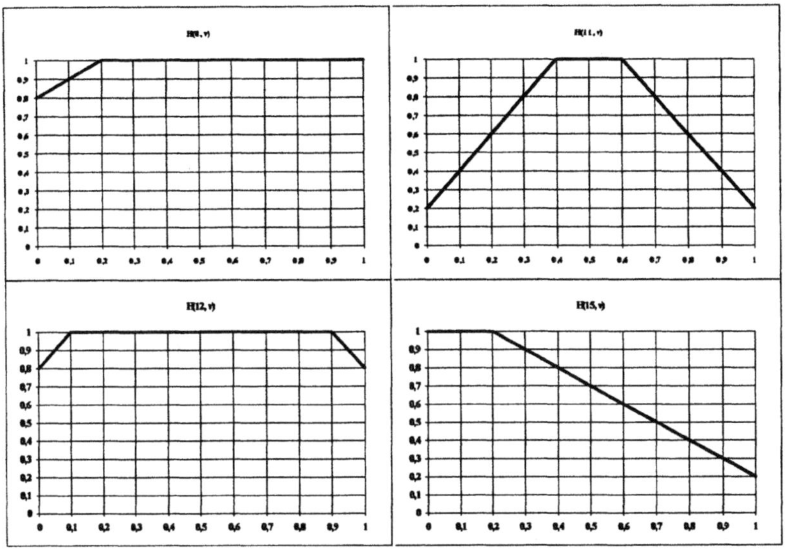

Figure 1.14. Decision functions $H(m,\nu)$ for all $A(m) \neq 0$

or, according to fuzzy-logic $D(\nu) = \min\left[H(m,\nu)\right] = \mu_{D(\nu)}$, where $\nu = [0.0, \ldots 1.0]$ for all *m*. *D(ν)* can be used to determine the linguistic description for a single risk. To that effect *D(ν)* is compared with given risk-preference functions F_p. The description looked for is then given by F_p with the smallest overall difference to *D(ν)*. Mathematically this process is described with the Hamming-metric equation[24]:

$$M_p := d(D, F_p) = \left|\mu_D(\nu) - \mu_{F_p}(\nu)\right| \tag{1.14}$$

for all *p* and $\nu(\mu_{F_p}(\nu)$, see *Figure 1.11*).

When differences $d(D, F_p)$ are regarded as fuzzy-sets M_p, the degree of membership of *D(ν)* to a linguistic risk can be calculated. The relative cardinality $\left\|M_p\right\|$ of M_p (see *Eq. 1.10*) is given by the following equation:

$$\left\|M_p\right\| = \frac{\left|M_p\right|}{|V|} = \frac{\sum\left|\mu_D(\nu) - \mu_{F_p}(\nu)\right|}{|V|} \tag{1.15}$$

To standardize the approach a fuzzy-logical set-complement is created:

$$f_p := 1 - \left\|M_p\right\| \tag{1.16}$$

This new value can be understood as the degree of membership of a linguistic risk-description where the single risk is given by the maximal value of f_p:

$$\max\,[f_p] \text{ for all } p \rightarrow \text{linguistic risk representation.} \tag{1.17}$$

As an example the fuzzy-set M_5 is calculated, based on the risk-function $D(v)$as related to hazard number 1. The index '5' indicates the comparison between the risk-function $D(v)$ and the risk-preference function F_5 which stands for "increased risk". TABLE 1.12 shows the computation of the nominator of equation (1.15).

TABLE 1.12. Computation of the counter of equation (1.15) related to hazard number 1.

No	v	$\mu_{D(v)}$	μ_{F_5}	$\lvert\mu_{D(v)} - \mu_{F_5}\rvert$
1	0.0	0.2	1.0	0.8
2	0.1	0.4	0.9	0.5
3	0.2	0.6	0.8	0.2
4	0.3	0.8	0.7	0.1
5	0.4	0.8	0.6	0.2
6	0.5	0.7	0.5	0.2
7	0.6	0.6	0.4	0.2
8	0.7	0.5	0.3	0.2
9	0.8	0.4	0.2	0.2
10	0.9	0.3	0.1	0.2
11	1.0	0.2	0.0	0.2
Σ				3.0

The cardinality $|M_5|$ is standardized with the cardinality of the appraisal space V where:

$$|V| = \sum_{v=0.0}^{1.0} \mu_v(v) = 11.0$$

because $\mu_D(v) = 1.0$ (see also explanations to equation (1.7) and (1.8)). $|V|$ indicates the amount of single values used to describe the risk-preference function. For relative cardinality it follows therefore that:

$$\| M_5 \| = \frac{|M_5|}{|V|} = \frac{3.0}{11} = 0.3$$

The so called f_p (i = 1)-value is used to determine the linguistic risk, whereby i = 1 represents hazard number 1:

$$f_5(1) = 1 - \| M_5 \| = 0.7$$

This value $f_5(1)$ is the degree of membership of the risk-function D_v to the linguistic risk-description "increased". The computation is done for all six risk-preference functions:

$f_1(1)$ = 0.5 "very small"
$f_2(1)$ = 0.6 "small"
$f_3(1)$ = 0.7 "decreased"

$f_4(1)$ = 0.9 "medium"
$f_5(1)$ = 0.7 "increased"
$f_6(1)$ = 0.6 "high".

The maximum value of $f_p(1)$ gives the linguistic description of a single risk. In this example max $[f_p(1)]$ = $f_{p=4}(1)$ = 0.9 results to a *linguistic answer* to the example-scenario:

"The risk based on hazard number 1 is '*medium*' with a membership 0.9."

Considering All Individual Hazards. To illustrate the extended ZHA by using a fuzzy sets method, reference has only been made to hazard number 1. It is obvious that all hazards (in this case 33) or risks have to be summarized to give an estimation of the overall plant risk. To do this, STEP 3 to STEP 7 have to be repeated for every individual hazard (see *Figure 1.8*). The results are given in

TABLE 1.13.

TABLE 1.13. "Fuzzy" Risk Assessments (max$[f_p(i)]$: max. degree of membership of hazard *i*).

Individual Hazard *i*	**Fuzzy Risk Assessment**	
	classification	max $[f_p(i)]$
20	high	1.00
21, 12	high	0.98
13	high	0.96
15	increased	0.98
30, 27, 16, 2	increased	0.95
25, 22	increased	0.94
29, 18, 5	increased	0.92
31, 17	increased	0.91
8, 33, 4	increased	0.87, 0.82, 0.82
26, 14	increased	0.80, 0.78
9, 10, 28	medium	0.99
11	medium	0.99
3, 24, 32	medium	0.98
1	medium	0.85
6, 23	medium	0.85
19	medium	0.83
7	reduced	0.82

Assessing the Overall Risk. Performing STEPS 8 to 10 is mathematically identical to STEPS 5 to 6 (see *Figure 1.8*) but the latter are used to establish the overall plant risk level. In STEP 8 the individual decision-function $H_{tot}(i, \nu)$ for every risk needs to be determined:

$$H_{tot}(i,\nu) = \min\left\{1, \left[1 - f_p(i) + F_p(\nu)\right]\right\} \tag{1.18}$$

where

$f_p(i)$ = degree of membership of the single-risk for the hazard i
i = hazard identification number (in this paper: i = 1 to 33)
ν = degree of risk preference, range (in this paper: 11 elements)
$F_p(\nu)$ = risk-preference function; see *Figure 1.11*
p = index of the risk-preference function, determined by $f_p(i)$.

In STEP 9 all individual decision-functions are connected by fuzzy intersection to compute $D_{tot}(\nu)$ as the overall decision-function:

$$D_{tot}(\nu) = \min\left[H_{tot}(i,\nu)\right] \tag{1.19}$$

at STEP 9 $D_{tot}(\nu)$ is compared with given risk-preference functions $F_p(\nu)$:

$$f_{tot,p} = 1 - \frac{\sum_{\nu=0.0}^{1.0} \left|\mu_{D_{tot}}(\nu) - \mu_{F_{tot,p}}(\nu)\right|}{|V|} \tag{1.20}$$

for all p, $F_p(\nu)$ with the smallest overall difference to $D_{tot}(\nu)$, represents *the desired linguistic risk description.* Finally, the risk-preference function $F_p(\nu)$ with e.g., the highest value of $f_{tot,p}$, indicates the overall-risk expressed in linguistic terms.

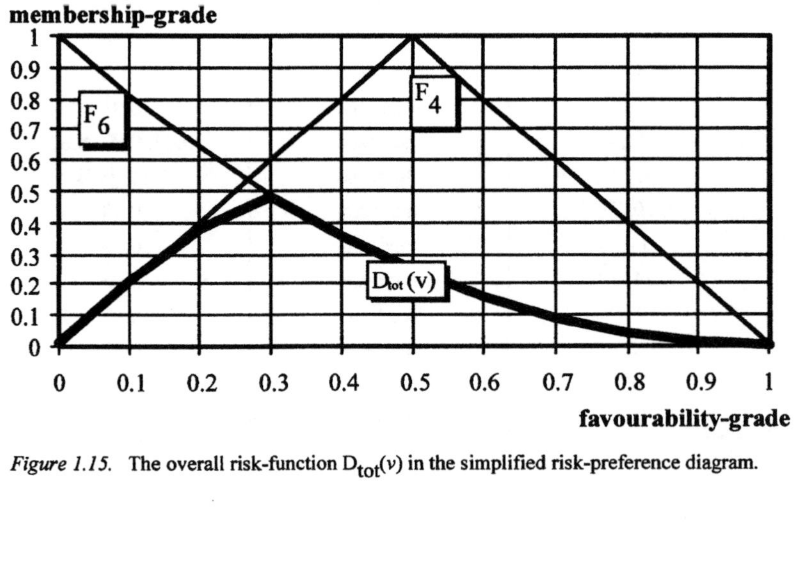

Figure 1.15. The overall risk-function $D_{tot}(\nu)$ in the simplified risk-preference diagram.

The analysis of all 33 hazards of the example leads to the overall plant risk. As illustrated in *Figure 1.15* a large section of the overall decision-function $D_{tot}(\nu)$ traces the risk-preference curve F_6. This is a geometrical representation of the following statement: ***"with the membership value of 0.83 the overall plant risk is high!"***

<u>*Remark*</u> The overall decision-function of *Figure 1.15* is given by:

$$D_{tot}(\nu) = \{0.01, 0.21, 0.38, 0.48, 0.36, 0.25, 0.16, 0.09, 0.04, 0.01, 0.00\}$$

$$\max\left[f_{tot,p}\right] = 0.83 \text{ for } p = 6 \tag{1.21}$$

This implies that the "linguistic result" will be here: "Plant risk is high".

1.6.4. PRIORITIZING HAZARD REDUCTION MEASURES

A catalogue of hazards and related risks can be used to develop risk reduction measures (see STEP 6 of *Figure 1.8*). It is obvious that the largest risks are considered first. By using fuzzy-logic as an extension of the classical ZHA, risks can be quantified and prioritized accordingly. As we can see in
TABLE 1.13, there are four individual risks classified as "high". By implementing the recommended safety measures these risks could be changed to 'medium'. On the assumption that these four risks were reduced, the analysis was repeated in order to calculate a new overall risk value $D_{tot}(\nu)$ (see STEP 3 to 10 in *Figure 1.9*). As a result $D_{tot}(\nu)$ is closer to the risk-preference function F_4, which represents "medium" risk:

$$\max\left[f_{tot,\, p=4}\right] = 0.84\,. \tag{1.22}$$

Considering that in the classical risk analysis most of the hazards are either in the 'medium' or 'increased' category (see *Figure 1.6*), from the above example it is somewhat surprising to find that the overall risk rating (after treatment by fuzzy-logic) is given as 'high'. Of course, it should be remembered that any system is only as good as its weakest element.

In the calculations, the fuzzy-logic intersections were used twice. Therefore, the 'lower' risks, which are more numerous, have no effect on the overall plant risk value. They would only be significant if they would have represented the highest risks in an analysis.

Quantifying risks by assigning degree of memberships could help the expert team conducting the analysis to settle differences of opinion in assessing risks. It is true that fuzzy-logic demands flexibility and a more differentiated and sophisticated approach than the classical ZHA.

The method described above could be used by risk management specialists to compare different plants and to help represent the effect of risk reduction measures. Fuzzy set theory in addition to ZHA can be a useful approach to solving qualitative problems in risk analysis for complex technical systems.

SUMMARY (CHAPTER 1)

The process of quantified regional risk assessment is probabilistic in nature. Hazard identification and analysis involves consideration of all situations in which the potential for harm may exist in order to identify those which are hazardous followed by a systematic analysis of the sequence of events which could transform this potential into an accident. This chapter deals with the objectives and methodological issues of the hazard identification process for regional risk assessment. The methodology involves performing a number of tasks and subsequent steps, prioritization of hazardous installations at the regional level and the use of specialized risk assessment models and tools together with a guide of implementation. Special attention is given to the use of new (hybrid) tools for estimating risks from technical systems by integrating fuzzy-logic into the Zürich Hazard Analysis (ZHA) method. Some practical examples are given.

References (Chapter 1)

[1] Belvisi, M; Boeri, G.C., Environmental Risk Assesment and Environmental Impact Assessment, ENEA, Ente per le Nuove Tecnologie L' Energia E L'Ambiente, Direzione Centrale Sicurezza Nucleare e Protezione Sanitaria. RT/DISP/93/01 (ISSN/0393-3016).
[2] Lees, F.P., Loss Prevention in the Process Industries, Hazard Identification, Assessment and Control, Vol. 2, Butterworths (Ed.), London, (1980).
[3] CONCAWE (1982). Reports are available from CONCAWE, The Hague, The Netherlands.
[4] AICHE (1965). Reports are available from AICHE (American Institute of Chemical Engineers, New York.
[5] IAEA-TECDOC-727 (1993). Manual for the Classification and Prioritization of Risks Due to Major Accidents in Process and Related Industries, IAEA, Vienna, (1993).
[6] HAZOP: Hazard and Operability Study ICI, (1974).
[7] Guide to Hazard and Operability Studies (The Hazop Study Guide), (1977 item 6) by the CISHC (Chemical Industry Safety and Health Concil), London, (1977).
[8] See also Heinrich, H.W., Industrial Accident Prevention, 4th ed. McGraw-Hill, New York, (1959).
[9] Risk Assesment for Hazardous Installations, compiled by: J.C. Consultancy Ltd., Pergamon Press, London, (1986).
[10] Aitken, A., Quantitative approach to Control and Instrumentation Systems, in F.R. Farmer (Ed.), Nuclear Reactor Safety, Academic Press, London, (1977).
[11] Nichols, R.W., John Player Lecture, The assessment and assurance of a pressure vessel reliability, Proc. Inst. of Mechanical Engineers, 189(47/75) (1975).
[12] Van de Putte, T., Purpose and framework of a safety study in the process industry, J. Hazardous Materials, 4 (1981) 232.
[13] Yager, R.R., Multicriteria decision with soft information - an application of fuzzy set and possibility theory, Fuzzy Mathematics, 2 (1982) 21-28.
[14] Kröger, W.; Mock, R., Risiko und Sicherheit, ETH lecture notes, Zurich, (1993).
[15] Zogg, H.A., "Zürich"-Gefahrenanalyse, Zurich Insurance Company, Risk Engineering, Zurich, (1987).
[16] Krause, J.P.; Mock, R.; Gheorghe, A.V., Assessment of risks from technical systems: Integrating fuzzy-logic into the Zürich hazard analysis method, Int. J. Env. and Pollution, 2 (1995).
[17] MIL-STD-882B, US Department of Defence, System Safety Program Requirements, (July 1987).
[18] Gillet, J.E., Rapid Ranking of Process Hazard , Process Engineering, 66(2) (1985) 19-22.
[19] Kleeli, A., Documents to ZHA, unpublished document deposited at Zurich Insurance Company, Risk Engineering, Zurich (1993).
[20] Schmucker, K.J., Fuzzy Sets, Natural Language Computations and Risk Analysis, in Computer Science Press, Rockville - MD, (1984) pp. 192.
[21] Zadeh, L.A., Fuzzy Sets, Information and Control, 8 (1965) 338-353.
[22] Chen, H.C.; Fang, J.H. A New Method for Prospect Appraisal, The American Association of Petroleum Geologists Bulletin, 77(1) (1993) 9-18.
[23] Zadeh, L.A., The concept of a linguistic variable and its application to approximate reasoning, Part I, Information Science, 8 (1975) 199-249.
[24] Zimmermann, H.J., Fuzzy Set Theory - and its Applications, 2nd Ed., Kluwer Academic Publishers, Boston, Dordrecht, London (1991) pp. 399.

CHAPTER 2

METHODS FOR ESTIMATING FREQUENCY AND MAGNITUDE OF ACCIDENTAL EMISSIONS

2.1. Introduction

Although there is no single authoritative source of acceptable definitions of the terminology used in Probabilistic Risk Assessment (PRA), it is widely accepted that the term risk implies the consideration of the measure of some form of loss in terms of both the likelihood and the magnitude of that loss. *This section presents the various methodologies and procedures used to calculate or estimate the unwanted consequences, effects, impacts or outcomes of severe accidents involving substances of a hazardous nature.*

These outcomes may cause death, injury, property damage or permanent damage to the environment and can be considered under the broad headings of fire, explosion, toxic effects and missiles. *The process of estimating quantitatively the effects of such outcomes is termed consequence analysis.*

Before presenting the various methodologies and procedures for the estimation and calculations of the consequences of severe accidents a number of past accidents will be described. These will be used to illustrate the type and effect of Major Hazard that can occur at installations handling hazardous materials.

2.2. Major Hazard Incidents

Major disasters are not new since natural disasters have been recorded throughout history. The potential for man-made disasters has grown with technological achievements. *In the context of this document a Major Hazard Incident has been taken to mean an accident involving one or more hazardous materials that has an impact in terms of death, injury or evacuation of people, damage to property or lasting harm to the environment.* This type of impact can be caused by an explosion, high levels of thermal radiation or by exposure to a toxic material. It is acknowledged that other (lesser) effects could be caused by ionizing radiations, suffocants, very cold substances (cryogenics) and corrosive substances, however it is not intended to consider these in the context of the guideline document.

For the reasons stated earlier, accidents at nuclear facilities are not considered in this section. Nevertheless the results of risk assessments carried out on any nuclear facilities in the study area are a basic component of the integrated risk assessment.

2.2.1. LIST OF COMPILED CASE HISTORIES

A list of compiled case histories for the period 1914-1979 can be found in *"Loss Prevention in the Process Industries" (Lees*[1]*, 1980)*. Switzerland has recently acquired an *Incident Reporting System PC-FACTS.*

> ***The Data Base PC-FACTS***
> *PC-FACTS is a accidents' events data base (including transportation of dangerous goods and processing industries and storage) developed by TNO Institute in Apeldoorn (Holland). The system has now been acquired by the Swiss Military Atomic and Chemical Laboratory ("AC-Laboratorium") and is located in Spiez (Switzerland). The information stored in PC-FACTS can be used to prepare reports on risk analysis based on accident/incidents reported on a National and international basis. Contact person is Dr. P. Roder (AC-Zentrum, 3700 Spiez, Switzerland; Tel: +41 33 55 11 11).*

Other reports and studies containing in some form lists of accidents/incidents involving leakage (gas/liquid), fire, or explosion are mentioned below:

- Davenport's list vapor cloud explosions for the period 1946-83.[2] (71 incidents, 30 where both release and deaths have been reported).

- Kletz's list[3,4] for the years 1970-80. It covers a whole range of worldwide incidents in the process industries (778 fires and explosions, involving 1196 fatalities). The repartition of deaths in the Netherlands is 22, in the UK 34, and in the USA 206. The ratios between these numbers are in line with corresponding data on general accidents in the chemical industry.

- Wiekema's list[5] (162 vapor cloud incidents in the USA for the years 1932-81, of which 62 were ignitions).

- Fawcett's list[6] (1959) is contained in a report which is internal to the UK Ministry of Works and served as a basis for a paper by Jarett[7] (1968) reporting and analyzing 74 accidental explosions up to the year 1938, and 24 explosions during World War II at munitions ships and stores.

- Cremer and Warner studies[8] (Rijnmond report, 1982). The report contains predictions for single installations contained in risk assessments that have been published. (Oxirane propylene plant at Seinehaven); AKZO chlorine plant at Botlek; UKF ammonia plant at Pernis.

- Smith and Warwick[9] (1974) of SRD (Safety and Reliability Directorate of the United Kingdom Atomic Energy Authority) have published a survey on "catastrophic vessel failures".

- Bush[10] (1975), in the USA, has published a survey on "pressure vessel reliability".

- Kellerman[11] (1982), in West Germany, has made an analysis of accidents related to nuclear technology.

Frequency and Cause of Evacuations Associated with Chemical Accidents
The frequency and cause of evacuations associated with chemical accidents have been reported by Sorenson[12] (1987) for the period from 1980 to 1984 for the U.S.A. The most frequent cause of evacuation were industrial accidents followed by train derailment. There were 300 evacuations. The average size of an evacuation was 1000 people and the largest involved 30,000 people. For every 1000 people evacuated, 8 were injured by exposure to chemicals. Injury occurred in 25% of the evacuations.

2.2.2. ACCIDENTAL ESCAPE OF TOXIC GAS

Toxic chemicals can cause harm to both animal and plant life. Effects from explosions and fires are usually confined to a relatively small area but toxic materials can be carried by wind or water over greater distances and can cause lasting damage to man and environment. Harm from toxic material is a function of the concentration of the toxic material and the duration of the exposure time. The process of calculating harm is inexact and is complicated by the fact that, as far as man is concerned, individual susceptibility varies considerably. The elderly, those in poor health, and the very young are those most at risk.

Two of the most important toxic chemicals produced in bulk are chlorine and ammonia.

Some accidental releases of toxic gas are related below:

Chlorine. This gas is produced at a rate of over 30 t per year. Therefore it is not surprising that there have been a number of accidental releases involving this material. Chlorine has also been used in warfare and some information concerning exposure to large releases has been obtained from World War I experience.

The most dramatic catastrophe involving chlorine in recent times involved a runaway train in Mexico. On August 1981 after a brake failure a train derailed, involving 28 chlorine tank cars, each of 50 t capacity. It has been estimated that over 100 t of chlorine gas was immediately released. In this incident 1000 people were known to be gassed fairly close to the source and had to be treated in hospital. Seventeen people died in this accident.

Example of a Fatal Accident due to a Chlorine Release
One of the worst industrial accidents involving chlorine occurred in December 1939 at Zarnesti in Romania. This disaster, probably caused by the rupture of a storage vessel, spilled 24 t of chlorine and killed 60 people. Many of those killed were close to the vessel but some were killed at a railway station about 250 m away. One person was killed about 800 m away - this is probably the greatest distance away from a peace-time chlorine release for a human fatality. It was fortunate that at the time the wind speed was low and therefore the rate at which the material dispersed enabled a number of people to escape to higher ground.

TABLE 2.1. List of ammonia releases (reproduced from Griffiths[15], 1982)

Year	Place	Quantity (t)	Source	Injured	Dead
1952	?	15	ST	20	15
1959	Ube, Japan	-	R(Ex)	40	11
1961	Creve Coeur, Illinois	350	Barge	-	-
1962	Brandenburg, Kentucky	-	R(Ex)	19	0
1965	Pasadena, Texas	-	R(Ex)	3	2
1968	Lievin, France	19	RdT	15	6
1969	Crete, Nebraska	90	RlyT	53	9
1969	Cumming, Iowa	-	-	-	0
1970	Belle, West Virginia	75	RlyT	30	0
1970	Blair, Nebraska	160	ST	3	0
1971	Floral, Arkansas	600	PL	0	0
1971	Texas City, Texas	-	PW	2	-
1973	Potchefstroom, South Africa	38	ST	65	18
1973	Conway/McPherson, Kansas	280	PL	2	0
1974	Hutchinson, Kansas	350	PL	4	0
1975	Texas City, Texas	50	PL	-	0
1976	Glen Ellyn, Illinois	52	RlyT	15	0
1976	Landskrona, Sweden	180	Ship to ST	-	2
1976	Enid, Oklahoma	500	PL	-	0
1976	Houston, Texas	19	RdT	200	6
1977	Mexico	-	-	102	2
1977	Cartagena, Columbia	-	-	22	30
1977	Pensacola, Florida	40	RlyT	46	2
1979	Vilvorde, Belgium	8	Rly or RdT	-	-

Legend (TABLE 2.1)

ST :	Storage tank	PL :	Pipeline
R :	Reactor	Cyl :	Cylinder
Ex :	Explosion	PW :	Pipework, including loading lines
RdT :	Road tanker	? :	Uncertain information
RlyT :	Railway tankcar	- :	No information available

Ammonia. It is a material of special importance in hazard control, because of its toxicity and its very widespread use. Ammonia is produced in similar quantities to that of chlorine but is much less toxic. It is transported by all methods, in bulk as well as in cylinders and drums. Nevertheless, there have been a number of accidents which have resulted in fatalities (see TABLE 2.1).

There are several examples of accidental releases of ammonia from pressurized containers in which the resulting gas cloud was denser-than-air, in spite of its low molecular weight relative to air.[13, 14] The production of dense gas mixture from ammonia releases has been reviewed by Griffiths[15] (1982).

Example of a Fatal Accident due to a Ammonia Release
One of the worst occurred at Potchetstroom in South Africa in 1973. A pressurized storage vessel was being filled from a rail tank when the vessel failed, possibly from being overfilled. About 38 t of ammonia were released more or less instantaneously. Exposure to ammonia resulted in the deaths of 18 persons, 6 of them outside of the works boundary. Five persons who died were at a distance of between 150 and 200 m from the release point.

The combination of factors determining the density of the mixture is highly dependent on the mode of containment and the details of the mode of containment failure. The meteorological conditions and the physical properties of ammonia also have an important influence. Static storage of large quantities of ammonia is usually in fully refrigerated tanks, but some systems utilize partially refrigerated containment at about 0°C in which there is an overpressure of about 4 atm. Pressurized ambient-temperature tanks are also in use. The vapor pressure at 20°C is about 9 atm. Ammonia-air mixtures are combustible in the concentration range 15.5% to 27% by volume. The frequency of accidental release of significant quantities is about one per year.

Over the last thirty years, there has been worldwide some 225 people killed, of which over one-third in North America. From 1971-77 there were 61 transport incidents causing injury or death in the USA alone (of which 21 injured/killed in one single accident in Cartagena, Columbia in 1977).

The crash at Houston in 1976 of a road tanker carrying 19 t of anhydrous ammonia under pressure has been reported by McMullen.[16] Another accidental escape of 38 t anhydrous ammonia occurred at Potchefstroom, South Africa in 1976.[17]

Railroad accidents with anhydrous ammonia have been reported. In November 1977, 40 t of anhydrous ammonia was accidentally released from a pressurized tank car after a train derailment at Pensacola, Florida,[18] killing two people and injuring forty six. Another accident occurred in Chicago, where 78 t of anhydrous ammonia were released.[19]

There are numerous reports of incidents involving chlorine and ammonia which have caused serious damage to the environment.

Damage to Environment due to a Ammonia or Chlorine Incidents
At an incident in La Barre, Louisiana, USA in 1961, in which between 27 and 35 t of ***chlorine*** *was released, there were reports of damage to animal life over an area of approximately 15 km^2.*
At an accident near Floral, Arkansas, USA in 1971, about 600 t of ***ammonia*** *were released from a pipeline. This ammonia reached a watercourse and killed thousands of fish.*

* * *

Dioxin Release at Seveso (Italy)
Approximately 2 kg of the chemical dioxin was released which affected an area of about 17 km^2. Although no persons died directly as a result of the release a number of persons were found to be victims of chloracne. There were a large number of deaths among the animal population and many other animals were slaughtered as a protection against dioxin entering the food chain. The dioxin released proved capable of sterilizing for agricultural use about 4 km^2 of land. The effects will last for several years. A large quantity of earth was removed from other areas in an attempt to return the land to agricultural use.

* * *

The Accidental Toxic Gas Release in Bhopal (India)
Due to reasons which have not been fully explained approximately two tons of water was added to 41 t of methyl isocyanate in a storage tank. Water and methyl isocyanate can react together in an exothermic reaction. The use of a refrigeration system to deal with this eventuality had been discontinued some six months earlier. The increase in temperature resulted in an increase in pressure which burst a rupture disc fitted to the tank and gases passed along a long line to a scrubber system. This system was inadequate to pass a large volume of gas (it was designed to pass process ventilation products not the full flow from a runaway reaction) and so the gases passed untreated to a flare which, at the time of the accident, was shut down for repair. A further possible safety feature was a pressurized water spray curtain - this failed due to insufficient water pressure. A major contribution to the high death rate was that many of the nearby population were asleep at the time in very high density accommodation and poorly constructed dwellings which offered virtually no protection. A large number of animals were also killed.

2.2.3. OTHER INCIDENTS INVOLVING CHEMICALS

In the UK there was a devastating cyclohexane explosion at the Nypro plant at Flixborough[20] in 1974. Other severe accidental releases of toxic gases have taken place during the past few years. A lorry carrying propylene exploded at proximity of a campsite near San Carlos de la Rapita, in Spain. A leak developed in the tanker as it approached the campsite, and a cloud of vapor started to form. This ignited, and the tanker broke into four pieces. The cloud of released polypropylene ignited as a low-level fireball which initially killed over a hundred and fifty people, and later on additional 61 people.

An accident which caused considerable damage to the environment occurred at Seveso, Italy in 1976 (see box on previous page).

The most horrifying incident involving a toxic gas release occurred in December 1984 in Bhopal, India, in which an escape of *methyl-isocyanate* killed at least 2500 people and may have injured 200,000 more. This disaster is possibly the worst industrial accident in the world's history (see box on previous page).

2.2.4. THERMAL RADIATIONS (FIRES)

An examination of the large financial losses incurred by the chemicals industry from major disasters suggest that some 30 per cent of the loss was caused by fire, 68 per cent by explosion and two per cent from other causes. However, many of the explosions were followed by fire which made a major contribution to the explosion loss. Fire causes death in two main ways, asphyxiation or radiation burns. The former is more likely to occur when people are trapped by fire in confined spaces, the latter in the open. Far fewer deaths are caused by blast than by fire; deaths from primary blast alone are very rare.

Following release of *flammable materials* there is the possibility (apart from the explosions described below) of the material igniting and burning in a manner which can give rise to high levels of thermal radiation. Depending on the physical properties (temperature, pressure, etc.), the mode of release and the time of ignition the material can be involved in a *pool, flash (vapor) or torch fire (flare).*

TABLE 2.2. Decision table for identifying fire types

Containment Failure	Pressure Storage	Instantaneous Release	Evaporation	Pool Formation	Ignition	Fire Type
Yes	Yes	Yes	No	No	Yes	BLEVE
Yes	Yes	Yes	No	No	No	EXPLOSIVE GAS CLOUD
Yes	No	No	Yes	No	Yes	FLARE
Yes	No	No	Yes	No	No	EXPLOSIVE GAS CLOUD
Yes	No	No	No	Yes	Yes	POOL FIRE
Yes	Yes	No	No	Yes	No	EXPLOSIVE GAS CLOUD
Yes	Yes	No	No	No	Yes	JET / TORCH FIRE

Based on the following decision table it is possible to identify the fire types that may arise from an accidental liquid or gas release from a damaged containment. The containment can be a vessel in which the material is stored, or a pipe and ancillary equipment such as pumps or gas compressors. Depending on the containment failure mode, respectively the mode of release, in case of intact containment, we can have the following situation. TABLE 2.2 gives a decision representation for identifying various types of fires.

Pool Fires. Liquid spilt onto a flat surface spreads out to form a pool. If the liquid is volatile, evaporation takes place and if the liquid is flammable then the atmosphere about the pool will be in the flammable range. If ignition takes place then a fire will burn over the pool. The heat from this fire will vaporize more liquid and air will be drawn in from the sides of the pool to support combustion. The system will then consist of a solid cylinder of flame burning above the pool. The principal hazard to people is from exposure to the high levels of thermal radiation generated. Whilst some of these fires can be spectacular, because the extent of injury depends on the proximity to the fire and the time of exposure, it is unusual for large numbers of people to be seriously affected and large accidents with multiple fatalities are rare. However plant damage and losses can be severe.

Flash Fires. A flash fire occurs when a cloud of a mixture of flammable gas and air is ignited. The shape of the fire closely resembles the shape of the flammable cloud prior to ignition but it also depends upon where within the cloud ignition occurred. In many cases the cloud extends back to the original point of release and can then give rise to a torch or pool fire dependent on the mode of release. When ignition occurs, the flame front races or "flashes" through the cloud very quickly. People or property close to or within the cloud are at risk from thermal radiation effects.

Example of a LPG Flash Fire
An example of a severe flash fire occurred at Meldrim, Georgia, USA in 1959 when LPG was released from a derailed train. The LPG spread over a wide area before ignition occurred. The resultant flash fire killed 23 people.

Jet or Torch Fires. A jet or torch fire usually occurs when a high pressure release from a relatively small opening (ruptured pipe, pressure relief valve, etc.) ignites. This gives rise to a torch which can burn with flame lengths several meters long. The flame is a hazard to persons nearby but the main hazard is generally its effect on adjacent vessels which may contain flammable liquids.

Example of an Accident Involving BLEVE
Example includes the jet-fire torching from a safety valve on top of a 50 t LPG tanker which was deflected onto its own unwetted surface and caused a BLEVE in Kingman, Arizona, USA in 1973. Thirteen firemen were engulfed in the ground level fireball and died.

2.2.5. EXPLOSIONS

These can be dense-phase explosions, confined or unconfined vapor cloud explosions, boiling liquid expanding vapor explosions (BLEVE) or dust explosions. All of these can lead to blast overpressures. Other causes of less destructive explosions are large vessel rupture through internal overpressure, runaway chemical reactions or explosions resulting from contact of a hot non-volatile body such as molten iron with water.

A "BLEVE" is a boiling liquid expanding vapor explosions

A number of *BLEVE's* have occurred as a result of flame impingement - a typical scenario being the *torch fire from the pressure relief valve* on an overturned rail tank-car impinging on an adjacent tank-car.

Dense-Phase Explosions. A dense-phase explosion occurs when a liquid or solid is suddenly converted to a gaseous form. The rapid increase in volume, results in a pressure wave which radiates from the source at a velocity greater than the speed of sound in air. This pressure wave can be very destructive. One of the most destructive explosions of this type which involved an industrial (rather than a military) hazardous material occurred at Oppau in Germany in 1921. The largest explosion to occur in the UK happened at a munitions store near Burton-on-Trent in 1944, when over 2,000 tons of TNT blew up (28 people were killed). Another accident took place at a Silvertown munitions' factory in 1917, where a 53 t explosion of TNT killed sixty-nine people, mostly within 250 m of the source. Beyond 650 m the damages to houses was described as slight.

More typical of the type of accident that happens in the chemicals industry is the explosion which took place in 1968 at the Shell refinery in Pernis, Rijmond, Netherlands, and involved the formation of a hydrocarbon vapor cloud (ca. 50 t). The TNT equivalent estimated from the resulting devastation was 20 tones. Two people were killed in the industrial area. Some damage occurred as far away as 8 km.

Example of a Dense-Phase Explosion at Oppau (Germany)
A mixture of ammonium nitrate and ammonium sulfate was stored in the open in a large heap before being made into fertilizer. This mixture tended to become somewhat solid and it became the practice to break it up using explosive charges - a technique that had been carried out a large number of times. On this particular occasion it appeared that some 4500 t of the material exploded with devastating effects. The explosion killed 561 people, 4 of whom were 7 km away in the town of Mannheim. 1500 persons were injured and 75% of the buildings in the nearby town of Oppau were demolished. A total of around 1000 houses were destroyed. All buildings within a range of between 250 and 300 m were demolished and a 10 meter deep crater roughly 100 m in diameter was formed. Damage to buildings up to 45 km away was reported.

Vapor Cloud Explosions. Since the Flixborough disaster in 1974, *Unconfined Vapor Cloud Explosions* (UVCE) have received much attention in the industrial, scientific, governmental, and insurance fields. The state-of the-art up to 1978 has been treated by Gugan[21], who also included a listing of known incidents up to that time. Further listing of UVCE can be found in the papers of Strehlow[22] and Davenport.[23, 24] Of 71 incidents up to the year 1978, 72% occurred within hydrocarbon processing plants, such as chemical plants and refineries; 23% were rail, truck or pipeline accidents; and the remaining 5% occurred in other places. *The requirement for a vapor cloud explosion is a large pre-mixed cloud of flammable vapor and air within the flammable range.* The combustion processes of large vapor clouds are still not fully understood, however the effects are strongly affected by the degree of confinement encountered, the size of the cloud and the degree of turbulence experienced. An example of a vapor cloud explosion was that which occurred at Flixborough in the UK in 1974.

Example of a Vapor Cloud Explosions at Flixborough (UK)
As part of its process the plant reacted cyclohexane (a flammable material with a boiling point of 81°C) with air at a temperature of 145°C at a pressure of about 8 bars gauge in a series of reactors. Due to a fault with one of the reactors it had been taken out of service and a temporary pipe in the shape of a "dog-leg" installed in its place. Some time later this temporary pipe failed and hot liquid was released which flashed into a mixture of vapor and entrained liquid. A large vapor cloud was formed which contained approximately 50 t of cyclohexane. Ignition occurred within a minute and a massive explosion resulted. The plant and on-site buildings were destroyed and 28 plant personnel, most of them in the nearby control room, were killed. Nearby houses suffered heavy damage and some windows were broken up to 15 km away.

Boiling Liquid Expanding Vapor Explosions (BLEVE). A *BLEVE* describes the sudden rupture of a vessel containing liquefied flammable gas under pressure due to flame impingement. The pressure burst and the flashing of the liquid to vapor creates a blast wave and potential missile damage. The immediate ignition of the expanding mixture of fuel and air leads to intense combustion and the creation of a *fire-ball.* The majority of BLEVES have occurred during the transport of pressurized liquefied gases but a number have occurred at fixed installations. Most probably the worst occurred at Mexico City in 1984.

LPG-BLEVE at Mexico City
A release of gas occurred during the early morning at a large LPG distribution plant. The initiating event was possibly a leak on a pipeline bringing LPG in from a refinery. A cloud of vapor was formed and ignited. There were several violent explosions (7 or 8) and numerous smaller ones. These explosions and the fires that followed killed at least 500 people, injured more than 7000 others and about 60000 persons had to be evacuated. Out of the original 48 'bullet' type storage vessels only 4 remained on their supports. One of these weighing about 20 t was found 1200 m away. There were also 6 spherical storage vessels on the site, the 4 smallest all exploded and large fragments of them traveled at least 400 m. The two larger spheres did not explode but collapsed through their legs buckling. Virtually all housing within a 300 m radius of the plant was severely damaged. It should be noted that when the plant was originally constructed some 25 years before the accident the nearest housing was about 300 meters away. However poor quality, flimsily constructed housing had been allowed to encroach to within 100 m of the site boundary.

Dust Explosions. These explosions are a hazard whenever combustible solids of small particle size are handled. A significant number of these explosions have occurred in flour mills or in buildings used for storing or discharging grain.

> ***Dust Explosions at Westwago***
> *A particularly large explosion occurred at Westwago near New Orleans, Louisiana, USA, in 1977. Over forty silos containing corn, wheat and Soya beans were involved and 35 on-site workers killed. Most of these were in an office block which collapsed when an 80 m tall concrete tower fell on it.*

2.3. Frequency and Magnitude of Accidental Releases of Hazardous Materials

Before preparing some consequence analysis it is necessary to obtain data on the frequency and magnitude of accidental releases of hazardous materials, classified according to an official classification scheme for dangerous products (usually based on physical properties or toxicity criteria). For fixed installations involving the processing or storage of dangerous products, information is needed on the type and quantities involved in a given process or in store. Then it is necessary to define possible release scenarios and to quantify the release of gases or liquids and to calculate their dispersion into the environment. This section considers the analysis of historical data on accidental releases and the estimation of the source term.

2.3.1. ANALYSIS OF HISTORICAL DATA

For the assessment of release of toxic material, it is possible to use "standard release data" compiled from case histories, based on the analysis of historical frequency and magnitude. A standard release pattern of the form:

$$\mathrm{Log}_{10} T = a \log_{10} f + b \tag{2.1}$$

may be used, where T is the magnitude of the release in tons, f is the frequency of the release in events per annum divided by 10'000, a and b are parameters. It is generally the case that only the reports of the more serious accidental releases have been recorded in the literature, and it is necessary to consider the extent of this under-reporting before any attempts is made to establish a standard release pattern such as might be represented by above equation.

There is a method for compensating for under-reporting regarding small accidental releases (Badoux[25], 1983). The method applies the *Pareto distribution*, using a linear regression analysis to a logarithmic transformation of all releases exceeding for instance 10 tons. Extrapolation of the straight line obtained from the linear regression so as to intercept the Y-axis gives a simply derived estimate for ln(N) and hence the total number of incidents.

Wiekema's List. Wiekema[26] (1983) gathered data on 162 vapor cloud incidents in the USA over the years 1932-1981. Sixty-two of these were ignitions. The compensated total number of releases for the full set (162) was calculated as described above and approximated 8000 releases. It is obvious that the method may introduce a measure of conservatism into the predicted outcome depending on the view taken of the degree of completeness in the original set.

Davenport's List. Another best known *list of vapor cloud explosions* is that collected by Davenport[2] (1983). From this list of 71 incidents, 30 have been selected, where both release size and deaths have been reported. The time-span for these 30 incidents was 37 years and the number of events for various categories over this period provide data for a frequency/magnitude plot. TABLE 2.3 shows the initial data for both the fatalities and the release size. Here there is a ratio 3 : 1 between tons released and the number of fatalities.

TABLE 2.3. Frequency/magnitude data based upon Davenport

Fatality Magnitude	Actual Cumulative frequency	Release Magnitude	Actual Cumulative Frequency	Best Fit Cumulative Frequency	Augmented Cumulative frequency	Augmented Finite Frequency
200	0.004×10^{-4}	500	0.004×10^{-4}	0.003	0.003	0.002
100	0.004×10^{-4}	200	0.007×10^{-4}	0.006	0.0006	0.003
50	0.007×10^{-4}	100	0.007×10^{-4}	0.010	0.010	0.004
20	0.011×10^{-4}	50	0.020×10^{-4}	0.017	0.017	0.007
10	0.027×10^{-4}	20	0.039×10^{-4}	0.033	0.033	0.016
5	0.045×10^{-4}	10	0.057×1^{-4}	0.054	0.054	0.021
2	0.091×10^{-4}	5	0.081×10^{-4}	0.078	0.093	0.039
1	0.115×10^{-4}	2	0.088×10^{-4}	0.097	0.168	0.075
-	-	1	0.105×10^{-4}	0.102	0.300	0.132
-	-	0.5	0.111×10^{-4}	0.112	0.528	-
-	-	0.2	0.116×10^{-4}	0.116	-	-

The finite frequency data for *vapor cloud explosions* gives the equation:

$$\log 500 = a \log 0.002 + b$$
$$\log 2 = a \log 0.075 + b$$
from which $a = -1.52$ and $b = -1.41$

Similar equations were derived for the reports and surveys mentioned below:

The Kletz's list. The Kletz's list[3] has been compiled at ICI, and spans the years 1970-81. It covers a whole range of worldwide incidents in the process industries and a ten-year total of 778 *fires and explosions* involving 1196 fatalities has been extracted. From the Kletz's list the total number of deaths worldwide from major accidents involving five or more fatalities is around 1500 (22 in the Netherlands, 34 in the UK, and 206 in the USA).

The finite frequency data for fatalities from major accidents gives the equation:

$$\log 300 = a \log 0.025 + b$$

$\log 3 = a \log 2.0 + b$
from which $a = -1.05$ and $b = +0.79$

The Fawcett's list. This list is contained in a report of the UK Ministry of Works, and the data have been published by Jarett.[7] The list reports 74 *accidental explosions* ranging in size from 500 lb. to 168,000 lb. up to the year 1938. This was supplemented in the report by 24 further explosions during World War II, at munitions ships and stores ranging in size from 300 lb. to over 5 million lb. The estimated peacetime finite frequency has been estimated to be according TABLE 2.4.

TABLE 2.4. Estimated frequency of explosions in peace time

Release Magnitude (tons)	Estimated Frequency (p.a. / 10,000)
2000	0.004
1000	0.004
500	0.007
200	0.014
100	0.015
50	0.030
20	0.050
10	0.070
5	0.085
2	0.195
1	0.220

The finite frequency data concerning *explosions in peacetime* gives the equation:

$\log 2000 = a \log 0.080 + b$
$\log 2 = a \log 3.90 + b$
from which $a = -1.76$ and $b = +1.36$

The Cremer and Warner Studies. It is of interest to compare the three previous examples from historic records of actual events taken from the *Rijnmond Report*[8] *(1982)* with the predictions for single installations contained in risk assessments that have been published.

For *chlorine*, the finite frequency data gives the equations:

$\log 100 = a \log 0.002 + b$
$\log 2 = a \log 2.6 + b$
from which $a = -0.55$, and $b = +0.53$.

For ammonia, the finite frequency data gives the equations:

$\log 500 = a \log 0.0005 + b$
$\log 2 = a \log 3.9 + b$
from which $a = -0.62$, and $b = +0.66$.

For propylene, the finite frequency data gives the equations:

$$\log 1000 = a \log 0.0035 + b$$
$$\log 1 = a \log 2.1 + b$$

from which $a = -1.08$, and $b = +0.35$.

The resulting slopes are compared in *Figure 2.1.*

The data reported and which can be calculated from the equations suggests that the frequency of the top event may be expected to be in the range 2 to 200 times in 100 million years. Such top events are frequently expected to be the consequence of the catastrophic failure of a large containment vessel, leading to the nearly instantaneous discharge of its total contents. Catastrophic vessel failures have been the subject of a number of studies.[27, 28, 29] From these historical statistics it is possible to make a probabilistic assessment of the likely annual failure rate of a containment vessel. In 1975 Bush suggested that there is a 99 per cent confidence at the upper bound for less than one failure in 100,000 vessel years when built to ASME Code, Section 1. This value may be considered conservative for new installations.

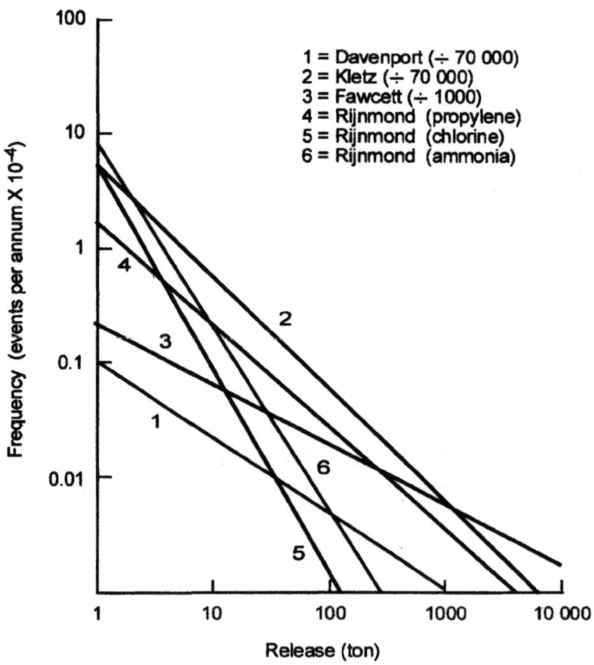

Figure 2.1. Comparison of the six slopes showing the events frequency per annum as a function of the tons released. *(Source: John Withers, Major Industrial Hazards)*

The concept of a standardized release pattern to provide the primary input to a risk estimate was explained previously. This can provide the basis for a simple estimating procedure, which uses as its prime independent variable a release pattern made up of discrete masses together with their associated release frequencies.

2.3.2. ESTIMATION METHODS FOR QUANTIFYING THE RELEASE AND DISPERSION

Estimation of the source term for the release of chemically toxic or flammable vapors is not simple. It may not even be possible to predict with certainty the possible size of a hole which may be formed by failure. This has proven to be the case especially for rail and road tankers accidents (real or simulated), but it is also the case for refineries and chemical installations. A superficial approach would be to postulate the worst possible case by assuming the total failure of a storage vessel and the immediate discharge of the entire contents into the surrounding atmosphere, but the likelihood may be so small that the risk from such an event becomes insignificant when compared to other risks, if the likelihood is more significant, the combined frequency of the event chain needed to complete the disaster may become so low as to be negligible.

The assessment of the consequences of an accidental release of a hazardous involves the sequences shown in *Figure 2.2*.

In order to perform some calculations and assess the consequences of accidental releases on men and property, considerations should be given to the choice of appropriate models and to the effects of mitigation measures:

- physical models
- effect models
- consideration of mitigating effects.

We present below a simple method for quickly estimating the dispersion, which is based on a scaling law relating the dispersion range to the mass released. Every escape or release can be considered equivalent to a discrete mass, and the dispersion of such mass can be related to the next appropriate dependent variable in the chain of events which has to be worked through. The independent variable that has been found most appropriate is the down-wind range to a given gas concentration. This in turn may relate to a lethality criterion in the case of a toxic gas, or to a flammability criterion in the case of a gas which may ignite.

For that purpose, the release of hazardous liquids can be classified broadly into two classes:

- near instantaneous release from a vessel which has suffered catastrophic failure (conveniently described by a mass release),
- slower release rate from a partial failure of the vessel, or from a pipe, or other device connected to the vessel system (described by a mass flow).

Since the development of a standard release pattern may rest upon an estimation of the sizes and frequencies of the larger possible events, *it is helpful to transform mass flows into equivalent mass release according to the scaling law which relates the downwind range to the mass released (method of Marshall[30]), when constructing and comparing release patterns with generic and plant-specific data.*

Determine Hazard Scenarios for Storage Vessel, Rail/Road-tanker

Pipe break, valve leaking, release valve damaged
Vessel failure (type: cooled, pressurized)
Pressure vessel failure mode (partial or total)
Fire, explosion, BLEEVE

Establish Release Mode of Hazardous Product

Release of toxic liquids/solids
Release of toxic vapors
Release of thermal/nuclear radiation's

Determine Mode of Action / Transmission to Target

Airborne dispersion (lighter/heavier than air)
Waterborne dispersion (water-soluble, ecotoxic)
High thermal radiation effect
Overpressure/blast effect
Projectiles/missiles

Calculate Effects on Target

Exposure to toxic vapors/liquids
Thermal radiation effects
Nuclear radiation effects
Blast overpressure effects

Take Allowance for Mitigating effects

Shelter availability
Evacuation / Rescue
First-aid / medical treatment

Assess the Consequences

Damage to people (probability of deaths, injury)
Damage to property (fire/structural damage)
Damage to environment
Assess financial losses and associated cost
(operating loss, remedial actions)

Figure 2.2. Overall approach to consequence estimations

Case 1: Instantaneous Release from a Pressure Vessel

We assume that most if not all of the vessel contents are released. The thermodynamic equation for standard release is:

$$\varnothing = 1 - \exp\left[-\frac{C_p}{\Delta H_v}(\theta_1 - \theta_2) \right] \tag{2.2}$$

where

$\varnothing$ = mass fraction vaporized
C_p = specific heat of the liquid
ΔH_v = latent heat of the liquid
θ_1 = storage temperature
θ_2 = boiling point of stored liquid

In the case of escape from a catastrophically ruptured container, however, turbulence caused by rapid boil will add spray to the flash fraction given by calculation from the formula. This may result in an ejection of the total content of the vessel as a spray of liquid droplets. The resulting turbulence will promote a cloud formation with air entrainment up to ten times the original mass.

Case 2: Slow Release from a Pressure Vessel

Where the release is not instantaneous, three kinds of release may be considered:

- non-flashing flow
- flashing liquid flow
- gaseous discharge

Case 2a: Non-flashing Flow

The following equation applies to the first case:

$$\dot{m} = Ka\sqrt{[2\rho(P_1 - P_2)]} \tag{2.3}$$

where

$\dot{m}$ = mass discharge rate (kg s^{-1})
K = discharge coefficient (-)
a = area of discharge (m^2)
ρ = liquid density (kg m^{-3})
P_1 = upstream pressure (Pa)
P_2 = downstream pressure (Pa).

Case 2a: Flashing Flow

The determination of flashing flow is problematic, but a simple approximate method has been described in the second Canvey report[31] and empirical methods have been described elsewhere.[32] Whilst empirical methods may not be applicable in all cases, it is often found in multiphase flow that the reduction factor between each of the three phases is 4. As a result this approximation gives for multiphase flow an addition of 25

per cent to the non-flashing flow for the case of flashing liquid flow, and 6 per cent for sonic gas release.
To transpose the mass release rates into equivalent cloud masses, the method of Marshall is followed.[30] The following empirical equation gives the equivalent mass of material in a cloud between flammable limits:

For Atmospheric Dispersion

$$Q_{FL} = \left(\frac{0.321}{D^{0.59}}\right)\left(\frac{m_o^{1.59}}{u^{-1.59}}\right)\left(\frac{1}{\chi_L^{0.59}} - \frac{1}{\chi_U^{0.59}}\right) \tag{2.4}$$

For Jet Dispersion

$$Q_{FL} = \left(3.40\rho_a^{1.5}\right)\left(\frac{m_o^{1.5}}{w_o^{1.5}}\right)\left(\frac{1}{\chi_L^2} - \frac{1}{\chi_U^2}\right) \tag{2.5}$$

where

- Q_{Fl} = quantity in the cloud (kg)
- D = a constant
- m = mass flow rate (kg s^{-1})
- u = wind velocity (m s^{-1})
- χ_L = concentration at lower flammability limit (kg m^{-3})
- χ_U = concentration at upper flammability limit (kg m^{-3})
- ρ_a = density of air (kg m^{-3})
- w_0 = jet velocity (m s^{-1}).

These relationship are illustrated in *Figure 2.3* and *Figure 2.4* (see Marshall[30]), which refers to a hypothetical hydrocarbon release with lower and upper flammability limits of 0.039 and 0.176 respectively.

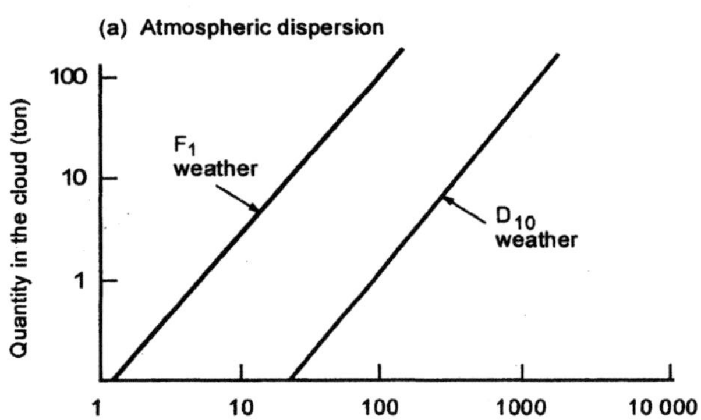

Figure 2.3. Transposition of mass release rates into equivalent cloud masses for atmospheric dispersion (after Marshall)

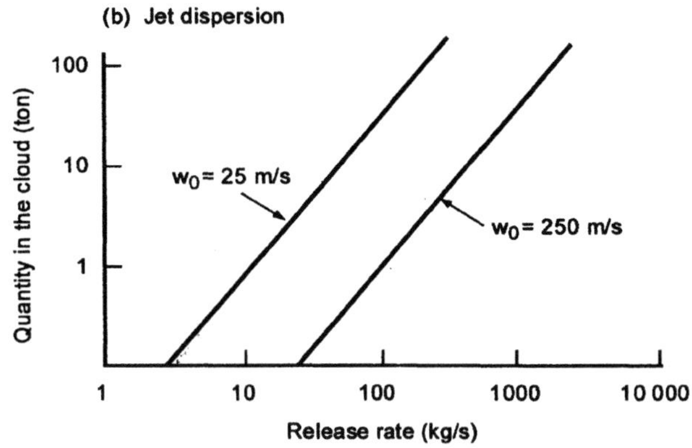

Figure 2.4. Transposition of mass release rates into equivalent cloud masses for jet dispersion (after Marshall)

At $u = 10$ m/s, neutral weather conditions (Pasquill category D):

$$Q_{FL} = 0.65m^{1.59} \tag{2.6}$$

At $u = 1$ m/s, stable weather conditions (Pasquill category F):

$$Q_{FL} = 60m^{1.59} \tag{2.7}$$

Pasquill weather classes

Convection and turbulence in the atmosphere are intrinsically very complicated processes. However by the use of stability classes, they can be included in dispersion models in a relatively simple way. Stability classes provide a scheme which relates the amount of turbulent energy in the atmosphere to general weather conditions such as wind speed or solar irradiation. Dispersion parameters (namely lateral and vertical dispersion coefficients) can then be defined depending on atmospheric stability. Pasquill[33] *has placed the stability of atmosphere into six categories (A-F) in terms of the time of year, of the day and the night-time, and the presence or absence of cloud. Category A is very unstable, category F (G) is very stable, D being neutral. Category G is sometimes added to this scale. Categories A and B can only occur at day time, and Categories E and F can only occur at night.*

We find that even with high leak rates the gas cloud will comprise up to 50 tons only, between flammable limits under the worst conditions. It has been remarked also, that for hydrocarbon clouds, the amount between flammable limits is 20 percent of the total quantity. Low-mass release rates are unlikely to impact the general population off site since:

- the relatively low release rates gives small equivalent masses and short hazard ranges
- many gases only form denser-than-air clouds under catastrophic conditions. At low release rates they drift harmlessly upwards.

According to Marshall the absolute minimum release rate of 10 kg per second is necessary for constitution of a major hazard.

The TNT concept of equivalent mass has been utilized by Davenport[2] in his survey concerning gas cloud explosions. This TNT equivalent is computed from a survey of the damage and relates to the estimated mass of explosive causing similar damage. An estimate of the yield can be made too, and this is expressed as the ratio of the TNT equivalent and the energy content of the release quantity (see TABLE 2.5).

TABLE 2.5. Estimated rates, masses, TNT equivalents, and yields for actual vapor cloud explosions (Adapted from Marshall's paper)

Location	Material	Release Rate (kg)	Cloud Mass (tons)	Equivalent TNT (tons)	Equivalent Yield (%)
Port Hudson	HC	23	55	45	7.5
Beek	HC	36	5	1	4
Pernis	HC	100	78	20	6
L. Charles	i-HC	100	9	11	10
E-St. Louis	HC	400	53	2	2
Decatur	i-HC	660	69	20-125	3-18
Flixborough	HC	1030	36	18	5

2.3.3. GAS DISPERSION MODELING

The basic concepts of atmospheric gas dispersion modeling (for neutral or positively buoyant gases) have been presented in *Integrated Regional Risk Assessment, Vol. I (Nicolet-Monnier & Gheorghe[34], 1995).*

Neutral Density Gas Dispersion Modeling

It is usual to express the variations in gas concentration in a plume under steady-state conditions as a function of wind speed, distance and time, using rectangular co-ordinates in three dimensions and using the concept of a diffusion coefficient.

The wind profile can be described by a logarithmic expression. It tends to be zero near the ground, whereas with increasing height, it grows first rapidly and then more slowly. The wind at any height is determined by the mean wind measured at 10 m above ground (reference height), times a logarithmic expression with the aerodynamic roughness length of the average obstacles along the horizontal air flow as parameter. For numerical purpose the logarithmic expression is approximated by an exponential form (see TABLE 2.6).

The down wind concentration (g/m) of a passive gas in a plume under steady-state conditions is a function of wind speed, distance and time. It is given by the Sutton and Pasquill formula:

$$C = \frac{2.8\ 10^{-3} Q}{u d h \theta} \tag{2.8}$$

where

- C = downwind concentration (g/m)
- d = downwind range (km)
- h = vertical spread (m)
- Q = mass rate of release (g/m)
- u = wind speed (m/s)
- θ = lateral spread (aperture in degrees).

TABLE 2.6. Pasquill stability (A-F) and lateral spread, θ

Weather Category	Weather Conditions	Frequency	θ-parameter (d = 100 m)	θ-parameter (d = 100 m)
A	Unstable		60	20
B	Unstable	22%	45	20
C	Unstable		30	10
D	Neutral	65%	20	10
E	Stable		15	5
F	Stable	13%	10	5

The vertical spread h varies from around 300 at 100 km to around 6 at 1 km in stable/neutral conditions, according to the following formulae:

$$\log h = 1.85 + 0.835 \log d - 0.010\ (\log d)^2$$

$$\log h = 1.48 + 0.656 \log d - 0.122\ (\log d)^2$$

The ground-level concentration at the center of a neutral density cloud or puff-type release is given by a similar type of equation, using three dispersion coefficients in the three directions (downwind, crosswind and vertical).

$$C = \frac{2Qi}{(2\pi)^{3/2}\ \delta_x \delta_y \delta_z} \tag{2.9}$$

This equation also represents the maximum concentration on the axis at a particular point. They effects of a toxic gas can be predicted using the total integrated dose at this point:

$$Dosis = \frac{Qi}{\pi\ \delta_x \delta_y u} \tag{2.10}$$

where: u = wind speed (m/s).

The values for δ_x ,δ_y ,δ_z are obtained by interpolation from the data in TABLE 2.7.

TABLE 2.7. Diffusion coefficients for neutral density gases

Pasquill Category	**x = 100 m** $\delta_x = \delta_y$	**x = 100 m** δ_z	**x = 4 km** $\delta_x = \delta_y$	**x = 4 km** δ_z
Unstable (A-C)	10	15	300	220
Neutral (D)	4	3.8	120	50
Stable (E-F)	1.3	0.8	35	7

There is some uncertainty about the relevance of the Pasquill categories to the atmospheric turbulence and diffusion factors affecting the dispersion of dense gas clouds. The ground level concentration at the center of a neutral density cloud or puff-type release is given by a similar type of equation, applying three dispersion coefficients (for downwind, crosswind and vertical). Winds speeds and direction are usually tabulated for certain regions.

The meteorological stations of the Swiss Meteorological Networks are distributed into 10 different networks types with different programs of observation. Part of the synoptic, climatological and agricultural-meteorological networks are automated and belong to the ANETZ (SMA-ANETZ automated network).

The basic processes that determines the dispersion of a passive cloud are complicated:

- the concentration distribution of the cloud is skewed due to advection by the mean wind velocity and wind direction (changing with height)
- the cloud as a whole is displaced by large horizontal eddies. Finally the cloud is subject to diffusion by small scale eddies (relative diffusion), both horizontally and vertically.

Of particular importance for a modeling approach is the interaction between windshear and vertical diffusion. With respect to gas dispersion in the atmosphere, advection has to be recognized as the main mechanism for downwind transportation, whereas convection and turbulence determines the vertical transport as well as the dilution process. In general we may expect that in stable atmospheric conditions the cloud is very skewed, because wind shear is large and vertical diffusion small. In unstable conditions the reverse is true. A proper description of wind profiles and of the characteristics of turbulence are required in terms of observable parameters. A satisfactory treatment of passive puffs is not yet available, but seems to be within reach. Scaling of the boundary layers in terms of similarity parameters forms an adequate basis for the description of dispersion in general and are discussed by Van Ulden and Holstag[35] (1985), and by Gryning[36] et al. (1987). A proper distinction between absolute and relative diffusion must be made (Csanady[37], 1973).

Dense Gas Modeling. Dense gas dispersion models have to take into account three distinct phases of the gas behavior. These are:

- initial mixing, which is calculated using source models
- gravity slumping, requiring special dense gas models,
- turbulent spreading, which may be modeled with neutral gas models

The basic concepts of dense gas dispersion modeling have been discussed in *Chapter 5*. In the next section we present a scaling law for estimating the concentration of a toxic gas cloud at a given distance from the source.

2.3.4. SHORT-CUT METHOD FOR ESTIMATING DOWNWIND TOXIC GAS CONCENTRATIONS

In the second Canvey report published by HSC an empirical equation for the radius of a vapor cloud has been given:

$$R = 30\, M^{1/3} \qquad (2.11)$$

where

R = radius of the gas cloud (km)
M = mass of the gas cloud (ton)

It is based on experimental data and makes use of the assumption that the ratio of the cloud to its height is 5 : 1. For neutral-density dispersion the equation of Pasquill (2.8) given previously, suggests that for a continuous release, the downwind concentration is directly proportional to the mass. For a puff-release the result will lie somewhere between these two. For an *average release* we obtain the following formula:

$$R = \left(1/C\right)^{0.76} \qquad (2.12)$$

whilst for low toxic gas concentration (C ranging from 1 - 5 %) the next formula applies:

$$R = k\,\left(Mass\right)^{n} \qquad (2.13)$$

where

n = 0.40 - 0.43.
k = factor depending on the physical properties of the gas and on the weather conditions

The values of the k-factor for different gases are given in TABLE 2.8. These values must be used with equation 2.11, taking a value for n equal to 0.42.

But the dependency of the concentration range upon density is likely to be a complex matter. Two important aspects of the weather which affects dispersion are speed and direction of the wind, and the stability of the atmosphere. Light wind and heavy gases allow the cloud to drift downwind without undue dispersion. Initial increase of wind speed or decrease of density increases the range, but as the wind becomes stronger and the gas gets lighter the dispersion rate increases so that the range decreases again.

TABLE 2.8. Values of the *k*-coefficient for different gases

	Values of the *k*-parameter as a function of weather Pasquill categories (C-F)				
Toxic Gas	**C5**	**D5**	**D2**	**E2**	**F2**
Ammonia	0.10	0.14	0.17	0.20	0.23
Chlorine	0.20	0.28	0.26	0.31	0.36
Propane	0.10	0.15	0.20	0.24	0.29
LNG	0.07	0.09	0.11	0.13	0.15
Butane	0.09	0.12	0.11	0.13	0.15
Hydrogen Fluoride	0.10	0.13	0.15	0.18	0.21

Influence Exerted by the Pasquill Categories

The individual risk due to drifting vapor clouds increases at night due to the weather conditions represented by the Pasquill categories E and F. The resulting dispersion is quite symmetric and in such a case the risk contours can be represented by a set of concentric circles. The risk contours for the daytime conditions cover a much smaller area and are conveniently represented by ellipses with the long axis oriented in the wind direction. For most practical purposes the combined average day and night contours may be represented by circles.

TABLE 2.9. Lethal concentration range (LC_{50}) for ammonia and chlorine

Toxic Gas	**Concentration**	**Range (R, in km)**	
		at D5	**at F2**
Ammonia (100-t release)			
LC_{50} (at 30 min.)	10,000 ppm	1.03	1.70
LC_{50} (at 10 min.)	17,000 ppm	0.69	1.10
Chlorine (50-t release)			
LC_{50} (at 30 min.)	500 ppm	1.45	1.86
LC_{50} (at 10 min.)	866 ppm	1.10	1.41

The hazard analysis for any particular gas container must consider all possible conditions of the ambient atmospheric turbulence, and the possibility of release occurring at any time of the day or night during any season of the year.[38]

When assessing the risk of toxic gas dispersion it is important to establish for a given region a table of weather probabilities distributed on the Pasquill categories A-F and also considering the period of the day (night/day time). The all weathers' probability is for a given place is equal to one and represents the sum of the weather probabilities of the different Pasquill categories for that location.

Using the scaling law some downwind ranges for chlorine and ammonia have been calculated for two different weather conditions, i.e., at D5 and F2. They correspond to a lethal concentrations (LC_{50}); some information is given in TABLE 2.9.

> *Caution is required when extrapolating to large catastrophic releases, using the above mentioned formulas. It must be emphasized that the formulas provide a simple basis for calculating the downwind range.*

2.3.5. PHYSICAL MODELS

The objectives of this section is to review the release or discharge models currently used in consequence analysis. Most accidents are the result of a hazardous material escaping from its containment. This may be from a crack or hole in a vessel or pipework, it may be from catastrophic failure of a pipe or vessel, it may be from a wrongly opened valve or it may be from an emergency relief system. *These leaks could be in the form of a gas, a liquid or a two-phase flashing liquid-gas mixture Figure 2.5.* It is essential at this stage to estimate the total amount of material involved. This may be greater or lower than the amount of material stored in any single vessel or pipework system due to interconnection with other vessels or pipework systems and also due to the relative position of the leak within the system.

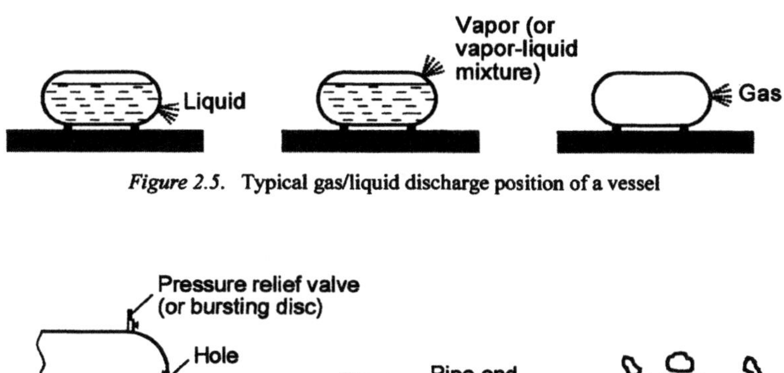

Figure 2.5. Typical gas/liquid discharge position of a vessel

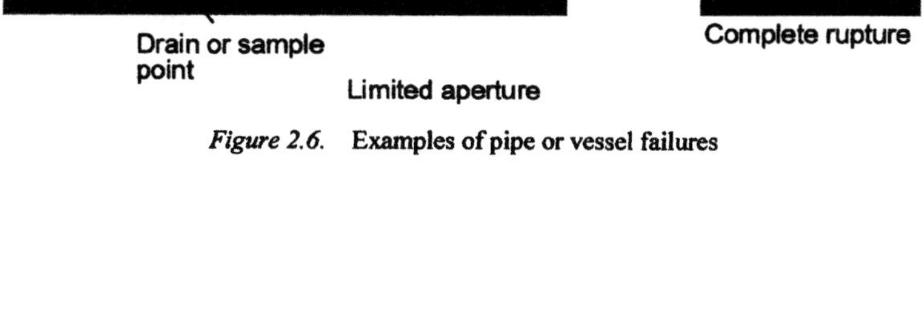

Figure 2.6. Examples of pipe or vessel failures

- *Vessels may catastrophically fail or leak from a crack, a hole or at a connection to pipelines (Figure 2.6).*
- *The behavior of the contents of the vessel depend on its initial conditions immediately before release - the main factors being the physical properties of the material and the temperature and pressure within the vessel.*

Scenarios for a Gas Release from a Container

The following cases are probable, supposing a tank containing pressurized liquefied gas at ambient temperature:

■ *Pressurized tank with a small hole in vapor space.* If the area of the whole is small as compared with that of the liquid surface, pure vapor is emitted at a rate which can be calculated making use of the pressure inside the vessel, the area of the hole and a knowledge of the thermodynamic properties of the gas in question. The gas will emerge as a jet of vapor alone (10-20%) into which air is entrained and its dilution and subsequent dispersion can be estimated. In such a case the rate of release is small.

■ *Pressurized tank with a large hole in vapor space.* This case is considered as a catastrophic failure of the vessel. If the area of the hole is large, the pressure is relieved suddenly and bulk boiling of the liquid occurs. The whole content of the tank may be ejected into air in finely fragmented form. In the same time considerable turbulence is generated and large quantities of air are entrained, which cools down while the droplets are evaporating. This is a particular mechanism of release.

■ *Pressurized tank with hole of intermediate size in vapor space.* At the present time the atmospheric dispersion of vapors escaping through a hole of intermediate size in a pressure vessel cannot be calculated, because it is not possible to define the range of hole sizes which qualify as intermediate.

■ *Pressurized vessel with hole in liquid space.* The analysis of the escape of a liquefied gas through a hole in the liquid space shows that different cases must be considered:

- if the whole is in the wall of the vessel itself, a jet consisting of unflashed liquid will be released and air will be entrained (two-phase jet).
- if the jet emerges from a long horizontal pipe, there is a possibility that the vapor will flash within the pipe.
- the same applies to a vertical pipe. Flashing depends on the critical length to diameter ratio of the pipe and on the thermodynamic properties of the gas.
- jet may emerge in such a way as to impinge on the ground or on another surface and this has an important effect on the airborne liquid fraction.

■ *Spillage of refrigerated liquid onto land.* If a tank containing refrigerated liquid should be pierced under the liquid surface, the liquid will spill into the surrounding bund and will evaporate gently, due to heat absorption from the ground or from the sun. Ammonia clouds will behave in a buoyant or perhaps passive fashion. For liquids with boiling points near or above the temperature of the ground, the influence of the wind is also important. It can be assumed that pure vapor escapes. *The SPILL code, for example, can be used for calculating the rate of evaporation.* If the escaping vapor is passive, a conventional Gaussian dispersion model can be used, whereas if the vapor is heavy a gravitational slumping model can be coupled with the atmospheric dispersion model for passive gases. If the spillage of refrigerated liquid is not confined by a bund, the boiling pool of liquid has a radius that increases with time.

■ *Spillage of refrigerated liquid onto water.* During spillage of refrigerated liquid onto or into water there is a rapid formation of a boiling pool that spreads in much the same way as it would on land. Ammonia will evaporate at its boiling point of -33 °C. The heat supply from the water and thus the evaporation rate remain constant, because the bulk of the water below remains at its ambient temperature.

- In the case of ammonia large concentration profiles closed to the ground have been observed at wind speeds higher than 12 m/s. At lower wind speeds the behavior is buoyant. For other gases a buoyant plume is formed.
- In the case of chemical reaction with water (for instance ammonia), some of the material will dissolve in the water (up to 65% for ammonia, but experimental values range from 30% to 98%).[39] If the spill takes place at a distance D below the water surface through a pipe of diameter d, most of the ammonia dissolves if $d/D \leq 10$.

■ *High velocity jet from a refrigerated vessel.* If there is a small hole in a refrigerated vessel below the liquid surface, so that the static head is high, a high velocity jet may emerge. Such a jet may well fragment into droplets. In such a case the droplet size distribution is very sensitive to the area of the orifice, its shape and its roughness. Predictions are not yet possible.

In the case of *liquefied gases stored under pressure,* the contents of the vessel which has catastrophically failed will rapidly flash off and form a vapor cloud, if unignited. If a source of ignition is found, then a large fireball will be formed. Other *materials in liquid form,* including many stored at reduced temperatures, will spill onto the ground below the vessel. The liquid will spread out to form a *pool* which will be confined in the event of the vessel being bunded (having a confining barrier around it). This pool will evaporate as a result of heat supplied from the air and the ground and form a vapor which will be dispersed in the atmosphere.

Holes and cracks will have discharge rates similar to pipe breaks of similar sizes. Depending on the position of the leak relative to the liquid level within the system, the discharge can be a *vapor* (discharge always above the liquid level), or a *liquid* (discharge always below the liquid level). However, a leak located between these two extremes can experience a range of conditions ranging from liquid phase to two-phase flow or vapor flow. Under each of these conditions the flow rate is varying as the pressure and static level within the tank changes. These effects can be summarized as follows (*Figure 2.7* and *Figure 2.8*):

A *gas/vapor discharge* results from:

- a hole in equipment (pipe, vessel, etc.) containing gas under pressure (*Figure 2.7*)
- a relief valve discharge of vapor only
- evaporation or boil-off from a liquid pool
- generation of toxic combustion products in fires.

A *two-phase discharge* results from:

- a hole in a pressurized storage vessel containing a liquid above its normal boiling point
- a relief valve discharge under certain conditions (possibly a foaming liquid, a runaway reaction or because the vessel it relieves has been moved and the valve is no longer at the top of the vessel).

Figure 2.7. Gas/vapor discharge from a hole in equipment containing gas under pressure

A *liquid discharge* results from:

- holes under liquid head in atmospheric storage tanks or other atmospheric pressure vessels or pipes (*Figure 2.8*)
- holes in vessels or pipes containing pressurized liquids below their normal boiling point.

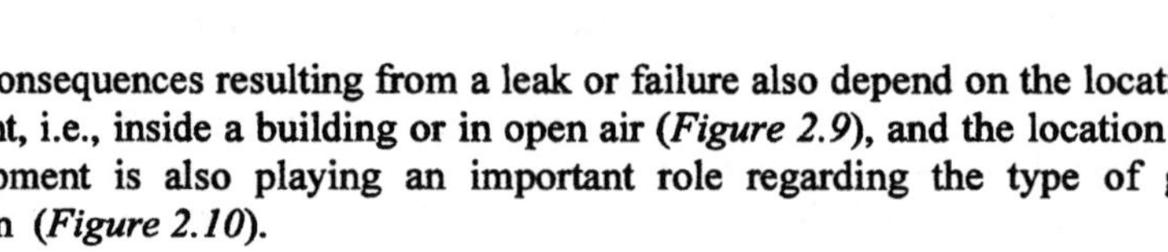

Figure 2.8. Liquid discharge from a hole in atmospheric storage tank or pressurized vessel

The consequences resulting from a leak or failure also depend on the location of the equipment, i.e., inside a building or in open air (*Figure 2.9*), and the location height of the equipment is also playing an important role regarding the type of gas/liquid dispersion (*Figure 2.10*).

Figure 2.9. Enclosure respectively equipment location influencing gas/vapor dispersion

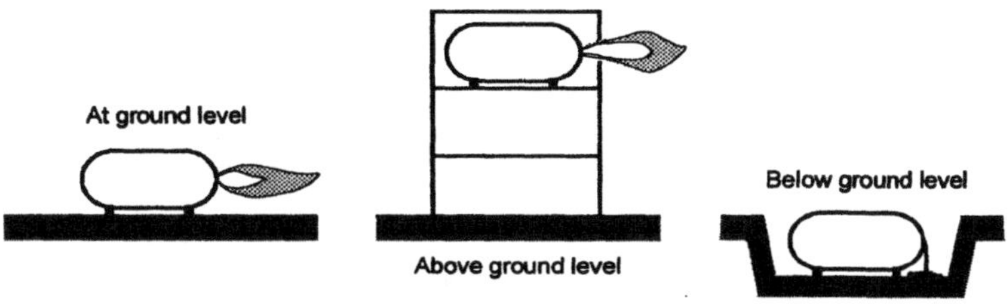

Figure 2.10. Influence of the location height of equipment on gas/liquid dispersion

Models for Calculating Release or Discharge Rates

There are a number of equations and models which deal with the release of liquids, two phase mixtures and vapors from various leak regimes. The most important are detailed in *Lees and Ang*[40] *(1989)*, which lists example base cases for a range of hole sizes, *Ramskill*[41] *(1987)*, *AIChE/CCPS*[53], *(1989)*, *Perry* and *Green*[42], *(1984)* and *CRANE Co.*[43] *(1981)*. Relief valve discharges can be determined by reference to the *AICheE/DIERS work Fauske et al.*[49] *(1986)*, and *Crozier*[44] *(1985)*.

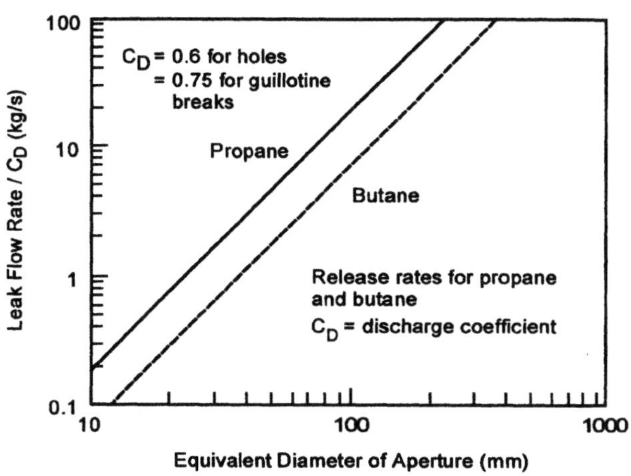

Figure 2.11. Estimated release rate for propane and butane, from apertures of different sizes

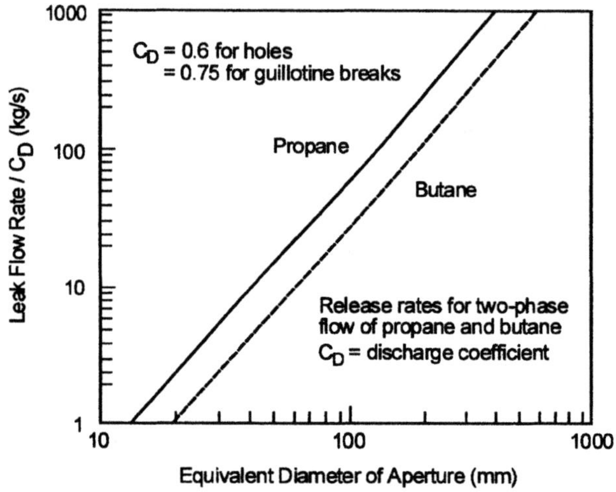

Figure 2.12. Estimated release rate for two-phase flow of propane and butane, from apertures of different sizes

Figure 2.11 and *Figure 2.12* show some curves which may be used to make an approximate estimate of the release rates of propane and butane from apertures of different sizes. These curves are derived from work carried out by the UK Safety and Reliability Directorate during the preparation of the Second Canvey Report, *Health and Safety Executive*[45] *(1981),* and show the leak flow as a function of the equivalent diameter of aperture (mm). *Figure 2.12* is to be used when dealing with a two-phase flow situation.

Discharge-Rate Calculation
There are a few computer codes which deal with discharge-rate calculations. These include the following:

DEERS	Two-phase flashing discharges (supplier: JAYCOR Inc.). See also *Klein*[46]*(1986)*
SAFIRE	AIChE, New York
PIPEPHASE	Supplier: Simulation Sciences Inc., Fullerton, California.

Computer packages for discharge rate calculations and consequence analysis
A few integrated computer packages for consequence analysis also include discharge calculation rate modules. In many cases the specific and detailed nature of the system under study may require manual calculations to be carried out. Apart from the specific references cited earlier, discharge rate calculation methods can also be found in the *TNO*[47] so-called *Yellow Book (1979)*, the World Bank Manual *Technica*[48, 51] *(1985, 1988)*, and *Safeti Technica.*

In the following, simple models are illustrated which let the user perform first hand calculations of the source term.

Gas Discharge

The calculation of the *gas flow rate through openings in a pressurized reservoir* (large vessel or large pipeline) is now described. The gas is assumed to behave as an ideal gas and the transformation is assumed to be a reversible adiabatic expansion. Two flow regimes are possible depending on the value of the critical pressure ratio:

$$r_{crit} = (p/p_a)_{crit} = [(\Gamma + 1)/2]^{\Gamma/(\Gamma-1)} \quad (2.14)$$

where

- p = absolute upstream pressure (N/m^2)
- p_a = absolute downstream pressure (N/m^2)
- Γ = gas specific heat ratio (C_p/C_v).

Depending on whether the ratio of the actual upstream and downstream pressures is lower or greater than r_{crit}, the flow regime is subsonic or sonic (choked). The gas flow is given by:

$$G_v = C_d \frac{Ap}{c} Y \quad (2.15)$$

where

- G_v = gas discharge rate (kg/s)
- C_d = discharge coefficient
- A = hole area (m^2)
- c = sonic velocity of gas at T = $(GRT/M)^{1/2}$
- T = absolute temperature in the reservoir (°K)
- M = molecular weight of gas (kg-mole)
- R = gas constant
- Y = flow factor.

The flow factor is dependent on the flow regime:

Subsonic Flow

$$Y = \left\{ \frac{2\Gamma^2}{\Gamma - 1} (p_a/p)^{2/\Gamma} \left[1 - (p_a/p)^{(\Gamma-1)/\Gamma}\right] \right\}^{1/2} \quad (2.16)$$

for $(p/p_a) \le r_{crit}$

Sonic Flow

$$Y = \Gamma \left\{ \frac{2}{\Gamma + 1} \right\}^{(\Gamma+1)/2(\Gamma-1)} \quad (2.17)$$

for ($(p/p_a) \ge r_{crit}$

Liquid Discharge

Using Bernoulli's equation, the liquid flow rate can be calculated with:

$$G_l = C_d\, A\delta \left[2(p - p_a)/\delta + 2gh\right]^{1/2} \quad (2.18)$$

where

G_l = liquid discharge rate (kg/s)
C_d = discharge coefficient
A = hole area (m^2)
δ = liquid density (kg/m^3)
p = storage pressure, absolute (N/m^2)
p_a = ambient pressure (N/m^2)
g = gravity constant (m/s^2)
h = liquid head above hole (m).

For fully turbulent flow at the discharge from small sharp edged orifices C_d assumes a value of 0.6 - 0.64.

If the liquid is superheated and if the diameter of the break is sufficiently small compared to the diameter of the pipeline or the dimensions of the tank (ratio of lengths lower than 12), the flow is assumed to remain liquid while it is escaping through the break. Immediately after, it flashes to vapor for the fraction:

$$f_v = \frac{C_{pl}\left(T_l - T_s\right)}{H_{ev}} \tag{2.19}$$

where

C_{pl} = specific heat of liquid (kJ/kg/°K)
T_l = liquid temperature (°K)
T_s = saturation temperature at atmospheric pressure (K)
H_{ev} = enthalpy of evaporation at atmospheric pressure (kJ/kg).

Non-flashing liquid is entrained in the vapor phase as aerosol. As a first approximation, it can be assumed that all the liquid is entrained if $f_v \geq 0.2$; none, of course, if $f_v = 0$; for values included in this range, a linear relationship could be considered.

Two-Phase Discharge

If a superheated liquid is discharged through a hole which has the equivalent diameter equal or greater than one tenth of the length of the pipe or the dimensions of the tank, or if the discharge is from the vapor space of a vessel containing a viscous or foamy volatile liquid, a two-phase critical flow develops. An empirical method by ***Fauske***[49] ***(1965)***, adapted by ***Cude***[50] ***(1975)***, and reported in the World Bank Manual ***"Technica"*** [51] ***(1988)***, is explained in the following.

It is assumed that the two phases form a homogeneous mixture in equilibrium; it is assumed also that the ratio of the critical pressure p_c at the throat to the upstream pressure p for water systems (0.55) can be applied to other substances.
The fraction of liquid flashing at p_c is:

$$f_v = \frac{C_{pl}\left(T_l - T_{s,c}\right)}{H_{ev,c}} \tag{2.20}$$

where

C_{pl} = specific heat of liquid (kJ/kg/°K)
T_l = liquid temperature (°K)
$T_{s,c}$ = saturation temperature at pressure p_c (°K)
$H_{ev,c}$ = enthalpy of evaporation at pressure p_c (kJ/kg).

The mean specific volume v_m of the two-phase mixture is:

$$v_m = v_g f_v + v_l (1 - f_v) \tag{2.21}$$

where

v_m = mean specific volume of mixture
v_g = specific volume of saturated vapor (m^3/kg)
v_l = specific volume of saturated liquid (m^3/kg)

The discharge rate of the mixture is:

$$G_m = C_d A_r \left[2(p - p_c)/v_m\right]^{1/2} \tag{2.22}$$

where

v_m = mean specific volume of mixture
G_m = discharge rate of the mixture (kg/s)
C_d = discharge coefficient (0.8 recommended)
A_r = effective hole area (m^2)
p = upstream pressure (N/m^2)
p_c = critical pressure (N/m^2).

The entrainment of liquid can be estimated as in the case of flashing immediately following the discharge (see above).

Discussion
Gas and liquid phase discharge calculation methods are well founded and are readily available from many standard references. However, many real releases of pressurized liquids will give rise to ***two-phase discharges*** *which must be taken into account. A simplified approximate method has been developed by Fauske and Epstein*[52] *(1986).*

Airborne Dispersion Models
One of the most important factors governing dispersion of a hazardous gas or vapor closely following release is the *density of that gas or vapor.* It is convenient therefore to classify clouds according to whether they are lighter than air, they have the same density of air or are denser than air (positively, neutral or negatively buoyant, respectively).

Evaporating Pool
Liquid spilled from a containment forms a pool which would then evaporate and become dispersed to the atmosphere. The vapor generation rate from an evaporating pool must be calculated before considering methods of estimating the dispersion of

gases and vapors. A liquefied gas can form a liquid pool if it escapes from refrigerated storage. Other liquids which boil above ambient temperatures can form slowly evaporating pools. The vaporization rate of a pool is the product of the average local vaporization rate and the pool area. However the local vaporization rate is in itself largely dependent upon the pool area. The final shape and size of the pool will be a function of the quality of material involved, the nature of the surface upon which it was spilt and whether or not the pool size is confined by a physical barrier such as a bund.

Pool vaporization rates therefore depend on a number of variables, the principal ones being:

- the spread of liquid on land or water;
- heat and mass transfer from the atmosphere; and
- heat transfer to or from the surface upon which the material has been spilt.

The way pools spread is also a very complex problem. This is very much dependent on the nature and type of surface involved and is difficult to model in a generic manner. The shear diversity and complexity of the physical phenomena which conspire to determine pool vaporization rates have made numerical solutions to the problem absolutely necessary. *Hand calculation methods* can be used *(AIChE/CCPS*[53]*, 1989)*, but accurate estimates need sophisticated computer models. The most recent and thorough of these is *GASP (Webber et al.*[54]*, 1990)*. This code makes predictions for a wide range of continuous and instantaneous liquid spills on land and water. Because the physical properties of the substances involved are so important in determining the evaporation rate, the code has been coupled to a data bank containing properties of a number of common hazardous substances. Other available computer codes include *Wu & Schroy*[55] *(1979)*, and *SPILLS (Fleischer*[56]*, 1980)*.

Estimating the impacts of LPG spills during transportation accidents
A model LPG-1 is described[57], which can assess the risks from transportation accidents related to LPG. The model's first module contains a representation of the spill mechanism following a transportation accident. It determines the amount of vapor flash-off and the size of any remaining pool based on the atmospheric conditions at the time of the accident, the amount of pre-heating of the container, and the thermodynamic properties of the material shipped. Subsequently, simplified fireball, vapor cloud explosion and BLEVE models are used to compute the extent to which pre-specified levels of shock-wave and heat radiation damage will be reached and the radius within which direct flame exposure will be experienced. During each phase of the analysis, the model considers relevant site-specific mitigating factors, such as the wind speed, air stability, combustion characteristics and the probability of encountering ignition sources.

Neutral and Positively Buoyant Clouds
Models for neutral and positively buoyant clouds are used to predict concentration and time profiles of flammable or toxic material downwind of a source. These models are almost always based on the concept of *Gaussian dispersion*. The models attempt to

determine the concentration of a hazardous gaseous material downwind of a release. The basic work is best described by *Pasquill*[58], *1974 and Gifford*[59] *(1976)*. Descriptions of neutral or positively buoyant gases and the way in which they disperse are given in *Hanna and Drivas.*[60] *(1987), Pasquill and Smith*[61], *(1983) and in the TNO Yellow Book*[47] *(1979)*.

Hand calculations to estimate the dispersion of neutral or positively buoyant clouds are still common in chemical process plant risk assessment but in other models do use computerized techniques. A good review of these models is given in *AIChE/CCPS (1987)*.

Such positively buoyant (lighter than air) clouds tend to naturally rise - in most circumstances this reduces the harm they can do, although hazardous situations can exist close to low-level releases. Unfortunately, many of the hazardous substances met in large quantities are either denser than air (e.g., LPG, or chlorine) or behave as though they are denser than air due to their storage temperature (e.g., LNG, or ammonia). Dense clouds can stay at a low level for a considerable distance downwind and can therefore pose a much greater hazard (indeed under relatively calm conditions, large releases of dense gases can travel upwind whilst under the influence of gravitational forces, such as slumping of large releases, or due to topographical features).

Negatively Buoyant Gases (dense gas dispersion)

The importance of dense gas dispersion has been recognized for some time. Attempts have been made to develop comprehensive computer models and a number of field experiments have been carried out which confirm the fact that dense gases behave in a markedly different manner with respect to neutral or buoyant gases. Probably the largest and most comprehensive field experiments were those carried out under the supervision of the UK Health and Safety Executive (HSE) at Thorney Island in the early 1980's *(McQuaid*[62]*, 1985, and McQuaid and Roebuck*[63]*, 1985)*. These were coordinated by the HSE and funded by a wide range of contributors from a number of different countries.

There are a number of mechanisms by which a *dense gas or vapor* can disperse in the atmosphere and become progressively diluted as it mixes with air. These mechanisms depend mainly on the *buoyancy and momentum* of the material involved. Momentum forces are associated with the early stages of release from pressurized equipment although gravitational forces can provide momentum following the *slumping stages* of large instantaneous releases. Whilst consideration of the momentum driven period of dispersion may satisfy relatively small releases of flammable gases which are diluted below the lower flammable limit during the momentum phase alone, *in many other situations dispersion beyond the transition to the buoyant plume dispersion must be considered. The point at which this transition occurs depends on the momentum and buoyancy forces acting on the dispersing material, although in certain situations gravity effects and collision with solid surfaces (buildings, trees, very rough ground, etc.) may become important before the momentum of the jet becomes negligible.* It is here not possible to discuss in detail the mathematics which describe this dispersion process. The solutions of the equations describing the gravity-slumping of a heavier-than-air gas cloud, the simultaneous movement in the wind and the entrainment of air into the cloud, together with heat effects, is sufficiently complex to require computer

modeling. Perhaps the most comprehensive *review of vapor cloud dispersion models* is that given by *Hanna and Divas/CCPS*[60] *(1987)*.
Publications which describe methods of calculating the dispersion of dense gas in the atmosphere are numerous.

Negatively Buoyant Gases (dense gas dispersion)
A number of codes are available, some of these deal only with instantaneous releases, others with only continuous releases, whilst there are others which are capable of dealing with both situations. At the moment, few codes can handle complex time-varying situations, although many codes are under development. These codes model the transition from a heavier-than-air cloud to a neutrally buoyant one, as the cloud dilutes and equilibrates with the temperature of the surrounding air. Therefore, they can also be used for neutrally buoyant releases, although the equations for this are generally simpler and, as stated earlier, can be, and often are, calculated by hand.

Dense gas dispersion *computer codes* which have been made available in substantial numbers are listed in TABLE 2.10.

TABLE 2.10. Computer codes for dense gas dispersion

CHARM	(Radian Corporation, USA)
DEGADIS	(US Coastguard)
HEGADAS	(SHELL)
DENZ/CRUNCH	(SRD, UK)
HASTE	(ERT, USA)
SLAB	(Lawrence-Livermore National Laboratory, USA)
SAFETI	(Technica, UK)
TRACE	(SAFER CORPORATION, USA)
DRIFT	(SRD, UK)

One of the most comprehensive is that by *Britter and McQuaid*[64] *(1987)*. Other recent publications worth referring to are listed below (see TABLE 2.11):

It must be appreciated by now that the subject of dense gas dispersion is a very specialized, technical one, and because of this it is important that calculations of the hazard ranges, due to the dispersal of dense gases, are carried out by those who have more than just a passing acquaintance with the topic. Even with the modern tendency to make codes easier and more attractive to use, caution must always be taken to ensure that the situations presented to the computer model is that which actually exists.

TABLE 2.11. Publications dealing with (dense) gas dispersion calculations

• Fryer and Kaiser (1979)	• Ermak (1985)
• Blackmore et al., (1982)	• Havens (1985)
• Britter (1982)	• Spicer (1986)
• Havens (1982)	• Journal Hazardous Materials (1987)
• Weber (1982)	• Deaves (1987)
• Bradley (1983)	• Havens (1987)
• Jagger (1983)	• Webber (1987)
• Hartwig (1984)	• Kukkonen (1988)
• Knox (1984)	• Spicer (1988)
• McQuaid (1984)	• Witlox (1988)
• Morgan (1984)	• Koopman (1989)
• Brighton (1985)	

There is no easy short-cut to carrying out dense-gas dispersion calculations, but for a few of the more common hazardous materials encountered in everyday life, *there are curves, derived from the use of modern codes, which calculate gas concentration as a function of distance and time for a range of release scenarios*. Examples of these for *flammable gases and chlorine* can be found in Chapters 8 and 14 of *Lees and Ang*[40] *(1989),* and in *Chemical Industries Association*[65] *(1987).*

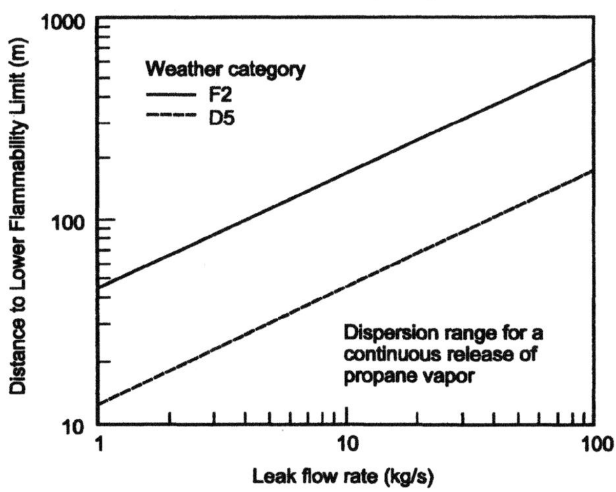

Figure 2.13. Dispersion range for a continuous release of propane

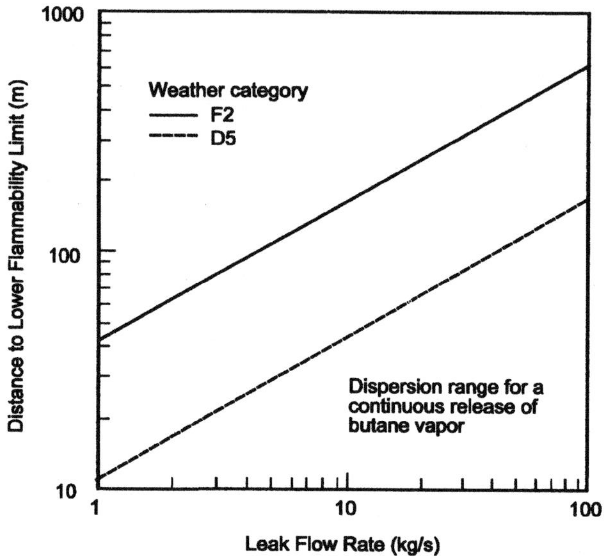

Figure 2.14. Dispersion range for a continuous release of butane

The *Figure 2.13* and *Figure 2.14* show curves for the dispersion of a continuous release of propane or butane as a function of distance to lower flammability limit against the leak flow rate for two weather stability classes (D and F) and related typical wind velocities (5 m/s and 2 m/s, respectively). *These curves were derived with the use of the SRD computer code CRUNCH.*

> ***Discussion***
> *The strength of most of the dense-gas dispersion models is their inclusion of the important mechanisms of gravity slumping, air entrainment and heat transfer processes. Their main weakness is the difficulties encountered with estimating the source term and the fact that a degree of skill is required by the user.*

Toxic Gas Risk Assessments

The use of Gaussian models for determining time-averaged concentrations in atmospheric calculations is well established. The maximum area affected by a severe toxic gas release will occur if the gas is released instantly, with neutral buoyancy, and at ground level. A simple method for determining the consequences from such a discharge has been reported by Thompson and Nightingale[66] (1988). The method consists of replacing the Gaussian cloud with an idealized hemispherical cloud with the same total mass of toxic gas and of uniform concentration.

The time-averaged concentration χ due to the instantaneous release of quantity Q of material is given by:

$$\chi = \frac{Q}{(2\pi)^{3/2}\sigma_x\sigma_y\sigma_z}\exp\left[-\tfrac{1}{2}\left(\frac{x^2}{\sigma_x^2}+\frac{y^2}{\sigma_y^2}+\frac{z^2}{\sigma_z^2}\right)\right] \tag{2.23}$$

where x, y, and z are the downwind, crosswind and vertical distances from the center of the cloud, and σ_x, σ_y, and σ_z are the respective dispersion coefficients. Q is the mass of the cloud.

The dispersion coefficients are a function of downwind distance from the source, atmospheric stability, and ground roughness. σ_x, and σ_y, are usually considered equal for instantaneous releases (i.e., radial symmetry), and σ_z is smaller. Maximum dose to persons open terrain, on the ground (z = 0) will be on the axis of the cloud (i.e., y = 0). The resulting dose will be given by:

$$D = \int_0^{\infty}\chi^n\,\mathrm{d}t = \frac{1}{\mathrm{u}}\int_{-\infty}^{+\infty}\chi^n\,\mathrm{d}x \tag{2.24}$$

where

u = wind speed
n = coefficient usually equal to 2.75 for ammonia or chlorine. (For carcinogens the dose is directly proportional to the concentration and n = 1.0).
χ = time-averaged concentration

For a person on the centerline the dose can be estimated according to:

$$D = \left[\frac{Q}{(2\pi)^{3/2}\sigma_x\sigma_y\sigma_z}\right]^n \frac{\sigma_x}{u}\left(\frac{2\pi}{n}\right)^{\frac{1}{2}} \tag{2.25}$$

This formula can be further simplified for an idealized hemispherical cloud of radius R and uniform concentration to give:

$$D = \left(\frac{Q}{\frac{2}{3}\pi R^3}\right)^n \frac{2R}{u} \tag{2.26}$$

Using the data of Hosker[67] (1974) for σ_x, and σ_y (as functions of downwind distance x and Pasquill stability category), and assuming radial symmetry of the cloud, it is possible, by setting Eq. (2.23) equal Eq. (2.24), to produce graphs of idealized cloud radius R against downwind distance x for given values of toxicity coefficient n, such that the dose to an individual from the idealized cloud would be the same as that for a Gaussian cloud. These graphs enable the dose D at distance x due to toxic release Q to be determined easily via Eq. (2.24). The individual risk may then be determined from a probit function. The exposure time for the idealized cloud is 2 R/u.

Nearly all existing methods of analyzing hazards associated with dispersing clouds of heavy gas (affected by turbulence) are based on the mean concentration, and ignore fluctuations about the mean. According to Chatwin[68] (1982), this approach is not correct, since the root mean square value of these fluctuations is not small compared to the mean. Assuming an instantaneous gas release at time t of a finite volume Q, it is possible to calculate a ***probability density function of concentration*** at position x and time t. Such functions should be particularly useful for hazard assessment of flammable gases. Estimating such probability density functions is in practice not easy, due to the lack appropriate field experiments. Another approach would be to estimate the ***mean square fluctuation of the concentration.*** This provides the simplest measure of the variability that can be expected between two gas releases. Calculation based on the mean concentration alone is not sufficient for proper hazard analysis. Approximation of the probability function by a Gaussian function near the center of the cloud is sufficient.

The hazard analysis for any particular gas container must consider all possible conditions of the ambient atmospheric turbulence, and the possibility of release occurring at any time of the day or night during any season of the year.[38]

Toxic Gas Release Taking Place Outside the Building

Being or going indoors provides considerable protection against toxic gas incidents. For residential population the mitigation provided, and hence the reduction in risk, depends upon the rate at which gas will enter the house and the behavior of the occupants once the gas cloud has passed. Davies and Purdy[69] (1986) have been reviewing the work that has been done on the rate of infiltration into houses and show how this rate, and the overall predicted risk, is affected by the behavior of the occupants.

The rate of change of indoor concentration (C_i) is equal to the effective ventilation rate (λ) times the difference between indoor (C_i) and outdoor concentrations (C_o):

$$\frac{dC_i}{dt} = \lambda\left(C_o - C_i\right) \tag{2.27}$$

Integrating with respect to time gives:

$$C_i = C_o\left(1 - e^{-\lambda(t-t_o)}\right), \text{ where } \left(t < t_c\right) \tag{2.28}$$

The gas concentration inside the building goes over a maximum value and starts decreasing exponentially once the cloud has passed:

$$C_i = C_d\ e^{-\lambda(t-t_c)} \tag{2.29}$$

where C_d is the peak indoor concentration given by:

$$C_d = C_o\left(1 - e^{-\lambda(t_c-t_o)}\right) \tag{2.30}$$

The toxic load to people, TL, for a gaseous exposure is a function of both the concentration (C) and the time of exposure (t). For a single exposure to a constant concentration the equation is:

$$TL = C^n\, t \tag{2.31}$$

For toxic irritant gases the value of n is greater than one (for chlorine[70], n = 2.75, and TL = $(3.2 \times 10^6 \text{ ppm})^{2.75} \times {}_{\text{min}}$).

Indoors, the concentration varies with time and we have the relation:

$$TL_{(indoor)} = \left[\left(C_i\right)_{av}\right]^n t, \text{ where } \left(C_i\right)_{av} \text{ is given by :}$$

$$(C_i)_{av} = \frac{C_o}{t_d}\left(t_p + \frac{e^{-\lambda t}}{\lambda}p - \frac{1}{\lambda}\right) + \frac{C_o}{\lambda\, t_d}\left(1 - e^{\lambda t}p\right)\left(1 - e^{-\lambda(t_d - t_p)}\right) \tag{2.32}$$

where

t_p = $t_c - t_o$, cloud passage time, and
t_d = $t_e - t_o$, time of exposure.

Such a model has been included in the code RISKAT (Pape and Nussey[71], 1985) and can be used to explore the protection afforded to people indoors.

The pattern of air flow through a house depends upon both the driving forces, wind and buoyancy, and the size and distribution of the openings of the house (doors, windows). The ventilation rates, λ, attributable to wind and buoyancy effects are:

$\lambda_w = A\ v$, respectively $\lambda_b = B\ |DT|^{1/2}$, where v is the wind speed and DT is the difference between the temperature inside and outside.

Different correlations (Dick[72], 1950; Warren[73] et al., 1980; Coblentz[74] 1973; Eisenbery[75], 1977) have been developed for gas infiltration into a housing on the basis of measurements made with tracer and pressurization techniques. Eisenbery's vulnerability model U.S.C.G. is used in American risk assessment studies together with the correlation of Coblentz. The latest infiltration's models are of the form:

$$\lambda = \lambda_s \left[N\Delta^{1.2} + MU_m^{2.4}\right]^{0.5} \tag{2.33}$$

where λ_s is the ventilation rate measured in standard test conditions and the constants M and N are characteristics of a particular type of house and its general environment.

For risk assessment purposes, the following equations giving the ventilation rate in closed houses do provide good estimates:

Exposed site: $\lambda = 0.87 + 0.13\ U_m$
Sheltered site: $\lambda_{(U_m<4.2\ m/s)} = 0.88$
$\lambda_{(U_m>4.2\ m/s)} = 0.22\ U_m$

where

U_m = average wind speed (m s^{-1})

The effect of wind direction and house orientation does not play an important role. The results of such correlation has been compared to field measurements:

- Swedish measurements have been reported by Kronvall[76], (1978), and
- Canadian measurements by Beech[77] (1979).

The ventilation rates found in kitchens and bathrooms/toilets are relatively high. It must be noted that is not always possible to individually switch off such ventilation in case of emergency. The ventilation rates to be used should always apply to the type of buildings for which the calculation is made. Compiled data may vary since the regulations concerning construction, isolation and ventilation rates may be different according to national standards.

The ventilation rates in occupied houses has been simulated by Dick and Thomas[78] (1951) as a function of the number of open windows:

$$\lambda_{(AL)} = 0.87 + 0.13\ U_m + 0.23\ (n + 1.4\ m) + 0.05\ (n + 1.4\ m)\ U_m \tag{2.34}$$

where

n = number of open top-vent windows
m = number of open casement windows
U_m = average wind speed (m s^{-1})

The random variation in wind direction and house orientation as well as the amount/number of windows being open has been taken into consideration. *If residents close all windows and doors, switch off the ventilation the individual risk is reduced by a factor 3 in the far field.*

Toxic Gas Release Taking Place Inside a Building
Many potential sources of accidental hazardous heavy gas releases are either located inside buildings or are themselves building-like structures. Chlorine gas is widely used for water treatment (i.e., for cooling water, drinking water, or swimming pool water preparation). It is frequently stored in drums, bottles or tanks inside buildings of many types. The presence of the building may restrict the initial mixing of a heavy gas release; it may also cause a delay before chlorine is emitted to the atmosphere. Outside the building dilution of the gas may be enhanced by the vigorous turbulence in the recirculating wake in the immediate lee of the building. If gravity effects are strong the dense gas may spread upwind and sideways. Thus these effects will influence the initial plume shape and possibly concentrations used by the dispersion models.

A simple analytical method has been provided by Brighton[79] (1986) for estimating the source terms for certain types of release of a heavy gas (chlorine, ammonia) or two-phase mixture inside a building or its wake.

2.4. Fire and Explosion

The production, transportation and storage of combustible fuels for industrial and domestic applications present a major potential hazard. Fire and explosion are indeed related to the working of most thermal power plants as well as to other transformation industries, to the production and storage of chemicals and to the transportation thereof. Fire in the process industries causes more serious accidents than explosion or toxic release, although the accidents in which the greatest loss of life and damage occur are generally caused by explosion.

Fire is therefore a serious hazard, but it is normally regarded as having a disaster potential less than explosion or toxic release. One of the worst explosion hazards, however, is usually considered to be that of unconfined vapor cloud which has drifted over a populated area and in this case the difference in the number of casualties caused by a flash fire rather than an explosion in the cloud may be relatively small. Uncontrolled large-scale burning is both life-threatening and potentially damaging to buildings and plants in the vicinity. Widespread fires are often envisaged as a major hazard in earthquake scenarios for urban areas.

In the present section we shall seek to distinguish the principal combustion phenomena which arise in hazard situations, in particular the jet flame, buoyant (pool) fire, fireball, BLEVE, and vapor cloud explosions; to characterize the hazard presented in terms of engulfment and flame spread, thermal radiation, pollutant emissions, overpressure or blast; and to present some theoretical models available for their prediction.

In the case of inflammable material it is necessary to provide an estimate of the chances of the material igniting. This may occur near the point of release or at some distance as the cloud drifts away. If the material does ignite, the ignition may be followed by an explosion. Such an explosion may occur or not. The chance of explosion and ignition is related to the mass released.

Methodology for Assessing the Risk of Fire and Explosions
The methodological approach to be followed is given below:

■ Collect information concerning the region to be assessed. This information should include the exact location of processing facilities, tank-farms and other fixed storage facilities for liquid and gas, etc., as well as type and quantities of dangerous materials stored; the tank temperature, pressure and dimensions. Known ignition sources should be recorded too. For calculation purpose the size and location of the rupture or vent should be known.

■ Collect for the region under consideration information on environmental conditions, such as air temperature, pressure, humidity, wind velocity, and stability class which are usually needed by the gas dispersion models. Use National meteorological statistics.

■ For fixed installations, collect data defining possible gas/vapor/liquid releases (type, size/quantity, toxicity, danger class, etc.), and location and size of spills.

■ Establish a list of toxic products classified according to their toxicity class, and store in a library file the physical and chemical properties of the products. The records should include: molecular weight and density at standard temperature and pressure; flammability; toxicity; flame temperature; thermal conductivity; heat of vaporization; and some other useful thermodynamic properties.

■ Determine the population density and type of housing using census data or other data available.

■ Prepare release scenarios for different classes of dangerous products and modes of release depending on storage type (atmospheric, cryogenic, pressurized).

Using a geographical information system (GIS) would allow to plot various information including population density and housing type on a grid (with a subdivision of 100 m x 100 m).

The assessment itself is taking place in two phases: (a) modeling/calculation of the physical effects, and (b) assessment of the damages to the vulnerable resources.

Phase (a): This phase, which consists in modeling and calculation of the phenomena of fire and explosion, proceeds in several steps:

- ❑ Based on a given release scenario and the type of storage (atmospheric, pressurized, cryogenic), estimate the size and type of the release (vapor/liquid phase, spill formation and evaporation).
- ❑ Calculate the gas dispersion, or the spill spreading and evaporation using an appropriate model (Gaussian plume, puff, or dense gas models) or a validated computer code (available commercially). Special models must be used for gases heavier than air (a/o., for LPG, ammonia, Chlorine, etc.), or when terrain/obstacles are to be considered for a three-dimensional dispersion simulation. Different models are usually used for long range dispersion.
- ❑ Decide when and where ignition may occur, based on the physical properties of the cargo, the gas concentration, the flammability limits, and the strength of the ignition source.
- ❑ Subsequent to ignition, either an explosion or a flash fire, torch fire or pool fire must be considered, depending on the ignition source and strength specified.

■ *Ignition:* There are three requirements for combustion, i.e., (a) fuel, (b) oxidizing agent (air), and (c) ignition source. The fuel is provided by the dispersed flammable vapor, whereas the oxidizing agent is usually provided by the oxygen in the air. Since the combustion will occur only over a certain range of fuel-air ratios, the vapor concentrations must be estimated, and for a given cell, considering the flammability limits and the presence of ignition sources, a decision can be made whether ignition could take place. Rural areas should have fewer ignition sources per unit area, whereas urban areas should have a greater concentration of ignition sources. The source strength should be greater for industrial areas (welding shops, smelters, petrochemical installations, etc.) than for residential and recreational areas.

The input data required for calculation are:

- Type of ignition source (capable of causing a fire or an explosion)
- Strength of the ignition source
- Flash point of the spilled substance
- Upper and lower flammability limits of the substance
- Concentration of the air-dispersed cargo for the time and grid cell location under consideration.

■ *Explosion:* Calculate the peak-overpressure and the dynamic impulse generated by the explosion of a flammable air-vapor mixture. The explosive yield is given by the product of the heat of combustion per unit mass and the total mass of fuel participating in the explosion, taking into consideration the flammability limits and the stoichiometry of the fuel-air mixture. The well-known scaling laws for explosion are assumed to hold.

The input data required for calculation are:

- Time of ignition
- Parametric values determining the gas/vapor concentration in space at the time of ignition

- Heat of combustion of the fuel (cargo)
- Location of the explosion epicenter
- Atmospheric temperature and pressure.

■ *Flash Fire:* Calculate the effective radiation intensity level and the effective radiation duration resulting from the flash fire. The flash fire is considered to be the rapid combustion without detonation of the fuel-air layer premixed within the flammable limits. Assume that the heat generated instantaneously by combustion is entirely radiated; thus the radiation loss as a function of time may be calculated. Assume the emissivity of the layer to be unity.

■ *Pool Fire:* Calculate the duration and magnitude of thermal radiation emitted by a burning pool of flammable air-vapor mixture. The radiation level must be calculated for any desired point in space.

The input data required for calculation are:

Flame size:

- Wind velocity
- Pool diameter
- Liquid burning rate
- Flame size (height, diameter, and angle of inclination of the flame from a burning pool under the influence of wind).

Thermal radiation from flame:
The flame from the burning pool is modeled as a cylindrical (inclined) radiator of uniform temperature. The atmospheric transmissivity and flame emissivity are assumed to be one. The position of the receptor with respect to the burning pool must be known.

- Thermal radiation from flames
- Radiation view factor between an inclined flame and an arbitrarily oriented surface in space (receptor). The view factor is a purely geometrical property of the spatial arrangement of the flame and receptor.
- Burning time
- Adiabatic flame temperature

Phase (b): *This phase consists to assess damages and injury to the vulnerable resource impacted by the physical events defined in the scenarios. Damage to vulnerable resources is conveniently discussed in terms of (a) toxic injury, (b) explosion damage, and (c) fire damage.*

- ❑ Estimate the percent of vulnerable resources affected within a given grid cell by a given event type, using the different probit equations or other methods described in chapter 3. Provision should be made to avoid double counting (i.e., inclusion of an element of some vulnerable resource in more than one category of damage or injury).

- ❑ Apply these percentages to the numbers of vulnerable resources present, giving the total numbers of vulnerable resources affected.

2.4.1. FIRES

The Combustion Process

Fire occurs when a source of heat comes in contact with a combustible material. If a combustible liquid or solid is heated it evolves vapor which may form a combustible mixture with the oxygen of the air. If this mixture is further heated to its ignition point, combustion starts. *Thus there are three conditions essential for a fire: fuel, oxygen, and heat. If on of the conditions is missing fire does not occur, and if one of them is removed, fire is extinguished.* Fire is sustained only if there is a net release of heat. The heat comes from the combustion of fuel. If this fuel is solid or liquid it must first be vaporized. Normally the heat required is initially supplied by a source of heat and is then provided by the combustion itself. Fire needs to be supported by the ignition source until it is self-sustaining. The amount of heat needed to cause ignition depends on the form of the substance. A gas or a vapor may be ignited by a small flame or a spark, while a solid may require a more intense heat source. Fire normally grows and spreads by direct burning. which results from the impingement of the flame on the combustible materials, by heat transfer, or by travel of burning material. The three main modes of heat transfer are conduction (through walls and ceilings), convection (75 %), and radiation (to near-by objects).

Depending on the country, there are different classification systems for fire:

- according to the combustible materials
- according to the size of the fire

Flammability

Combustion of a flammable gas-air mixture occurs if the composition of the mixture lies in the flammable range and if the conditions exist for ignition, i.e., if the bulk gas is heated up to its auto-ignition temperature, or if an ignition source with sufficient energy to ignite it is present. Collections of flammability data can be found in *"Limits of Flammability of Gases and Vapors* (Coward and Jones[80], 1952) in *"Flammability Characteristics of Combustible Gases and Vapors"* (Zabetakis[81], 1965), and, among others, in the *"Handbook of Industrial Loss Prevention"* (FMEC,[82] 1967). Such tables usually indicate the flammability limits (upper/lower), auto-ignition temperature, and flash point (closed/open cup).

Flammability limits are affected by pressure, temperature, direction of flame propagation, gravitational, field and surroundings (obstacles). Flammability limits are also affected by the addition of an inert gas such as nitrogen, carbon dioxide or steam4. The flammability of a substance depends strongly of the partial pressure of oxygen in the atmosphere. The oxygen concentration affects both the flammability and the other flammability parameters.

Ignition Temperature and Energy
Some ignition temperatures for hydrocarbons are presented in TABLE 2.12.

TABLE 2.12. Ignition temperature of paraffinic hydrocarbons

Paraffinic Hydrocarbon	**°C**	**Paraffinic Hydrocarbon**	**°C**
Methane	537	n-Hexane	223
Ethane	515	n-Heptane	223
Propane	466	n-Octane	220
n-Butane	405	n-Nonane	206
n-Pentane	258	n-Decane	208

The flash point of a flammable liquid is the temperature at which the vapor pressure of the substance is such as to give a concentration of vapor in the air which corresponds to the lower flammability limit. The *open cup flash point* is usually a few degrees higher than the *closed cup flash point.* Flash point is a main parameter in hazard classification of liquids and in government regulations based on these.

Sources of Ignition
Sources of ignition include the following:

- flames, direct heat and hot surfaces
- welding and cutting
- mechanical sparks
- chemical energy
- vehicles
- arson
- self-heating
- static electricity
- electrical equipment and instrumentation

Many potential sources of ignition are associated with activities. These need to be controlled by a permit-to-work system.

Static Electricity. In liquids, static electricity is generated by electrical potential differences between equipment during unloading operations (hose connections, etc.) or by friction of the flowing liquid during pumping operation. Different pieces of equipment should be earthen. The charge generated by a liquid flowing in a pipeline is greatly increased if the liquid contains water. The accumulation of a charge is a hazard, because it is accompanied by an increase in the strength of the electrical field. If the latter exceeds a certain value the atmosphere becomes ionized and this may lead to a spark discharge. Most industrial dust and powders are poor electrical conductors and tend to generate static electricity. Operations in which this occurs include micronizing, grinding, mixing, sieving, gas filtration, pneumatic conveying and mechanical transfer.

Static electricity is also generated when a gas containing particles, either liquid or solid, issues from an orifice. There is a static electricity hazard when steam or carbon dioxide is used for cleaning or purging (inerting) of equipment which has contained flammable products. The human body can also become electrostatically charged by contact with or induction from charged objects or by friction of clothing.

The ignition energy of the source is also important TABLE 2.13. The ignition energy is dependent on the composition of the gas mixture.

TABLE 2.13. Minimum ignition energy for selected substances in air

Substance	Minimum Ignition Energy (mJ)	Substance	Minimum Ignition Energy (mJ)
Carbon disulfide	0.01- 0.02	n-Butane	0.25
Hydrogen	0.019	n-Hexane	0.25
Acetylene	0.02	Ethylene	0.12
Methane	0.29	Benzene	0.22
Ethane	0.24	Ammonia	up to >100
Propane	0.25	-	-

The rate of combustion of a flammable mixture is affected by the burning velocity, which is a property of the mixture, and of temperature and pressure. The actual flame speed may be much greater than the burning velocity, and the speed depends on other factors such as turbulence and pressure waves. If detonation occurs, the flame speed is greater by order of magnitude than the burning velocity.

Probability of Ignition

Regarding gas cloud fire or explosion, it is possible to use a number of different assumptions regarding ignition sources. The most conservative of these is to assume that the cloud always ignites at the furthest possible distance away from the release point. This assumption can be refined by assuming a constant ignition source probability density and using one the equations below.

Ignition Source Probability. The ignition probability can be taken into account by assuming a uniform ignition source density μ, which is defined as the probability that the flammable region of the cloud encounters an active ignition source for each square meter it covers.

The probability of ignition of the cloud $\delta\psi$ in traversing between points s and $s+\delta s$ downwind can be given as:

$$\delta\psi = \mu s \tan\theta \; \delta s \tag{2.35}$$

This can be integrated to determine the probability, ψ, that a cloud will have ignited by the time a distance, s, has been reached:

Sectoral Cloud. We have the following formula:

$$\Psi_s(s) = 1 - e^{-\theta_s s^2 \mu} \tag{2.36}$$

with

$$\theta_s = \tan^{-1} \frac{a}{2b}$$

Elliptical Cloud. We have the following formula:

$$\Psi_s(s) = \left| -\exp\left\{ -\frac{\mu a b}{2} \left[\pi + 2 \sin^{-1} \frac{s-b}{b} + 2 \frac{s-b}{b} \left(1 - \left(\frac{s-b}{b} \right)^2 \right)^{1/2} \right] \right\} \right| \tag{2.37}$$

with

$$\theta_E = \tan^{-1} \left\{ \frac{a}{s} \left[1 - \frac{(s-b)^2}{b^2} \right]^{1/2} \right\} \tag{2.38}$$

> *The use of sectoral (directed) clouds is likely to give larger risks far away from the plant (i.e., off-site locations where the public may be present) whereas the use of elliptic clouds is likely to give larger risks for locations on or near the plant.*

The question now arises as to which form is the most realistic. reference to incident data and to photographs confirms that the shape of the area which the cloud covers at any time is highly variable and appears to depend on such factors as the wind speed and the density of the cloud. The use of elliptic cloud footprints are more realistic for light gases as methane or materials where low concentrations are of interest such as toxics; sectoral footprints are more realistic for clouds that are denser than air.

The increase in source density makes ignition at some point during the cloud travel more likely. Secondly the relative probability that the cloud reaches a certain point downwind before ignition occurs is reduced.

A more detailed approach is to use individual point ignition sources. Calculation shows an increase in risk between the hazardous sites and the ignition sources; and a reduction in risk in areas between the sources. Care must be taken when using this approach with sectorial clouds.

The risk can vary with direction. This can occur as a result of preferential wind direction or by consideration of an anisotropic ignition source distribution, either using different values for the ignition source probability density in different areas or directions, or by using point ignition sources. Another potential source of anisotropy is caused by the variation of the surface roughness and the presence of obstacles around the site. This will change the dispersion characteristics of a vapor cloud depending on the wind direction.

Radiation from a Point Source

In this case the flame is considered as a point in which part of the energy released in a fire is emitted in the form of radiation. The heat radiation load is proportional to that radiation and depends on the distance from the source to the receiver.

The heat load at a distance x from a point source is given by:

$$q = \left(\frac{Q_R}{4\pi\, x^2} \right) \tau_a \tag{2.39}$$

in which Q_R (J/s) is the part of the combustion energy sent out as radiation, and is given by:

$$Q_R = \eta\, m\, h_c \tag{2.40}$$

where

η = radiation fraction (-)
m = burning rate (kg/s)
h_c = heat of combustion (J/kg)

Four different categories of fires can be considered:

- Pool fire
- Jet fire
- Flash fire, and
- Fire-balls

Pool Fire. It occurs when an accumulation of flammable liquid as a pool on the ground or on a different liquid surface is ignited. A steadily burning fire is rapidly achieved as the fuel vapor to sustain the fire is provided by evaporation of the liquid by heat from the flames. For liquefied gases, significant heat transfer from the surface on which the pool is formed usually contributes to the vaporization of the fuel. The rate of consumption of fuel is a function of the properties of the fuel such as latent heat, heat of combustion etc., which results in typical rates of regression of the pool depth of 6-13 mm/min.

There are three methods of calculating the thermal radiation fluxes from a pool fire:

- Point source method,
- Solid flame method, and
- Volume emitter method.

Point Source Method. In the point source method it is assumed that the heat is radiated from the vertical axis at the center of the pool. The radiation flux is given by the formula:

$$I = \frac{f H_c}{4\pi R^2} \tag{2.41}$$

where

I = incident radiation per unit area

R = distance from the source

f = fraction of the heat of combustion assumed to radiate in all directions from the notional counterpoint source

H_C = heat of combustion per unit time.

Solid Flame Method. The solid flame method has the advantage over the simple point source method as it takes account of the actual shape and volume of the flame, although it is reduced to a simple geometrical shape for ease of manipulation. It is however a simplification to assume that a flame emits thermal radiation solely from its surface.

Volume Emitter Method. The volume emitter method takes account of the fact that the sources of radiation are hot molecules and particles distributed throughout the whole volume of the flame. The radiation is determined by factors like the path length, concentration and temperature of the molecules and particles. However, it is extremely difficult to do this; this is the reason why the normal procedure is to use the point source method.

The portion of the thermal radiation from a source which is incident upon a nearby target is given by the relationship:

$$Q_t = Q_s F_{ts} \tau \tag{2.42}$$

where

τ = atmospheric transmissivity (a function of the path length and the physical characteristics of the atmosphere) (Simpson[83], 1984);

Q_t = thermal radiation received at distance d (W m^{-2});

Q_s = total heat radiated (W);

F_{ts} = geometric view factor (or form factor, or configuration factor).

The geometric view factor is the ratio between the received and emitted radiation energy per unit area (of the receiver/emitter), for complete transmission. This factor is determined by the dimension and shape of the flame, and by the location and orientation of the receiving object. The calculation can be difficult but fortunately tables are available which give the view factors for a large variety of shapes and orientations (*Considine*[84], 1984; *TNO*[47], 1979; *Mudan*, 1984, and *Institute of Petroleum*[85], 1987).

The heat radiation from a fire burning in a catchment area to a pressure storage can be estimated according to the following method (*Figure 2.15*):

- The heat radiated and the intensity of heat radiation from the flame burning on the liquid pool (enclosed by the catchment) are calculated from the equations

$$Q = k_2 L W \rho C$$

$$I = \frac{k_3 L W \rho C}{(L+W)h + L W} \tag{2.43}$$

with

k_2 = 5.1 x 10^{-5}

k_3 = 2.5 x 10^{-5}

where

C	=	net calorific volume of liquid in the pool (KJ kg^{-1})
h	=	height of flame (m); it is equal to 2 x width W of the pool (wake)
I	=	intensity of heat radiation from the flame envelope (kW m^{-2})
L	=	length (m) of the pool (wake)
Q	=	heat emitted by the fire (kW)
W	=	width of the pool (m)
ρ	=	density of the liquid (kg m^{-3})
k_2	=	a constant (m^{-1} s^{-1})
k_3	=	a constant (m^{-1} s^{-1}).

It is assumed here that the heat is radiated from a radiating area (wake) which is a vertical rectangular plane through the center line of the flame envelope parallel to the equator of the cylindrical pressure vessel receiving the radiation. The heat flux received by this vessel is obtained from the view factor method.

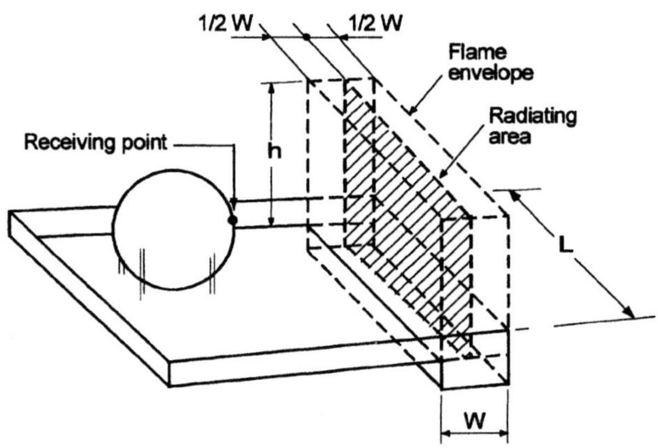

Figure 2.15. Heat radiation from a fire burning in a catchment area to a spherical pressure storage (Source: Imperial Chemical Industries Ltd. - ICI/RoSPA 1970 IS/74)

The burning time of a pool fire after ignition can be determined as follows:

$$t_b = \frac{V_p}{A_p\, r_b} \tag{2.44}$$

where

t_b	=	Pool burning time (s)
V_p	=	Volume of fuel remaining in the pool at the time of ignition (m^3)
A_p	=	Area of the pool at the time of ignition (m^2)
r_b	=	Burning rate of the fuel (m/s)

Discussion
Pool fires have been studied for many years and the empirical equations used are well validated. The treatment of smoky flames is still difficult, also flame impingement effects are not simulated.

Jet Fire

A jet fire occurs when a flammable liquid or gas, under some degree of pressure, is ignited after release from a hole or crack in a pressure vessel, from the end of an open pipe or from the orifice of a pressure relief valve. The pressure behind the liquid or gas tends to generate a fairly long stable flame. This jet flame can be extremely intense and can impose high heat loads on nearby plant and equipment.

Jet fire modeling is not as well developed as pool fire modeling. However, there are a number of publications which describe the various approaches (Bagster[86], 1986; API 521[87], 1982; and Hustad and Sonju[88], 1985). The API method is relatively simple. An example of its application to an LPG jet flame is given below. *Figure 2.16* shows the flame length, and *Figure 2.17* the distance to a given level of thermal radiation against the flow rate. As in pool fires flame impingement effects are not simulated.

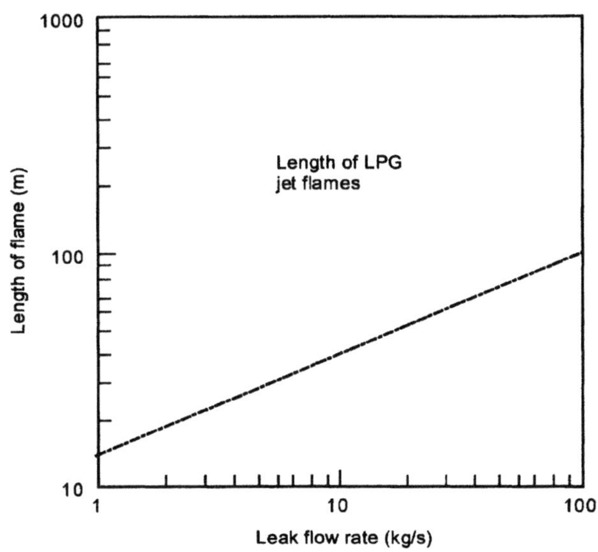

Figure 2.16. LPG jet flame length as a function of distance

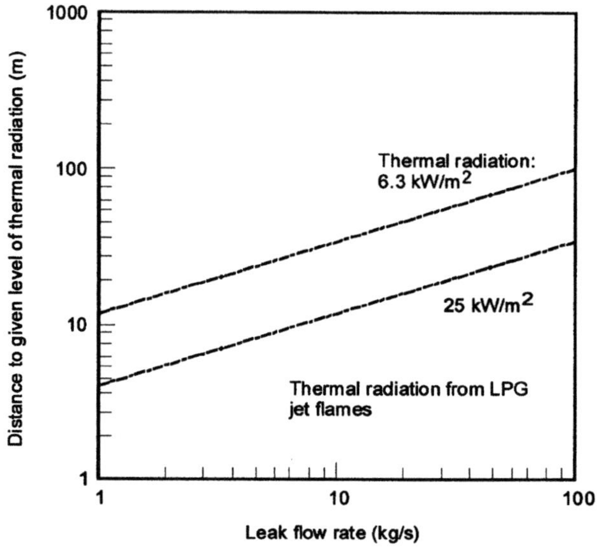

Figure 2.17. Estimation of the thermal radiation from LPG jet flames as a function of distance

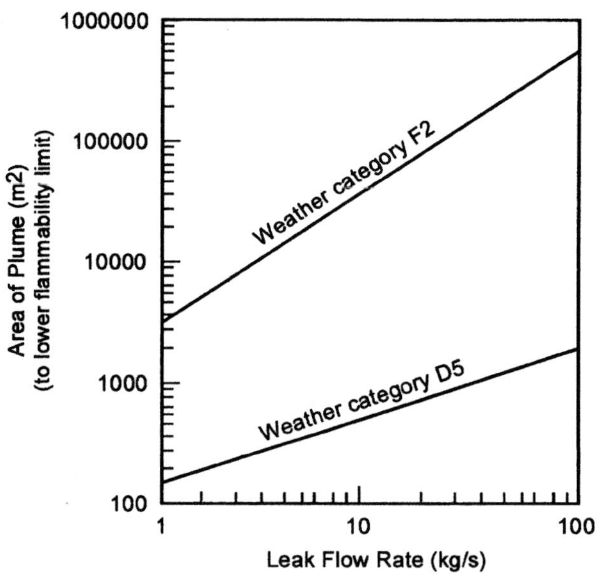

Figure 2.18. Area of the plume to the lower flammability limit

Flash Fire

Flash fire occurs when a cloud of a mixture of flammable gas and air is ignited. The shape of the fire closely follows the shape of the cloud prior to ignition but also depends upon the position within the cloud where the ignition took place. The speed of burning depends on the concentration of the flammable material in the cloud and, to a lesser extent, on the wind speed. Ignition of the cloud may take place whilst the cloud still extends to the release point (under these circumstances this can give rise to a pool or a jet fire, depending on the nature of the release). It is also possible that the flame may accelerate to a sufficiently high velocity for an explosion to occur. *Figure 2.18* shows the area of the plume to the lower flammability limit against leak flow rate for plumes of LPG for two weather stability classes (D and F) and related typical wind velocities (5 m/s and 2 m/s, respectively).

Fireball

Fireball occurs when there has been a release of considerable violence and vigorous mixing and rapid ignition takes place. The initial flammable cloud is often hemispherical before ignition but rapidly approximates to a rising sphere, due to thermal buoyancy. If the release of fuel is directed upwards, such as when a vessel suddenly ruptures, then a spherical shaped fireball forms immediately.

An important source of a fireball is due to the phenomena known as a *"Boiling Liquid Expanding Vapor Explosion", or BLEVE. These usually occur with flammable liquids stored under pressure at ambient temperature,* liquids such as liquefied petroleum gas, propylene or ethylene oxide. The event starts with an external fire, possibly fueled by a spillage or leak from the vessel itself, which has flames impinging on areas of the vessel which are in contact with the liquid contents. Boiling of the liquid increases the vapor pressure but keeps the wetted vessel surface relatively cool. However, where the flames impinge on areas of the vessel blanketed by vapor, heat transfer is poor and the metal surface temperature rapidly rises. At these high temperatures the metal weakens and, with increasing internal pressure, ruptures. As a result of the vessel failure the pressurized contents rapidly escape and expand forming a large cloud of vapor and entrained liquid. The cloud is ignited by the original flames and a huge fireball is formed.

Casualties from Fire Balls and BLEVE
Casualties can be due to not only thermal radiation but also to the effects of the blast and to missiles. Fireballs tend to produce large heat fluxes for a short period of time.

Some useful formulas for fireballs as a result of a BLEVE are given in *TNO*[47]. Both can be handled the same way.

Peak diameter D (m) of the fireball:

$$D = 29M^{1/3} \tag{2.45}$$

Fireball duration t (s):

$$t = 4.5M^{1/3} \tag{2.46}$$

or

$$t = 8.2M^{1/6} \text{ for } M > 30 \text{ tonnes.} \tag{2.47}$$

where M is the initial mass of flammable liquid (tons).

The constants used in the above equations depend on the nature and amount of the material involved. Where some degree of precision is required, it is recommended to use the methods described in *TNO*[47], *Marshall*[89], *(1987) and Lees/Ang*[40], *(1989)*.
Recent formulas given in TNO[90] (1992) are shown below.

The first general formula is valid for different combustible mixtures (average BLEVE fireball):

$$d_v = 6.48m^{0.325}, \quad \text{and} \quad t = 0.852\, m^{0.26} \tag{2.48}$$

where

d_v = fireball diameter (m)
m = mass (kg)
t = duration (s)

and the second formula below gives an average value for methane, ethane, and propane.

$$d_v = (7.00 \pm 0.30)m^{1/3}, \quad \text{and} \quad t = (2.63 \pm 0.30)m^{1/6} \tag{2.49}$$

> ***Discussion***
> *BLEVE dimensions and duration have been studied by many authors and the empirical basis consists of several well-described incidents, as well as many smaller laboratory trials. The use of a surface emitted flux estimate is the greatest weakness, as this is not a fundamental property.*

There is a voluminous literature on fire and fire protection and only a few selected items were mentioned here. Further references can be found in Loss Prevention in the Process Industries (Lees[91], 1980).

2.4.2. EXPLOSIONS

The second of the major hazards is explosion. In the process industry explosions cause fewer serious accidents than fire but more than toxic release. *An explosion is a process involving the production of a pressure wave resulting from a very rapid release of energy*. In the case of an explosion in air, the air will become heated locally due to its compressibility. This will increase the velocity of sound causing the front of the disturbance to steepen as it travels through the air, thereby increasing the pressure and density of the air until a peak pressure wave is developed at some nominal distance. *The magnitude of this pressure wave will govern the loading and therefore the damage to structures, people, etc., nearby.*

Some Definitions

Explosions are of two kinds: deflagration and detonation. In a deflagration the flammable mixture burns relatively slowly. For hydrocarbon-air mixtures the deflagration velocity is typically of the order of 1 m/s.

In a detonation the flame front travels as a shock wave which releases the energy to sustain the shock wave. At steady state the detonation shock front reaches a velocity equal to the velocity of sound in the hot products of combustion and thus much greater than the velocity of sound in the unburned mixture. For hydrocarbon-air mixtures the detonation velocity is typically of the order of 2000-3000 m/s. For comparison the velocity of sound in air at 0°C is 330 m/s.

A detonation generates greater pressures and is more destructive than a deflagration. A deflagration may turn into a detonation in the presence of obstacles or when traveling down a long pipe. A basic distinction is made between confined explosions and unconfined explosions:

- Confined explosions are those which occur within some sort of containment (vessel, pipework, buildings).
- Unconfined explosions are those which occur in the open air.

Explosions may cause death and injury in a number of ways. It is common practice to relate the blast overpressure and duration time to the quantity of explosive material first, and second, to try to relate the pressure time history of the blast wave at a specified distance from the explosion source to a damage criterion such as percentage fatalities.

This section will consider the prediction of blast overpressure effects from vapor cloud explosions, condensed phase explosions and catastrophic failure of large vessels under pressure.

Prediction of Blast Overpressure Effects

There is a considerable literature on explosions. Much of the information available relates to chemical, nuclear explosions, or hydrogen-air explosions in a containment, and to military applications. The data given in literature relates to point source TNT explosions, since it is convenient to conduct damage experiment on this basis. The usual approach to estimating blast effects from industrial chemical explosions is to form a TNT equivalent. To obtain a TNT equivalent the estimated weight of the released material has to be multiplied by two factors: the first provides a ratio between the energy per unit weight of TNT and the released material; the second is an arbitrary efficiency factor.

The idealized structure of a blast wave can be described as follows: before the arrival of the front of the shock wave the pressure is at the ambient level. The time taken for the front to travel from the source of the explosion to the point at which the blast is measured is known as the *arrival time*, t_a.

At the arrival time the pressure rapidly rises to a peak value which is known as the *peak positive overpressure*. This pressure then decays back to the ambient pressure in a time known as the *positive phase duration*. This is followed by a further decline to produce a pressure lower than ambient and eventually returns back to the ambient pressure. The period from the end of the positive phase to the final return to the ambient atmospheric pressure is known as the *negative phase duration.* The parameters of most interest are the peak positive overpressure and the area enclosed by the positive overpressure time curve.

The characteristics of a blast wave produced from a TNT explosion are well understood and full account is given in a book by Baker[92] et al., (1983). The magnitude of the key parameters such as overpressure, duration time and impulse at known distances from the source can be related to the mass of explosives by means of a scaling law.

Scaling Law for Explosion Parameters

If a given positive pressure is caused by a given amount of explosive at a given distance, the same pressure will be experienced at twice that distance only when the amount of explosive is increased eight times.[93, 94] *Figure 2.19* shows the overpressure from a condensed phase explosion as a function of explosive mass and distance.

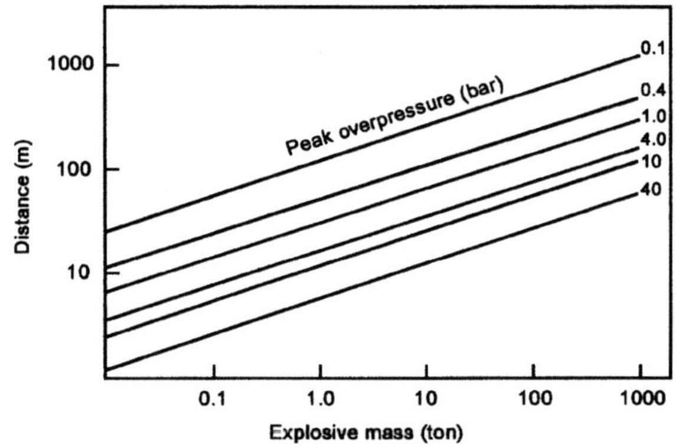

Figure 2.19. Overpressure from a condensed phase explosion as a function of explosive mass and distance.

It is important to take time into account as well as overpressure when estimating the damage that can be caused by a blast wave, and as a first approximation it is convenient to formulate the possible damage in terms of impulse, simply expressed as half the product of pressure and time. The experimental work as demonstrated a significant scatter, due to differing chemical explosives, differing containers, differing geometrical aspects etc. (Baker[92], 1983). Death due to lung damages are very rare. Lung damages are caused by external pressure on the body.

A vapor cloud explosion occurs when a release of gas mixes with air and is ignited. The mixture must be within a limited ***flammability range*** *for an explosion to occur. The effects of a vapor cloud explosion depend to a large extent on the* ***degree of confinement.***

For hazard assessment the casualties are usually categorized under just two headings: primary blast effects, respectively secondary blast effects.

■ *Primary blast effects:* It concerns casualties being directly affected by impact of the blast wave or from fragments of the explosive's casing. From the experimental data of Bowen et al.[95] on animals it is possible to derive a set of curves showing the relationship between impulse and duration time for 50% lethality of the different species at standardized body weight (*Figure 2.20*). From the data in *Figure 2.19* and *Figure 2.20*, and the scaled relationship between duration time and explosive mass, it is possible to derive a relationship between mass of explosive and distance for 50% lethality to a 70 kg man, and this is illustrated in *Figure 2.21*. A similar relationship for primary deaths caused by whole body translation is also shown on this graph.

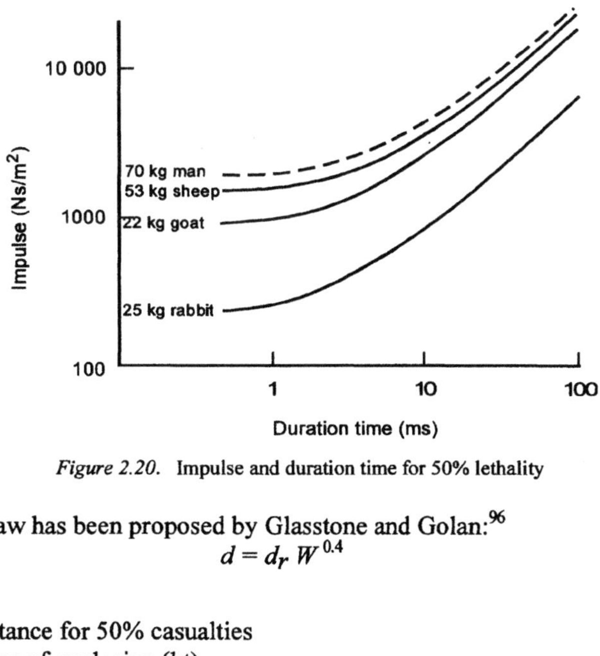

Figure 2.20. Impulse and duration time for 50% lethality

A scaling law has been proposed by Glasstone and Golan:[96]

$$d = d_r \, W^{0.4} \qquad (2.50)$$

where

d = distance for 50% casualties
W = mass of explosive (kt)
d_r = distance for 1-kilotonne explosion

The 50% lethality contour for primary blast deaths, given the distance in terms of primary independent variable, i.e., the explosive mass, can be obtained from:

$$L = \frac{2\,10^{-2}\,T^{1/3}}{\left[1 + (20/T)^2\right]^{1/6}} \qquad (2.51)$$

where

T = explosive mass (t)
L = distance (km)

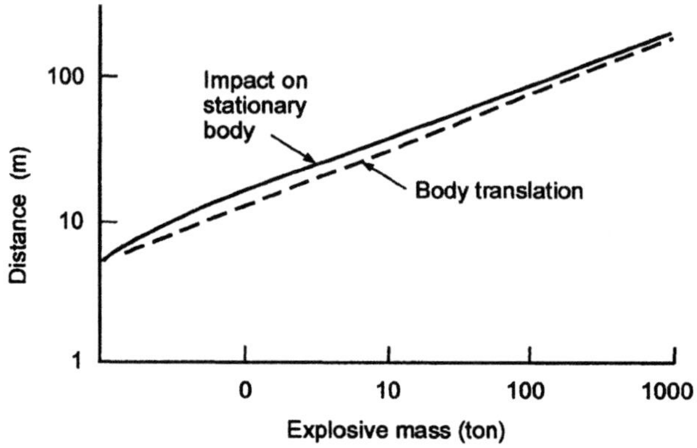

Figure 2.21. Primary blast deaths caused by a point source explosion (assuming 70 kg man; 50% fatality)

It is also possible to derive fatality relationship in probit from animal data (*Figure 2.22*). Short-duration time pressure waves (associated with small explosions) require a much higher pressure for lethal effect than do the longer duration times (associated with larger explosions).

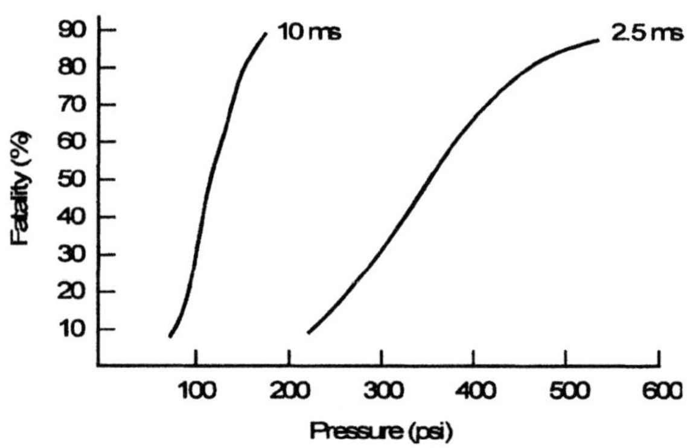

Figure 2.22. Estimated percentage fatality as a function of overpressure and duration time

■ *Secondary blast effects:* It concerns casualties resulting from being hit by secondary missiles, walls, ceilings, trees, tiles, roofs, glass etc., or falling due to a collapsed floor, or being buried. It is not possible to identify cause and effect relationship for secondary categories in the same way as for primary categories. Instead the findings of the UK Explosive Storage and Transport Committee[97] have been used to provide a risk contour which gives 50% chance of home destruction from various size of TNT explosion. The relationship between home destruction and explosive mass is subject to considerable variation because of:

- differences in the blast wave overpressure
- asymmetry of the wave due to variation in terrain and in obstructions
- differences in the types of buildings

For residential areas in the UK a contour was calculated for different categories of housing, where as a result of specified explosion, the number of inhabitable houses outside the contour are balanced by the number of houses inside the contour (50% damage contour). A similar formula as above can be given:

$$L = \frac{13.2\ 10^{-2}\ T^{1/3}}{\left[1+(3.175/T)^2\right]^{1/6}} \qquad (2.52)$$

To form an estimate of the number of people likely to be affected by secondary blast action, an estimate is required of the ratio of the deaths and injuries to homes made inhabitable (for that it is only necessary to know the number of homes inside the contour, see above).

An estimate of the relationship between uninhabitable homes and casualties has been formed from a review of the following categories of event:

- natural disasters (including tornadoes, and earthquakes)
- man-made disasters (including chemical explosions, bomb explosion, and domestic gas explosions in the UK).

From the review of the man-made disasters, the ratio of deaths to uninhabitable houses is of the order of 1 : 10 - 15. In the case of natural disasters the death toll is rather less than this (due to warnings). The average figure for home occupancy in the UK is 2.5, so that the number of casualties in a 45 ° sector is then given by the formulae:

$v = \pi R^2$ where R is the distance from the hazard source to the population boundary (v is the clear space)

$q = (x - v) / 8$

$s = (z - v) / 8$ where x and z are the primary and secondary damage zones around the hazard source

$n = D\,(q + (s / 25))$ where D is the population density and n the number of fatalities.

Because of the much larger 50% damage radius there will be more fatalities arising from secondary effects than from primary even though only one fatality is to be expected for every ten homes destroyed. The deaths of primary and secondary effects are added together. From LPG incident in Mexico City it can be concluded that there was only minor damage from the blast wave, but that the greatest damage was probably caused by fire (direct flame contact, fumes and lack of oxygen) and from explosions of the gas that accumulated inside the houses.

Unconfined Vapor Cloud Explosions.

Open-air, so called *unconfined vapor cloud explosions* (UVCE) have been thought to be impossible and are very difficult to theoretically understand. However, the presence of relatively minor turbulence producing obstacles with the requirement for a certain critical mass may explain the fact that a large number of so-called unconfined vapor cloud explosions have occurred.

An unconfined vapor cloud explosion is one of the most serious hazards in the process industries. Although a large toxic release may have a greater theoretical casualty potential, it is a very much rarer event.

> ***Unconfined Vapor Cloud Explosions (UVCE)***
> *The blast wave from an unconfined vapor cloud explosion is characterized by a relatively slow rise to peak pressure and a relatively long duration (typically a few tenths of a second). Vapor cloud explosions produce levels of overpressure of the order of 1 bar and do not produce craters.*

A theoretical estimate of the hazard from an unconfined vapor cloud explosion has been done by assuming a particular scenario, calculating the emission and dispersion of the gas and determining the explosion effects. Studies on these lines have been described by Decker[98] (1974) and by Eisenberg[99] et al., (1975).

TABLE 2.14. Estimated frequencies of unconfined vapor cloud explosions (after Kletz, 1977)

	Cause of Explosion	Frequency (Explosion/-plant-year)
(1)	**Caused by Failure of:**	
	Pressure vessel	10^{-5}
	Special pipeline	10^{-5} - 10^{-4}
	Normal pipeline	10^{-4} - 10^{-3}
	Normal pump	10^{-2}
	Severe duty pump	10^{-1}
	Reciprocating compressor	10^{-1}
(2)	**Caused by leak from:**	
	Batch reactor	10^{-2} - 10^{-1}
	Tanker filling hose	10^{-2} - 10^{-1}

In the case of unconfined vapor cloud explosion the energy in the blast wave is generally only a small fraction of the energy theoretically available from the combustion of all the material which constitutes the cloud. The ratio of the actual energy released to that theoretically available is referred to as the *explosion efficiency.* The explosion may occur at some distance of the point of release and thus threaten a considerable area. The material escaping may be a gas, a volatile liquid, a superheated liquid, or a refrigerated liquid. Flashing superheated liquids tend to give rise to the largest vapor clouds. Most UVCE have been caused by such flashing liquids. The estimated frequencies of unconfined vapor cloud explosions is given in TABLE 2.14.

With regard to the cause of release Strehlow[100] (1973) states that an evaluation of spills in the chemical industry showed that 40% were due to component malfunction and 60% due to human error. The probability of ignition has been estimated to 0.1-0.5 for leaks larger than 10 tons. For small leaks the probability of ignition is much less. Kletz[101] (1977) quotes a value of 10^{-4} for polyethylene plants.

The distance drifted by a cloud of flammable vapor before exploding depends on the situation. Kletz suggests that the assumption of no drift is probably good enough for approximate calculations, but that a drift of 100 m may occur in 1 case in 5 or 10. The drift determines the explosion center and the blast intensity further away. In open situations with few sources of ignition the cloud may drift much further. The time delay before ignition may be as long as 15 minutes and the quantity of material which can accumulate in this time may be very large.

The relations between explosion effects and fatalities are those given by the probit equations. Usually, the majority of deaths due to an explosion are caused by lung hemorrhage. The corresponding probit equation is shown below.

Confined Gas Explosions.
Under total confinement, most gases will, when mixed with air at atmospheric pressure, produce *a maximum pressure of 8 bars when ignited* (Harris[102], 1983, and Marshall[89], 1987).

In most practical confined situations there will be a vent, or a weak point (sometimes deliberately inserted) within the structure which will relieve some of the explosion gases and cause a reduction in peak overpressure.

> ***Confined gas explosions***
> *Confined gas explosions may occur in equipment (such as storage tanks), amongst groups of plant items and/or buildings (partially confined explosions) or inside buildings.*

Condensed Phase Explosions
They arise as a result of detonation of high explosives such as TNT or RDX or materials such as some organic peroxides which are used as propellants for military purposes.

> ***Condensed phase explosions***
> *Condensed phase explosions are the type which most closely approximates that of the idealized blast wave structure described earlier in that it is characterized by an abrupt pressure rise, a short (1 to 10 ms) positive phase duration and a very high (10000 bar) peak positive overpressure. For confined or semi-confined explosions a further phase of the blast wave exists as a result of reflections from surrounding structures.*

Failure of a large vessel under pressure results in a blast wave which is similar to the ideal blast wave structure during its positive phase but has a larger negative phase and is followed by multiple shocks. The stored energy released from the vessel is transferred to fracture energy, blast wave energy and kinetic energy of missiles. *Generally something between 40 and 80% of the total energy is transferred to the blast wave.* This depends on the amount of energy spent in fragmenting the vessel.

Estimating the Effects of Explosions

The most common method of estimating the effects of explosions is to determine the mass of TNT which would cause an equivalent amount of damage. This is based on the assumption that, in the far field at least, a blast wave from any source of explosion will tend towards that of a TNT explosion. *This method is known as the TNT equivalence technique, and has been outlined by the UK Advisory Committee on Major Hazards, ACMH-2 (1979).*[103] This method is under review, to take account of improved understanding of the underlying mechanisms of flame acceleration in partially confining structures. A number of people also use a *model developed by TNO (Wiekema*[104]*, 1979),* which is based on actual unconfined vapor cloud explosions and employs one of two defined explosion yields. The model is limited to flammable materials of medium reactivity. Both methods are strictly empirical and are not based on solid theory.

For condensed phase explosions the TNT equivalent mass is evaluated by using a *TNT efficiency factor* which is an estimate of the proportion of the available energy of the explosion transferred to the blast wave. This efficiency factor is then multiplied by the total stored energy to determine the energy in the blast wave. The mass of TNT required to produce an equal energy blast can then be calculated using 4520 kJ/kg as the mass specific energy for TNT. The efficiency factor for high explosive varies from about 60% to 130%, however as a first approximation it can be assumed that 1 equivalent ton of high explosive will produce the same blast energy as 1 ton of TNT. Explosives used as propellants generally transfer only up to 25% of their available energy to the blast wave on explosion.

Having obtained the TNT equivalent mass for the scenario under consideration, it is then possible to estimate the *blast parameters of an explosion* at any distance from the source. A number of sources publish plots of blast parameters versus scaled distance Z for high explosives (Baker et al.[92], 1983):

$$Z = \frac{R}{W^{1/3}} \tag{2.53}$$

where

R = distance (m) from the source of the explosion
W = charge weight (kg TNT).

In the case of the *rupture of a pressurized vessel,* it can be assumed (as a first approximation) that the resulting pressure profile will be close to that of a TNT explosion and can be modeled accordingly using the TNT equivalence technique. It is generally accepted that this is more valid the further one is away from the source. The equivalence model tends to over-estimate the blast wave parameters close to the source and other methods need to be used *(Baker et al.*[92]*, 1983).* The total stored energy of a gas in a pressure vessel is given by:

$$E = \frac{p_1 V_1}{\Gamma - 1}\left[1 - (p_a/p_1)\,(\Gamma - 1)/\Gamma\right] \tag{2.54}$$

where p_1 and V_1 are the pressure and volume of the vessel; p_a is the atmospheric pressure and G the ratio of specific heats (C_p/C_v).

For a first approximation it should be assumed that 50% of the stored energy is transferred to the blast wave.

> *The TNT equivalent mass of a gas cloud explosion is difficult to estimate with any real accuracy. A large number of factors affect the magnitude of the blast wave energy. These include turbulence, the volume of gas, the composition of the cloud, the location of the ignition source relative to the cloud, the shape of the cloud and the proportion of the total energy transferred to the blast wave.*

The complexity of this problem led to the production of a number of models such as ACMH-2 [103], 1979 and Wiekema[105], 1984. The range of efficiency factors obtained from such models can be as low as a fraction of one per cent up to a few tens of percent. The UK Advisory Committee on Major Hazards recommends that an approximate value of 3% of the total available energy should be assumed to have been transferred to the blast wave. It should however be noted that the TNT method should not be used to predict blast wave parameters for gas explosions at a distance of less than 10 cloud diameters from the source of the explosion.

> *The TNO multi-energy method (Wingerden[106], 1989) is now considered to give results which are much more representative of those observed in actual explosions.*

Computer models do exist which attempt to model the basic physical principles of explosion behavior. These models are generally neither simple nor easy to use. Probably the best known and most widely used is the code *FLAX* which was developed by the *Christian-Michelson Institute at Bergen, Norway.*

> ***Discussion***
> *The TNT equivalent model as described is relatively easy to use. Neither it, nor the TNO model, is solidly based on theory, but they both predict well the observed UVCE incidents. One difficulty is that in the TNT approach some expertise is required in selecting the explosion yield. The other weakness of the TNT model is the substantial physical difference between detonations and UVCE deflagration. The TNO correlation model is based on actual UVCE incidents and employs one of two defined explosion yields, but it is limited to flammable models of medium reactivity.*

Missiles

The consideration and prediction of the effects of fragments of pressure components which fail under incident conditions is important as there have been many deaths and cascade damage effects due to such fragments. Most of the events seem to be associated with the storage of flammable liquids such as liquefied petroleum, often resulting in the projection of missiles (sometimes still containing liquefied gas) to distances much greater than the thermal hazard range from the initial event. *The effect of these missiles is to cause physical damage to property and people and to act as an initiating event for further incidents due to damage to plant and also as a result of starting secondary fires.* A number of studies have been carried out into the cause, likelihood and effect of missiles. These include *Baker et al.*[92], *1983, Association of American Railroads*[107], *1972* and *1973,*[108] and by *Holden*[109], *1989.*

> ***Probability of Missile Projection from Cylindrical Liquefied Gas Vessels***
> *The comprehensive study by Holden confirmed the assumption of others that the probability of missile projection from cylindrical liquefied gas vessels which fail when affected by fire is almost 0.8. Major fire engulfment events usually produce up to about 4 fragments; non-fire events tend to produce slightly more - this is for cylindrical vessels. The generally larger spherical vessels tend to produce more fragments, a useful mean being around 10.*
>
> ***Distance Traveled by the Missiles***
> *The distance traveled by the missiles depends on the shape of the fragment produced. Cylindrical end tube fragments, which are closed at one end, tend to act like rockets and can travel anything up to 1 km. However, as a rough guideline, it can be assumed for cylindrical vessels that about 80% of the fragments will travel less than 200 m. For cylindrical vessels, the fragments are generally projected in directions roughly axial to the vessels orientation at the time of rupture. For spherical vessels there is a tendency of a non-random directional distribution.*

When assessing the hazards from missiles, it should be particularly noted that nearby pipework and thin walled tanks are very vulnerable to impact from vessel fragments. Large thick-walled pressure components can also be susceptible.

2.4.3. PROBABILITY OF FIRE AND EXPLOSION

In the case of inflammable material it is necessary to provide an estimate of the probability of the material being ignited. This may occur on the site, near to the point of release, or it may occur off-site as the cloud drifts away; finally it may not occur at all. If the material does ignite, the ignition may be followed by an explosion. Here again, such an explosion may occur on site close to the ignition or off-site or not at all. The chance of ignition can be related to the mass of the material released. The chances of ignition and explosion should the cloud drift off site have been estimated in the Canvey report. Should an unignited drifting cloud reach the edge of a populated area, it is suggested that there will be a 70% chance of ignition occurring at the edge, a 20% chance at the center, and a 10% chance of non-ignition. If the cloud is smaller than 5 tons, any prospect of ignition off-site is to be ignored. Off-site ignition followed by explosion may occur in some 7% of cases. TABLE 2.15 and
TABLE 2.16 show the chances of ignition and explosion of inflammable clouds for different release sizes and "on-site" respectively "off-site" conditions. These data were obtained from Wiekema's list, respectively, Davenport's list of reported vapor cloud explosions. The chances of ignition/explosion were finally estimated by Withers.[110]

TABLE 2.15. Probability of ignition and explosion "on-site"

Release Size (tons)	Chances of Ignition	Chances of Ignition and Explosion
5,000	1.00	0.70
2,000	0.94	0.50
1,000	0.80	0.40
500	0.60	0.32
200	0.45	0.20
100	0.38	0.15
50	0.27	0.11
20	0.17	0.06
10	0.11	0.04
5	0.06	0.025
2	0.04	0.012
1	0.022	0.006
0.5	0.013	0.003
0.2	0.006	0.001
0.1	0.001	0

TABLE 2.16. Probability of ignition and explosion "on-site"

	Chances of Edge		Chances of Central		
Release Size (tons)	Ignition	Ignition and Explosion	Ignition	Ignition and Explosion	Chances of no Ignition
5,000	0	0	0	0	0
2,000	0.05	0.04	0.01	0.01	0
1,000	0.14	0.13	0.04	0.03	0.02
500	0.07	0.06	0.03	0.02	0.3
200	0.05	0.03	0.01	0.007	0.49
100	0.05	0.02	0.01	0.005	0.56
50	0.04	0.01	0.01	0	0.68
20	0.03	0.01	0.01	0	0.79
10	0.02	0.01	0.01	0	0.86
5	0.01	0	0	0	0.92

The procedural steps to be followed for assessing the probability of fire and/or explosion are indicated in *Figure 2.23*.

A typical estimate of off-site fatalities involves the following steps (see box):

Step 1: *Calculate maximum downwind range in each of eight 45-degree sectors, starting with the north (i.e., sector 1 = North)*
Step 2: *Assume that no more than 20% of the released quantity will be within flammable limits and that the maximum attainable cloud size is only 50 tons due to dispersion effects.*
Step 3: *Assume that the boundary limit for fatalities due to fire alone will correspond to a radiation level of 12.6 kW/m^2.*
Step 4: *Calculate fatalities due to blast from an explosion alone following the damage relationship.*
Step 5: *For ignition and explosion, the consequential fatalities per release are based on the firestorm data (i.e., 36% of the exposed population (obtained by multiplying the population density by the area derived from application of the previous boundary limits*
Step 5: *For fire only, the fatalities are more simply assumed to be one-fifth of the casualties caused by fire and explosion.*
Step 7: *Obtain the number of fatalities per annum for a given mass release by multiplying the derived frequency of release by the ignition/explosion chance, by the chance of the weather being suitable for the downwind range to overlap the population edge, and by the number of consequential fatalities per release.*

Figure 2.23. Method for assessing the probability of fire and explosion

Summary (Chapter 2)

The chapter summarizes methodologies and procedures used to calculate or estimate the negative consequences, effects, impacts or other types of outcomes of severe accidents involving substances of a hazardous nature. A number of worldwide accidents involving various types of chemical substances are presented and described. Models to calculate frequency and magnitude of accidental releases of hazardous materials as well as fire and explosions risks are included.

References (Chapter 2)

[1] Lees, P., Loss Prevention in the Process Industries, Hazard Identification, Assessment and Control, Vol. 2, Butterworths, London, (1980), pp. 863-928.

[2] Davenport, J.A., A study of vapour cloud incidents - an update, I. Chem. E. Fourth International Symposium on Loss Prevention, Harrogate, September (1983).

[3] Kletz, T.A., Turner, E., Is the number of serious accidents increasing?, ICI Safety Note 79/2B, Chem. Ind. Assn., London, (1979).

[4] Kletz, T.A., Second International Symposium on Loss Prevention, I. Chem. E. Symposium Series, (1968) 50.

[5] Wiekema, B.J., Analysis of vapour cloud accidents, Proceedings of the Fourth Euredata Conference, Venice, (1983)

[6] Notes on the basis of outside safety distances for explosives involving the risk of mass explosion, ESTC, 3(7) Explosion, (1959) 43.

[7] Jarett, D., Derivation of the british explosives safety distances, Annals of the New York Accademy of Sciences, 152(1) (1968) 18-35.

[8] Rijnmond Report, Reidel, Dordrecht Netherlands, (1982).

[9] Smith, A.; Warwick, Second Survey of Defects in Pressure Vessels, SRD R30, London: HMSO, (1974).

[10] Bush, S., Pressure Vessel reliability, Trans. ASME 54-70, Febr. 1975.

[11] Kellerman, O., Unfallanalyse in der Kerntechnik, Technische Ueberwachung, 13(11), (1982).

[12] Sorensen, J. H., Evacuations due to off-site release from chemical accident: Experience from 1980 to 1984, J. Hazard. Mater., 14 (1987) 247-257.

[13] Canvey Report: An investigation of potential hazards from operation in the Canvey Island/Turrock Area, Her Majesty's Stationary Office, London (1975). See also the Annex "Ammonia release Incidents": Griffiths, R.F., Observations of non-buoyant behaviour of ammonia following accidental discharges to the atmosphere.

[14] MacArthur, J.G., Ammonia Storange Tank Repair, Ammonia Plant Safety, Symposium 1-3, (1971).

[15] Griffiths, R. F., Production of dense gas mixtures from ammonia releases - A review, in R.E. Britter and R.F. Griffiths (eds.), Dense Gas Dispersion, Chemical Engineering Monographs 16, Elsevier, (1982), pp. 197-212.

[16] McMullen, G., A review of the 11th May ammonia truck accident, City of Housten, Health Department Report (unnumbered) (1976).

[17] Lonsdale, H., Ammonia tank failure - South Africa, Ammonia Plant Safety, 17 (1975) 126-131.

[18] Railroad Accident Report: Louisville and Nashville Railroad Company freight train derailment and puncture of anhydrous ammonia tank cars at Pensacola, Florida, 9 November 1977, US National Transportation Safety Board, Report No. NTSB-RAR-78-4, (1978).

[19] Rail Road Accident Report: Chicago, Burlington and Quincy Railroad Company train 64 and train 824 derailment and collision with tank car explosion, Crete, Nebraska, February 18, 1969, US National Transportation Safety Board, Report No. NTS-RAR-71-2, (1971).

[20] Flixborough, Report of the Court Inquiry into the Flixborough disaster, Her Majesty's Stationary Office, London, (1975).

[21] Gugan, K., Institution of the Chemical Engineers, Gulf Publishing Co., Houston, (1979).

[22] Strehlow, R.A., Unconfined Vapor Cloud Explosions - An Overview, in 14th Symposium (International) on Combustion at Pensylvania State University, (August 1972).

[23] Davenport, J.A., A Study of Vapor Cloud Incidents, (Sept. 1977), 83rd National Meeting of the AIChE in Houston, Texas, (March 1977).

[24] Loss Prevention and Safety Promotion in the Process industries, IChemE 4th International Symposium on, 12-16 Sept. Harrogate, England, Vol. I - Safety in Operations and Processes, EFCE Publication Series No. 33, Vol. I, (1983).

[25] Badoux, R.A., Some experiences of a consulting statistician in industrial safety, Proceedings of the fourth Nat. Rel. Conferenc 3B, (1983).

[26] Wiekema, B.J., Analysis of vapour cloud accidents, Proc. of the fourth Euredata Conference, C1, Venice, (1983).

[27] Smith, A.; Warwick, Second Survey of Defects in Pressure Vessels, SDR R30, London: HMSO, (1974).

[28] Bush, S., Presure vessel reliability, Trans. ASME, p. 54-70, February 1975.

[29] Kellerman, O., Unfallanalyse in der Kerntechnik, Technische Überwachung 13, Nr. 11, (Nov. 1982).

[30] Marshall, J.G., I. Chem. E. Symposium Series 58(11), (1980).

[31] Canvey Second Report, London: HMSO, (1981).

[32] Wallis, J.B., Int. J. Multiphase Flow, 6 (1980) 97; see also: Akker, H.E. et al., Discharge of pressurized liquefied gases through apertures and pipes, I. Chem. E. Symposium Series 80, 1, E23, (1983).

[33] Pasquill, F., Atmospheric diffusion, Ellis Horwood, Chichester, (1974).

[34] Nicolet-Monnier, M.; Gheorghe, A.V., Integrated Regional Risk Assessment, Vol. I, Continuous and Non-Point Source Emissions - Air, Water, Soil, Kluwer Academic Publishers, Dordrecht, The Netherlands, (1995).

[35] Van Ulden, A.P.; Holstag, A.A. M., Estimation of atmospheric boundary layer parameters for diffusion applications, J. Climate and Appl. Met., 24 (1985) 1196-1207.

[36] Gryning, S.E.; Holstag, A.A.M.; Irwin, J.S.; Sivertsen, B., Applied Dispersion Modeling, (1987).

[37] Csanady, G. T., Austr. J. Phys., 8(4), (1955) 545.

[38] Chatwin, P.C., The influence of basic physical processes on the statistical properties of dispersing heavy gas clouds, Proc. 7th Biennial Symposium on Turbulence, Rolla, MO, (1981).

[39] Raj, P.K.; Hagopian, J.; Kalelkar, A.S., Prediction of Hazards of spills of anhydrous ammonia onto water, Report NTIS AD 779400, (1974).

[40] Lees and Ang, Safety cases - Within the Control of Major Industrial Accident Hazards (CIMAH) Regulations (1984), Butterworth & Co., London, (1989).

[41] Ramskill, Discharge Rates calcualtions Methods for Use in Plant Safety Assessments, UKAEA/SRD. SRD Report No. R352, HMSO London, (1987).

[42] Perry and Green, Perry's Chemical Engineering Handbook, 6th Edition, (1984).

[43] Crane and Co., Flow of Fluids through Valves, Fittings and Pipe - metric edition, Technical Paper No. 410M Crane, New York, (1981).

[44] Crozier, Sizing relief valves for fire emergencies, Chemical Engineering, 92 (22 Oct. 1978).

[45] Health and Safety Executive (1981). CANVEY - A second report, A Review of potential hazards from operations in theCanvey Island/Thurrock area three years after publication of the Canvey Report, HMSO London, (1981).

[46] Klein, H. H., Analysis of DIERS venting tests: Validation of a tool for sizing emergency relief systems for runaway chemical reactions, Plant/Operations Progress, 5 (1 January) ,(1986) 1.10.

[47] TNO (1978). Methods for the Calculation of Physical Effects of the Escape of Dangerous Materials Liquids and Gases. (The Yellow Book, 2 volumes). See the new edition: (1992).

[48] Technica. Manual of Industrial Hazards Assesment Techniques, Office of Environmental and Scientific Affairs, World Bank, Washington DC, (1985).

[49] Fauske, H. K., The Discharge of saturated water through pipes, CEP Symposium Series 59 (1965).

[50] Cude, A. L., The Generation Spread and Decay of Flammable Vapour Clouds, I.ChemE Course " Process safety - Theory and Practice", Teeside Polytechnic, Middlesborough, (July 1975).

[51] TECHNICA Ltd., Techniques for Assessing Industrial Hazards - A Manual World Bank Technical Paper No. 55, The World Bank, Washington DC, (1988).

[52] Fauske and Epstein, Emergency relief vent sizing for fire emergencies involving liquid filled atmospheric storage vessels, Plant Operation Progress 5, 4th October, (1986).

[53] AIChE/CCPS, Guidelines for Chemical Process Quantitative Risk Analysis, Center for Chemical Process Safety, American Institute of Chemical Engineers, New York, (1989).

[54] Webber, A Users' Guide to GASP on Microcomputers, UKAEA, SRD. SRD Report R521, (1990).

[55] Wu-Schroy, Emissions from Spills, Air Pollution Control Association and WPCF Joint Conf. on Control of Specific (Toxic) Pollutants, Air Pollution Control Association Florida, Section Gainesville, Florida, USA, (1979).

[56] Fleischer, SPILLS - An evaporation/air dispersion model for chemical spills on land, Westhollow Research Centre, Shell development Centre, Houston, Texas, USA, (1980).

[57] Van Aerde, M.; Stewart, A.; Saccomanno, F., Estimating the impacts of LPG spills during transportation accidents, J. Hazard. Mater., 20 (1988) 375-392.

[58] Pasquill, Atmospheric Diffusion: The Dispersion of Windborne Material from Industrial and Other Sources. 2nd Edition Wiley, New York (1974).

[59] Gifford, Turbulent diffusion - Typing schemes: A review, Nuclear Safety, 17 (1 Jan./Feb.), (1976).

[60] Hanna; Drivas, CCPS Guidelines for Use of Vapour Cloud Dispersion Models, Center for Chemical Process safety of the American Institute of Chemical Engineers, New York, (1987).
[61] Pasquill and Smith, Atmospheric Diffusion, 3rd Edition, Halstead Press - John Wiley, New York, (1983).
[62] McQuaid, Heavy gas dispersion trials at Thorney Islands - Journal of Hazard. Mater., 11 (June), (1985).
[63] McQuaid and Roebuck, Large Scale Field Trials on Dense Gas Vapour Dispersion, Commission of the European Communities, Report EUR 9903, (1985).
[64] Britter and McQuaid, Workbook on the Dispersion of Dense gases, HSE, UK Research Report 17/88, (1987).
[65] General Guidance on Emergency Planning within CIMAH Regulations for Chlorine Installations, Chem. Ind. Assoc., London, (1987).
[66] Thompson, J.R.; Nightingale, A.P.M., A simple method for determining the maximum consequences of notional toxic and radiotoxic gas discharges, J. Hazard. Mater., 17 (1988) 239-245.
[67] Hosker, R.P., IAEA-SM-181-19, Vienna, (1974) pp. 291-309.
[68] Chatwin, P.C., The use of statistics in describing and predicting the effects of dispersing gas clouds, J. of Hazard. Mater., 6 (1982) 213-230.
[69] Davies, P.C., Purdy, G., Toxic gas risk assessment - The effects of beeing indoors, in Refinement of Estimates of the Consequences of Heavy Toxic Vapour Releases, Institution of Chemical Engineers, North Western Branch, Manchester, 8th January 1986, Symposium papers No. 1, (1986).
[70] Davies, P.; Hymes, I., Chlorine toxicity criteria for hazard assessment, Chemical Engineer, (June 1985).
[71] Pape, R.P.; Nussey, C., A basic approach for the analyses of risks from major toxic hazards, The Assessment and Control of Major Hazards, Manchester, 22-24 April, 1985, London, The Institute of Chemical Engineers, IChemESymposium Series No. 93, (1985), pp. 367-388.
[72] Dick, J.B., The Fundamentals of Natural Ventilation of Houses, J. Int. Heat. Vent. Eng. (JIHVE), (June 1950), pp. 133-134.
[73] Warren, P.R.; Webb, B.C., Ventilation measurements in housing, Proc. of the Conference "Natural ventilation by design", Garston UK, (December 1980).
[74] Coblentz and Aschenbach, Residential Energy Consumption: Single family housing, Housing and Urban Development P72, (March 1973).
[75] Eisenbery et al., USCG Vulnerability Model, Report No. CG-D-53-77, (Feb 1977).
[76] Kronvall, J., Testing houses for air leakage using a pressurization method, ASHRAE Transactions, 84(I), (1978).
[77] Beech, R.K., Relative Tightening of New Housing in the Ottawa Area, National Research Concil of Canada, Division of Building Research, Building Research Note No. 148, (June 1979).
[78] Dick, J.B.; Thomas, B.A., Ventilation research in occupied houses, JIHVE, (June 1951), pp. 306-326.
[79] Brighton, P.W.M., Heavy gas distribution from sources inside buildings or their wakes, in Refinement of Estimates of the Consequences of Heavy Toxic Vapour Releases, Institution of Chemical Engineers, North Western Branch, Manchester, 8th January 1986, Symposium papers No. 1, (1986).
[80] Coward and Jones, Limits of flammability of gases and vapors, BM 1952 Bull. 503, (1952).
[81] Zabetakis, Flammability characteristics of combustible gases and vapors, BM 1965 Bull., 627 (1965).
[82] Handbook of Industrial Loss Prevention, FMEC, (1967).
[83] Simpson, I.C., Atmospheric Transmissivity - The Effects of Atmospheric Attenuation on Thermal Radiation, UKAEA Safety and Reliability Directorate, Culcheth, UK. Report SRD R-304 (1984).
[84] Considine, M., Thermal Radiation Hazard Ranges from Large Hydrocarbon Pool Fires, UKAEA/SRD, SRD-Report R297, HMSO, (1984).
[85] Liquefied PetroleumGas - (Model code of safe practice in the petroleum industry, pt.9), Institute of Petroleum, Wiley, UK.
[86] Bagster, Pool and Jet Fires - Major Industrial Hazard Projects, Warren Centre for Advanced Engineering, University of Sidney, NSW, Australia, (1986).
[87] API 521. Guide for Pressure-Relieving and Depressurizing Systems, API Recommended Practice 521, American Petroleum Institute, API, Washington DC, second edition, (1982).
[88] Hustad and Sonju, Radiation and Size Scaling of Large Gas and Gas-oil Diffusion Flames 10th International Colloquium on Dynamics of Explosion and Reactive Systems, American Institute of Aeronautics and Astronautics, New York,(1985).
[89] Marshall, Major Chemical Hazards, Ellis-Horwood Series in Chemical Engineering, John Wiley and Sons, (1987).

[90] Methods for the calculation of physical effects - Resulting from releases of hazardous materials (liquids and gases), CPR 14 E, Committeee for the Prevention of Disasters, TNO, The Hague, second edition, (1992), ISSN 0921-9633/2.10.014/9203.

[91] Lees, F.P., Loss Prevention in the Process Industries, Hazard Identification, Assessment and Control, vol. 1 and 2, Butterworths, London, (1980).

[92] Baker, W.E.; Cox, P.A.; et al., Explosion Hazards and Evaluation, Elsevier, Amsterdam, (1983).

[93] Hopkinson, B., Britisch Ordonance Board Minutes 13565, (1915).

[94] Zuckermann, S., Experimental study of blust injuries to lungs, the Lancet, 219, (August 1940).

[95] Bowen, I.G., .

[96]Glasstone, S.; Dolan, P. J., The Effects of Nuclear Weapons, Castle House Publications, London, (1980).

[97] Jarrett, D.E., Derivation of British explosive safety distance, Annals of the New York Academy of Sciences, 152 (1968) 18.

[98] Decker, D. A., An Analytical Method for Estimating Overpressure from Theoretical Atmospheric Explosions, Annual Meeting of Nat. Fire Prot. Ass., (1974).

[99] Eisenberg, et al., Vulnerability Model, A simulation system for assessing damage resulting from marie spills, Enviro Control Inc., Prepared for the US Coastguard. Report AD-A015-245, (June 1975).

[100] Strehlow, R. A., Unconfined vapour cloud explosion - An overview, Fourteenth Symp. on Combustion, Combustion Inst., Pittsburgh, Pa, (1973).

[101] Kletz, T. A., Unconfined vapour cloud explosions, An attempt to quantify some of the factors involved, Eleventh Loss Prevention Symp., Am. Inst. Chem. Engineers, New York, (1977).

[102] Harris, R. J., Gas Explosions in Building and Heating Plant, E & F.N. Spon., London, (1983).

[103] ACMH-2 (1979). Advisory Committee on Major Hazards, Second report, HMSO, London, (1979).

[104] Wiekema, Vapor Cloud Explosions - Chapter 8, TNO Yellow Book, Appeldoorn, The Netherlands, (1979).

[105] Wiekema, Vapour Cloud Explosions - An Analysis Based on Accidents, J. Hazard. Mater., Vol. 8 and 9, (1984).

[106] Wingerden, Vapour Cloud Explosion Blast Protection - Plant Operation Progress, Vol. 8(4), (Oct. 1989).

[107] Report on summery of ruptured tank cars involved in past accidents, Association of American railroads, AAR Chicago Res. Center, Railroad Tank Car safety Res. and Test Proj. RA-01-2-7, Chicago, Illinois, (1972).

[108] Summary of ruptured tank cars involved in past incidents, Association of American railroads, AAR Report R130, Washington DC, (1973).

[109] Holden, Assessment of Missile hazards: Review of Incident Experience Relevant to a Major Hazard Plant, UKAEA/SRD, London, SRD Report No. R477, (1989).

[110] Withers, J., Major Industrial Hazards - Their appraisal and Control, Gower Technical Press Ltd., (1988), pp. 127-137. (ISBN 0-291-39725-5).

CHAPTER 3

EFFECT MODELS

3.1. Introduction

The physical models discussed in the previous chapter considered the dispersion of airborne flammable or toxic materials, the creation of high levels of thermal radiation from various types of fires, the production of overpressures from explosions and the generation of missiles. This section will now consider the effects of these on people, property and the environment. TABLE 3.1 summarizes the elements involved when considering the assessment of damages to people and property.

3.2. Effects of Hazardous Material Dispersion (Toxicity Effect)

There are two main outputs from calculations of the way in which hazardous materials are dispersed in the atmosphere. The first is the determination of the concentration of flammable materials with a view to establishing the hazard ranges of these substances to some pre-determined concentration such as the *Lower Flammability Limit (LFL)*, or *Lower Explosive Limit (LEL)*. The results of these calculations are then used as inputs to the modeling and determination of the characteristics of fires and explosions. The effects of these will be considered under the heading of fires and explosions and so will not be discussed here. The main group of substances to be dealt with are therefore those which have toxic effects on plant and animal life.

The objective of using toxic effect models is to assess the consequences to man, animals and plants as a result of exposure to toxic materials. Considering first the effects on man it is difficult, for a variety of reasons, to evaluate precisely the toxic responses caused by acute exposures to toxic substances. Humans experience a very wide range of adverse effects which can include irritation, neurosis, asphyxiation, organ system damage and death. In addition the scale of these effects is a function of both the magnitude and duration of exposure.

There is also a high degree of individual response among different persons in a given population, due to factors such as general health, age and susceptibility. A further cause of difficulty is that there are known to be thousands of different toxic substances and there is by no means enough data (on even some of the more common ones!) on the toxic response of humans to permit a precise assessment of the hazard potential of a substance. In most cases the only data available are from controlled experiments with animals, under laboratory conditions.

TABLE 3.1. Damage assessment

Event	Vulnerable Resource	Injury / Damage Type		Injury / damage Cause	Factors for People (being outdoors, unsheltered)
Toxicity	People	- Death - Non-lethal injury - Irritation		Toxic vapor: - Concentration , or - Cumulative dose	- Concentration and time - Concentration
Explosion	People	- Death		- Direct blast - Impact	- Peak overpressure - Impulse
		- Non-lethal injury	- Ear drum rupture - Bone fracture - Puncture wounds - Multiple injury	- Direct blast - Impact - Flying fragments - Combination	- Peak overpressure - Impulse - Impulse - Impulse or peak overpressure and impulse
	Structures	- Structural damage - Glass breakage		- Direct blast - Direct blast	- Peak overpressure - Peak overpressure
Pool Burning	People	- Death - First degree burn		- Thermal radiation	- Duration/magnitude of thermal radiation
	Structures	- Ignition		- Thermal radiation	- Duration/magnitude of thermal radiation
Flash Fire	People	- Death - First degree burn		- Thermal radiation	- Effective duration/ magnitude of thermal radiation
	Structures	- Ignition		- Thermal radiation	- Effective duration/ magnitude of thermal radiation

The extrapolation of the effects observed in animals to the effects likely to occur in humans or indeed in other animals is not easy and is subject Frequency and Magnitude of Accidental Releases of Hazardous Materials to a number of judgments. The methods described in Vol. I of *Integrated Regional Risk Assessment - Continuous and Non-point Source Emissions*[1] to calculate the effect of toxic releases on the environment from continuous releases are also applicable to calculate the effects of accidental releases. Regarding the dispersion of gases heavier than air the relevant methods and models are described in the present volume II (Chap. 5).

3.2.1. PROBABILISTIC APPROACH TO TOXICITY

For most of the assessment the percentage of vulnerable resource, e.g. people injured or property damaged, can be related to the causative factors, using probit functions.[2,3]

There are a large number of references which give useful information on the methods of predicting the likelihood that a release event will result in serious injury or death. A number of substances in common have been examined in depth. In the UK, chlorine was considered by a sub-group of the UK I. Chem. E Major Hazards Assessment Panel, and associated publications - *Withers*[4] *(1985), Major Hazards*

Assessment Panel[5] *(1987), Withers & Lee*[8]*(1985)* have given an extensive review of the animal data for man. The same group has also reviewed ammonia - *Withers*[6] *(1986)* and a study is nearing completion of phosgene.

Definition of a Probit Function
A probit (P_T) is a probability unit lying between 1 and 10, which is directly related to the % fatalities. A probit function takes the following form:

$$P_T = a + b \log_e V$$

where the dependent variable P_T is a measure of the percentage of vulnerable resource affected and the independent variable V is some function of the factor that causes injury or damage to the resource. The coefficient, a, is the intersection with the y-axis and the coefficient, b, is the slope parameter. Both coefficients are computed by maximum likelihood estimation similar to the computation of coefficients for classical regression equations. The variable P_T is referred to as a probit (probability unit). The logarithm is conventionally used in probit equations, because it transforms the relationship between causative factor and response into a Gaussian function. This is frequently the case when the variable to be computed is a measure of percent.

If an attempt is made to estimate the proportion of the population which may suffer a defined degree of injury it is necessary to have information on the statistical distributions relating the probability of injury to the dose (total intake). Typically this is a log-normal distribution, but for these purposes it can take the form of a probit equation. Here the mean of the Gaussian normal distribution has been shifted in order to avoid negative values and takes a value of 5 instead of 0, with the same standard deviation of 1. The following two tables give the values corresponding to the transformation of percentage figures *p* (%) into probits (

TABLE 3.2a/b). The probits corresponding to the probabilities $\pi = 0$ *and* $\pi = 100\,\%$ are infinitely large.

Note: Probits are the five digit numbers in the table. Percent are read along the top and side margin of the table. The vertical column of percents gives the decade and the horizontal column gives the unit.

Definition: *The dose is the product of concentration level and duration*

The category of injury sustained by exposed people depends, in general, upon both the duration of exposure and the concentration level experienced. This dependence is non-linear. *This means that it is not appropriate to use the dose, as a variable, to assess response to irritant gases.*

The probit equation may be used, for instance, to relate the effect of an exposure to a given concentration and duration. The quantity ($C^n t$) is known as the *toxic load.*

The dependence of toxic gas lethality on concentration and time was found to be described by a non-linear function of the form ($C^n t$), where *C* is the concentration of the toxic gas, *t* is the time duration of the exposure, and *n* is an exponent depending on the specific gas. The corresponding probit equation takes then the form:

$$P_T = a + b \log_e\left(C^n t\right) \qquad (3.1)$$

where

P_T = The variable P_T (probability unit) is a measure of the percentage of people/resources affected (probits)

a, b, n = constants

C = concentration (ppm)

t = exposure time (min.)

TABLE 3.2a. Transformation of percentage figures into probits (after Fisher and Yates[7])

p (%)	0	10	20	30	40
0	...	3.7184	4.1584	4.4756	4.7467
1	2.6737	3.7735	4.1936	4.5041	4.7725
2	2.9463	3.8250	4.2278	4.5323	4.7981
3	3.1192	3.8736	4.2612	4.5601	4.8236
4	3.2493	3.9197	4.2937	4.5875	4.8490
5	3.3551	3.9636	4.3255	4.6147	4.8743
6	3.4452	4.0055	4.3567	4.6415	4.8996
7	3.5242	4.0458	4.3872	4.6681	4.9247
8	3.5949	4.0846	4.4172	4.6945	4.9498
9	3.6592	4.1221	4.4466	4.7207	4.9749

TABLE 3.2b. Transformation of percentage figures into probits

p (%)	50	60	70	80	90
0	5.0000	5.2533	5.5244	5.8416	6.2816
1	5.0251	5.2793	5.5534	5.8779	6.3408
2	5.0502	5.3055	5.5828	5.9154	6.4051
3	5.0753	5.3319	5.6128	5.9542	6.4758
4	5.1004	5.3585	5.6433	5.9945	6.5548
5	5.1257	5.3853	5.6745	6.0364	6.6449
6	5.1510	5.4125	5.7063	6.0803	6.7507
7	5.1764	5.4399	5.7388	6.1264	6.8808
8	5.2019	5.4677	5.7722	6.1750	7.0537
9	5.2275	5.4959	5.8064	6.2265	7.3263

If the gas concentration is varying with time the function ($C^n t$) must be replaced by the quantity:

$$\int C^n \, dt \tag{3.2}$$

which can be approximated by a finite sum. To evaluate the probit, the toxic load, ($C^n t$), must be calculated at positions of interest. At a given location the concentration will vary over time as the cloud passes and dilutes. The total toxic load for the location is obtained by considering different time steps and the average concentration during those time steps. Then for *m* time steps the *total toxic load* is given by an approximation to this integral:

$$\text{Total toxic load} = \sum_{i=1,m} \left(C_i^n \, t_i \right) \tag{3.3}$$

where

C_i = Duration of a time step
t_i = Concentration during that time step at a given location

The probit equations for assessing the lethal toxicity are given for:

<u>Chlorine inhalation</u>

$$P_T = -8.29 + 0.92 \log_e \left(C^2 \, t \right) \text{(Withers and Lees[8], 1985)} \tag{3.4}$$

$$P_T = -17.1 + 1.69 \log_e \sum_{i=1,m} C_i^{2.75} \, t_i \text{ (Eisenberg[2] et al., 1975)} \tag{3.5}$$

<u>Ammonia inhalation</u>

$$P_T = -35.9 + 1.85 \log_e \left(C^2 \, t \right) \text{ (Withers[9], 1986)} \tag{3.6}$$

$$P_T = -30.57 + 1.385 \log_e \sum_{i=1,m} C_i^{2.75} \, t_i \text{ (Eisenberg[2] et al., 1975)} \tag{3.7}$$

The assessment algorithms (probit equations) for toxic damage are highly dependent on the type of toxic substance released, whereas those for fire and explosion are independent of the substance type, although the variables used in the algorithm do depend on the type of substance. The toxic damage caused by irritant gases in general falls into three categories:

- Death
- Sub-lethal injury
- Irritation

TABLE 3.3 gives the constants for the lethal toxicity probit equation for a number of the more common chemicals.

The important factor in the determination of the effects of toxic material is to clearly study the known data about the material in question. These include the *MHAP monographs* for chlorine, ammonia and phosgene, publications by *NIOSH / OSHA*[10] *(1978)* and *Haber*[11] *(1986)*. In any case, before interpreting the results of an assessment involving toxic materials, agreement should be reached with those concerned about the concentration of toxic material which should be considered as various action levels or hazard indicators. Major sources of toxicity information are *Bridges*[12] *(1984), AIChE /*

CCPS[13] *(1988)*, there are also databases many of which are now computerized and some of which are on "Compact Disc-Read Only Memory" (so-called CD-ROM). These include *RTECS-NIOSH*[14] *(1987)* and *TOXLINE*[15] *(1990)*.

TABLE 3.3. Constants for the lethal toxicity probit equation

Substance	Constant *a*	Constant *b*	Constant *n*
Acrolein	-9.931	2.049	1.00
Acrylonitrile	-29.420	3.008	1.43
Ammonia	-35.900	1.850	2.00
Benzene	-109.780	5.300	2.00
Bromine	-9.040	0.920	2.00
Carbon monoxide	-37.980	3.700	1.00
Carbon tetrachloride	-6.290	0.408	2.50
Chlorine	-8.290	0.920	2.00
Formaldehyde	-12.240	1.300	2.00
chloride	-16.850	2.000	1.00
Hydrogen cyanide	-29.420	3.008	1.43
Hydrogen fluoride	-35.870	3.354	1.00
Hydrogensulfide	-31.420	3.008	1.43
Methyl bromide	-56.810	5.270	1.00
Methyl isocyanate	-5.642	1.637	0.653
Nitrogen dioxide	-13.790	1.400	2.00
Phosgene	-19.270	3.686	1.00
Propylene oxide	-7.415	0.509	2.00
Sulfur dioxide	-15.670	2.100	1.00
Toluene	-6.794	0.408	2.50

3.2.2. TOXICOLOGICAL CRITERIA

It should be noted however, that the information on dose-response relationships necessary to apply a probit function is only available for the chemicals shown in TABLE 3.3 (information on a few other chemicals is slowly becoming available). In circumstances where chemicals other than those listed in TABLE 3.3 are involved, or where a rapid result is required, it may be necessary to use of established toxicological criteria. Some are those listed below:

■ *Emergency Response Planning Guidelines (ERPG's) for Air Contaminants* issued by the American Industrial Hygiene Association (AIHA). Over 40 ERPG's have been developed.

■ *Immediately Dangerous to Life or Health (IDLH) Levels* established by the National Institute for Occupational Safety and Health (NIOSH). NIOSH has published IDLH concentrations to be used as acute toxicity measures for common industrial gases.

■ *Emergency Exposure Guidance Levels (EEGLs)* and *Short-term Public Emergency Guidance Levels (SPEGL's)* issued by the American National Academy of Sciences/National Research Council.

■ *Threshold Limit Values (TLV's)* established by the American Conference of Government Industrial Hygienists (ACGIH) including Short Term Exposure Limits (STEL's) and ceiling concentrations. The STEL's are designed to protect workers from acute effects resulting from exposure to chemicals. They are however designed as limits on exposure excursions lasting up to 15 minutes.

The use of established toxicity measures such as ERPG's, IDLH, EEGL's, SPEGL's and TLV/STEL is usually a simpler approach than the probit model - and is also applicable over a wider range of chemicals. However the results are more difficult to interpret when the release is over a different time period to that published.

Definitions
An EEGL is defined as a concentration of a gas, vapor or aerosol that is judge to be acceptable and that will allow exposed individuals to perform specific tasks during emergency conditions lasting from 1 to 24 hr.

SPEGL's are defined as acceptable concentrations for exposures of members of the general public. These are generally set at 10 to 50% of the EEGL and are calculated to take account of the effects of exposure on sensitive populations. Their advantage over IDLH values are that they have been developed for a range of exposure times and consider the effects on sensitive populations.

An IDLH level represents the maximum airborne concentration of a substance to which a healthy male worker can be exposed for as long as 30 minutes and still be able to escape without loss of life or irreversible organ system damage. Because IDLH's were developed to protect healthy worker populations they must be adjusted to account for sensitive populations.

3.2.3. THE TOXIC EFFECT MODEL (TEM)

In Switzerland, the *Toxic Effect Model (TEM)*[16] is used on a regulatory basis for assessing the toxicity risks to the population resulting from the dispersion of toxic gases. The documentation and tools (TEM model) can be purchased from the Swiss Federal Printing Office, EDMZ ("Eidgenössische Drucksachen- und Materialzentrale) in Bern.[17] A computerized version is also sold in the United States. TEM is a puff model which simulates the dispersion of toxic gases, taking into consideration the different weather conditions (Pasquill category scheme), and can cope with various types of intoxication. The model (*Figure 3.1*) requires as input a source term (total quantity of toxic gas released suddenly as a puff), thus due to this assumption, the results will be of conservative nature. Further the molecular mass of the gas, atmospheric pressure and temperature are required. There are different modules, of which one calculates the dispersion and concentration of the gas; a second module calculates the doses at different distances, on the basis of Concentration data, and a third module estimates the acute toxic effects to people, coupling dose and effects.

Dispersion. The assumptions for calculating gas dispersion are as follows: it is assumed that the gas is fully mixed with air. The dispersion is taking place in the form of a half-sphere. The diameter of this half-sphere is changing proportionally to the distance from the source. The possibility and influence resulting from a fire is also taken into consideration, as well as the weather conditions. The model indicates the same concentration overall inside the half-sphere. The diffusion parameters are assumed to be the same in all the three directions. It is therefore not necessary to know the temperature of the gas, air, and soil. The model is using an average concentration of the gas cloud for its calculation.

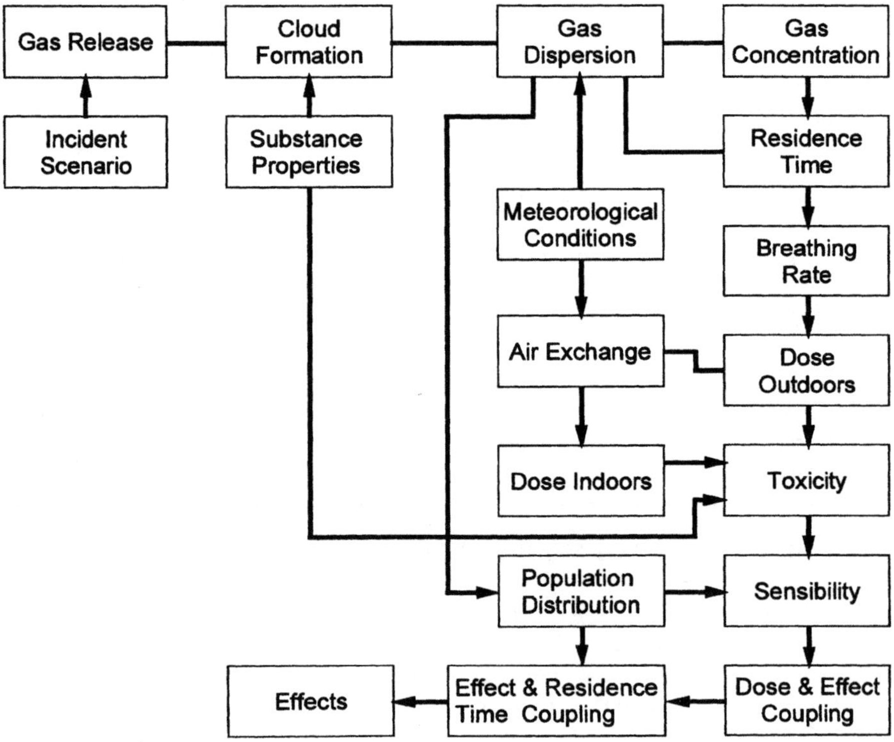

Figure 3.1. Description of the Toxic Effect Model (TEM) and associated sub-modules

Dose Inside and Outside Buildings.

Dose Calculation. In order to estimate the acute effect from a gas intoxication it is necessary to know exactly the accumulated dose, which was inhaled, after a certain concentration level was reached. Such a doe can be estimated using either the *rule of Haber* or the *concentration - time product*:

$$Dose = k\,c\,t \tag{3.8}$$

where

k = constant
c = concentration
t = inhalation time

In above equation (3.8) we substitute for k the breathing rate, and for the concentration c, we correct this concentration by subtracting a value corresponding to the lower concentration limit:

$$Dose = B\,t\,\left(\left|(c-c_l)\right|+(c-c_l)\right)/2 \tag{3.9}$$

where

B = breathing rate
c_l = lower concentration limit
$|..|$ = absolute quantity

Using formula (3.9) allows to conveniently average concentration errors affecting the calculation of the dose. This means, that a slow gas release will produce low concentrations, and the residence time in the toxic gas cloud will be larger for the same dose. On the contrary large concentrations will reduce the possible residence time. In that sense the calculation of the dose makes the model independent of the incident scenario and eliminates most of the uncertainties of the usual models.

The calculation of the dose cannot, of course, always follow exactly above equation, since lipophilie and gas transfer in the lungs, and other body compartments, plays a role. The physiological factors can be taken into consideration by means of the toxicological data, e.g., the coupling of doses and effects.

Doses Inside Buildings. If we consider the air exchange rate between the open air and a closed building, it is then possible to calculate the indoor gas concentration and hence the dose inside the building. For that purpose we need to know two parameters, namely the gas exchange rate and the temperature difference between indoors and outdoors. The model does also consider to a lesser extent the wind dependency of the gas exchange rate.

Dose-Effects

The coupling of *dose - effects* is, when considering the dispersion of toxic gases, nothing new, but it was up to now based on the use of *probit functions* (Bowen[18], 1978; Poblete[19], 1984), which must be determined empirically for each substance. Two factors are required, at minimum, but these are only known for a few substances, and the values found in the literature have a broad dispersion range. Another problem is due to the fact that between the factors of the probit function, that correspond to a normal distribution of the population, and the toxicological data, there is hardly any correlation involving the molecular functional groups. In the TEM model the approach is different, since use is made of a correlation between doses and effects, which is based on the mass action law followed by a linear coupling of the effects (Ariens[20] et al., 1978; Lüpke[21] et al., 1981). Simultaneously, it is assumed that the receptor /drug ratio may be non-linear, whereby the slope of the curve, i.e., the sensitivity, will be considered. It is sufficient to only consider the relative effect:

$$E_R = E_{obs} / E_{max} = D^{SF} / \left(D^{SF} + k_d\right) \tag{3.10}$$

where

E_R = relative effect
E_{obs} = observed effect
E_{max} = maximal effect
SF = Stoichiometric factor
k_d = dissociation constant (LD_{50}, or LC_{50})

If we take for the dissociation constant a value corresponding to LC_{50}, then the transfer factor from the air to the corresponding organ is considered, and the coupling between concentration, dose, and effects is fairly good. On the basis of these data, it is possible to evaluate the maximal distance, at which lethality may be expected. By taking some simplification into account, it is also possible to estimate the maximal distance, at which the irritability threshold will be reached.

The dose-effect - coupling does also reckon with the distribution of the population's sensibility. Conservative prognostics can be obtained for sensitive population groups, by setting the SF-factor (Eq. 3.10) to be smaller than one.

A comparison of the results obtained using TEM, with the *risk contours* as they were deduced mathematically by Cox and Comer[22] (1982), who applied a *Parametric Correlation Method (PAM),* shows in both cases the well known sigmoid-type of curves.

3.3. Effects of Thermal Radiations

Fire is a main cause of loss in the process industries. Most accidents involving large loss of life which are reported in the loss prevention literature are explosions. Explosions are frequently followed by fires, but it is usually the former which are most lethal. The hazard potential of fire is generally judged to be less than that of explosion or toxic release.

A large fire may give rise to effects both on man and on buildings and equipment. Although all modes of heat transfer are involved to some degree, the most important on open plant is radiation. Two of the most important types of fire are flash fires and pool fires. The models given in Chapter 2 describe the fire in terms of the thermal radiation intensity and the time duration. It is the same two parameters which are required in the relations used to estimate injury and damage.

The modeling of high thermal radiation effects which are likely to cause injury or damage to people and property is much more straightforward than for toxic effects. The relations between thermal radiation dose and fatalities are those given by the probit equation given below. ***Both explosion and flash fire must be considered when estimating a population at risk.*** The population at risk is defined in terms of the cell model (i.e., taking a risk corridor and subdividing the area in squares of 100 m with the corresponding population density). The results are sensitive to the location of the population cells relative to the wind direction (especially for a vapor cloud drifting away).

3.3.1. EFFECTS ON PEOPLE

A large fire may give rise to the following effects on man: (a) death, and (b) severe burns. In order to estimate the effects of fire on man it is necessary to know the relationship between the thermal radiation intensity-time profile, or dose, and the degree of injury. Firemen as well as people trapped in a building in fire may also be intoxicated by the development of noxious vapors or asphyxiated by lack of oxygen or the presence of inert gases such as nitrogen, carbon dioxide or halon.

TABLE 3.4. Estimated relations between heat radiation intensity and burn injury (after Eisenberg et al., 1975)

Burn Injury Effects	**Heat Radiation Intensity (cal/cm^2s)**		
Explosion yield	**20 kt**	**1 Mt**	**20 Mt**
First degree burn	1.75	0.297	0.0886
Second degree Burn	3.14	0.643	0.221
Slightly clothed: few, if any injuries	1.75	0.297	0.0886
Significant injury threshold	2.80	0.594	0.210
Lethality threshold	3.50	0.792	0.243
• near 50%	6.30	1.385	0.442
• near 10%	14.0	3.07	0.952
Effective time Duration, *s*	1.43	10.10	45.20

A large amount of experimental data exists and a large number of simple tabulations, charts and theoretical models are available. Most of these charts, models etc. refer to bare skin. The effects can be considerably modified due to the presence of such factors as clothing (which most probably will protect but in a few cases may make the situation worse), instinctive responsive (to turn and run away) and the existence of solar radiation exposure in sunny climates.

TABLE 3.4 gives an indication of the burn injury effects on people as a function of the radiation intensity. The data show the relation between thermal radiation, which is the time integral of the radiation intensity, and burn injury for ***nuclear explosions*** of different yields (expressed as the kilo-ton (kt) equivalent of TNT-explosive, i.e., tri-nitrotoluol, taken as reference).

TABLE 3.5. Consequence effects of heat radiation on people and equipment

Incident Flux (kW/m^2)	Damage to Equipment	Damage to People
37.5	• Damage to process equipment • Cellulosic equipment will pilot ignite within one minute exposure	• 100% lethality in 1 min. • 1% lethality in 1 s
25.0	• Spontaneous ignition of wood after long exposure • Unprotected steel will reach thermal stress temperatures which can cause failure • Pressure vessels need to be relieved or failure will occur	• 100% lethality in 1 min. • 1% lethality in 1 s
12.5	• Minimum energy to ignite wood with a flame • Melts plastic tubing • Thin steel wit insulation on the side away from the fire may reach a thermal stress level high enough to cause structural failure	• 1% lethality in 1 min. • 1st degree burns in 10 s
4.7		• Causes pain if duration is longer than 20 s, but blistering is unlikely. • Possible injury after 30 s exposure
2.1		• Minimum to cause pain after 1 min.
1.6		• Causes no discomfort for long exposure

Eisenberg, et al.[2] *(1975)* developed a probit model to estimate the injury levels for a given thermal radiation dose from pool and flash fires based on data from nuclear tests:

$$Y = -14.9 + 2.56 \ln\left(t\, I^{4/3} \cdot 10^{-4}\right) \tag{3.11}$$

where

Y = Probit function
t = exposure time in sec.
I = thermal radiation intensity (W/m^2 or $J\ m^{-2}\ s^{-1}$)

TABLE 3.5 shows the effects and consequences from heat radiation on people and equipment, whereas *Figure 3.2* taken from *Mudan (1984)* shows a simple relationship between incident thermal flux, time and damage (injury/fatalities).

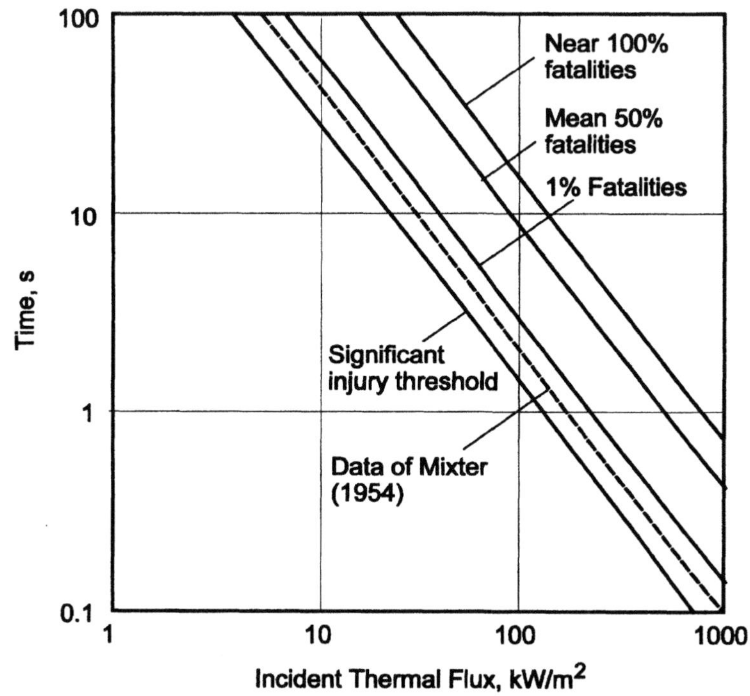

Figure 3.2. Fatality and injury levels for thermal radiations

3.3.2. EFFECTS ON BUILDINGS

The effect of a large fire on buildings and equipment is to cause ignition. The problem is frequently treated by considering the ignition of wood as a reference value.

For the purpose of injury, a lower heat radiation level (relative to that level which may cause fatality) is appropriate. A heat radiation level of 4.7 kW/m^2 is considered high enough to trigger the possibility of injury for people who are unable to be evacuated or seek shelter. That level of heat radiation would cause injury after 30 seconds' exposure. Accordingly, a risk injury criteria of 50 in a million per year (50 x 10^{-6} per year) at the 4.7 kW/m^2 heat flux is suggested for residential areas (TABLE 3.6). International experience with the implementation of that criteria indicates that it is achievable and appropriate.

TABLE 3.6. Effects of heat radiation

Heat Radiation (kW/m^2)	Effect
1.2	• Radiation received from the sun at noon in summer
2.1	• Minimum to cause pain after 1 minute
4.7	• Will cause pain in 15-20 seconds and injury after 30 seconds exposure (at least second degree burns will occur)
12.6	• Significant chance of fatality for extended exposure. High chance of injury • Causes the temperature of wood to rise to a point where it can be ignited by a naked flame after long exposure. • Thin steel with insulation on the side away from the fire may reach a thermal stress level high enough to cause structural failure.
23.0	• Likely fatality of extended exposure and chance of fatality for instantaneous exposure. • Spontaneous ignition of wood after long exposure. • Unprotected steel will reach thermal stress temperatures which can cause failure. • Pressure vessels needs to be relieved or failure would occur.
35.0	• Cellulosic material will ignite within 1 minute's exposure • Significant chance of fatality for people exposed instantaneously to the fire

Source: Department of Planing of the State of New South Wales, Advisory Paper 4.

The approach taken by Eisenberg for ignition by a flash or pool fire is to use the relations given by Lawson and Simms[23] (1952) in the way indicated below:

Pool Fire

Use the spontaneous ignition equation given by Lawson and Simms[23] for pool fire:

$$(I - I_s)t^{4/5} = k_1 \tag{3.12}$$

with I_s = 25,400; and k_1 = 6,730

where

I = Thermal radiation intensity (W/m^2)
I_s = Critical intensity for spontaneous ignition (W/m^2)
t = Time (s)

k_2 = Constant (J/m^2 s$^{1/5}$)
The radiation intensity I must exceed the critical value I_s and the duration of burning must exceed the time t given by the equation above.

Flash Fire
Use the pilot ignition equation given by Lawson and Simms[23] for flash fire:

$$\left(I - I_p\right)t^{2/3} = k_2 \tag{3.13}$$

with I_p = 13,400; and k_2 = 8,050
where k_2 = Constant (J/m^2 s$^{1/3}$)

The thermal radiation intensity I must exceed the critical value I_p and the effective exposure time t must exceed the value resulting from above equation for ignition to take place.

Discussion
Thermal effect models are simple and are based on extensive experimental data. The main weakness arises if duration of exposure is not considered. Thermal effect data relates to bare skin and it is necessary to take account of the effect of clothing and sheltering.

3.4. Effects of Explosions

The second of the major hazards is explosion. Explosion in the process industries causes fewer serious accidents than fire but more than toxic release. When it does occur, the resultant loss of life and damage are greater than for fire.

An explosion involves a process whereby a pressure or blast wave is generated in air by a rapid release of energy. The front of this wave can cause damage to the objects it impacts as it passes through the air. An explosion may give rise to the following effects: (a) blast damage; (b) thermal effects; (c) missile damages; (d) ground shock; (e) crater, and (f) personal injury. Not all of these effects are given by every explosion. An explosion at some distance from the ground will hardly cause a crater.

A survey of blast effects caused by accidental explosions has been made by Robinson[24] (1944). Information on the effect of air blast on a wide variety of objects have been given by Glasstone[25,26] (1962 / 64). There is a considerable amount of data on the effects of air blast on buildings (Clancey[27], 1972). A description of the effects of the overpressure in the Flixborough explosion has been given by Sadée, Samuels and O'Brien[28] (1976-77).

The objective of explosion effect models is to predict the impact of blast overpressure on people and structures.

3.4.1. BLAST DAMAGES

Damages Due to Overpressure Effects

It is common practice to relate the blast overpressure and duration time to the quantity of explosive material and to try to relate the pressure-time history of the blast wave at a specified distance from the explosion source to a damage criterion such as percentage facilities.

There have been a number of different approaches to determine the response of structures to a given blast load. A number of these draw a comparison between the magnitude of the predicted blast wave and existing data from explosions of a similar scale. Other approaches attempt to model the response of a structure to an applied load. Much of the data on explosions comes from military experience but a number of large industrial explosions have been investigated in depth.

TABLE 3.7. Effects of explosion overpressure on buildings

Explosion Overpressure (kPa)	**Effects**
3.5	• 90% glass breakage • No fatality and very low probability of injury
7.0	• Damage to internal partitions and joinery (repairs are possible) • Probability of injury is 10%. No fatality.
14.0	• House inhabitable and badly cracked
21.0	• Reinforced structures distort • Storage tanks fail • 20% chance of fatality to a person in a building
35.0	• House inhabitable • Wagons and plant items overturned • Threshold for eardrum damage • 50% chance of fatality for a person in a building and 15% chance of fatality for a person in the open air.
70.0	• Threshold for lung damage • 100% of chance of fatality for a person in a building or in the open air • Complete demolition of the house.

Note: 1 Pa = 1 N/m² → *Note:* $1\ Pa = 1\ N/m^2$

A probit equation has been derived by Fugelso et al.[29], (1972) for *damage to frame structures* from an explosion equivalent to 500 ton of TNT. This probit equation relates structural damage to peak overpressure.

$$Y = -23.8 + 2.92 \ln p^o \tag{3.14}$$

where

p^o = peak overpressure (kPa)

A 50% damage contour has been defined for residential houses (English type). This can be calculated from the following formula:

$$L_{50} = \frac{13.2\ 10^{-2} T^{1/3}}{\left[1 + \left(3.175/T^2\right)\right]^{1/6}} \tag{3.15}$$

where

T = Explosive mass (tons)
L_{50} = Distance for 50% damage to residential constructions (km).

As a first approximation this contour can be considered as equivalent to a 50 per cent damage contour, where as a result of a specified explosion, the number of inhabitable houses outside the contour are balanced by the number of inhabitable houses inside the contour.

The *shattering of glass* is an important damage effect, since flying glass can cause severe injuries, but the strength of window glass is very variable. The following probit equation relates glass breakage to peak overpressure.

$$Y = -18.1 + 2.79 \ln p^0 \tag{3.16}$$

TABLE 3.7 outlines consensus correlation between residential building damage and blast overpressure.

Using a similar analysis to that adopted in establishing a heat flux injury level, it can be suggested that an explosion overpressure level of 7 kPa be the appropriate cut-off level above which significant effects to people and property damage may occur. Accordingly, an injury/damage risk criteria of 50 in a million (50 x 10^{-6} per year) at the 7 kPa explosion overpressure level is suggested for residential areas. International experience with such implementation confirms this level as appropriate.

It should be noted that this correlation is applicable to standard European or North American brick built dwellings and much more severe damage would be experienced by less strongly constructed buildings.

The damage to industrial buildings is less easy to correlate since these range from buildings with strong reinforced concrete walls to lightly constructed buildings with large wall and roof areas.

Determination of Damage Circles for an Unconfined Vapor Cloud Explosion

The relation below indicates the damage radius for certain types of damages (TNO[30], 1992):

$$R = C(\eta E)^{1/3} \tag{3.17}$$

where

R = Radius of a particular damage circle (m)
C = Limit value for a characteristic damage ($mJ^{-1/3}$)
E = Energy content of the explosive part of the gas cloud (5×10^{-9} J < E < 5×10^{-12} J), i.e., the cloud has a mass equivalent to 100 kg - 100 t
η = Yield factor, and $\eta = \eta_c \, \eta_m$
η_c = Yield loss on account of the non-stoichiometry of the cloud ($\eta_c = 30\%$)
η_m = 33% for isochoric combustion, or respectively 18% for isobaric combustion.

The value of E (Joule) is obtained by multiplying the mass of the gas cloud times the heat of combustion (J/kg), listed in the *"Perry, Chemical Engineers' Handbook"*, after a unit conversion (cal/gmol ↔ J/kg). The different radii corresponding to the maximum distance for various expected damages are finally obtained by substituting the corresponding C values in the formula. The values of C are given in TABLE 3.8 for UVCE:

TABLE 3.8. Limit values for characteristic damage types

C ($m.J^{-1/3}$)	Type of Damage
0.03	Heavy damage to buildings and processing equipment
0.06	Repairable damage to buildings and facades damage to dwellings
0.15	Glass damage causing injury
0.4	Glass damage (~10% of panes broken)

Missile Damage. If the explosion occurs in a closed system, fragments of the container may form missiles. In addition, objects may also be turned into missiles by the blast from either a confined or unconfined explosion (e.g., the top cover of a canalization).

On certain assumptions it is possible to make approximate estimates of the behavior and effects of the fragments from a container in which an explosion has occurred. The problem may be considered under the following aspects: *(a)* energy, *(b)* velocity, *(c)* traveling range, and *(d)* penetration. Most of the fragments do not travel the maximum distance, but fall at distances between 0.3 and 0.8 of the maximum. Clancey[27] (1972) has proposed a correlation for calculating the maximum horizontal range of fragments from a cased charge on the ground surface. The correlation is based on experimental determination with TNT explosions in lightweight containers.

Ground Shock. The ground shock produced by an explosion may be regarded as a sinusoidal disturbance and is characterized by its frequency and amplitude, and also by its maximum velocity and acceleration. For charge exploding at the ground surface the ground shock depends mainly on the quantity of explosive and on the nature of the ground. The damage done by ground shock to normal houses has been correlated by

Langefors and Kihlstrom[31] *(1963)*. For an explosion at or near the surface damage by blast will extend to a much greater distance than damage by ground shock.

Crater Formation. The factors which affect the crater produced by an explosion are the position of the charge relative to the ground surface, the nature of the ground, and the type and quantity of explosive (high/low brisance explosive). The explosion at Flixborough did not make a crater.

Overpressure Effects on People

A large explosion can cause injury to man mostly through the following effects: (a) heat radiation, (b) blast, and (c) combustion products. The effects of heat radiation have been described previously.

Injury from blast includes:

- Direct blast injury
- Indirect blast injury (due to overpressure), second blast injury (due to missiles), and tertiary blast injury (as a result of whole body translation).

Injury may also be caused by hot, toxic and dusty gases produced by the explosion.

It so happens that people are much more resilient to blast overpressures than structures. The major threat to people is produced by missiles, structural collapse or whole body translation. Death or injury to humans which arises directly from the blast overpressure alone varies with the position of the body and its relationship to possible pressure reflecting objects. The effect of blast overpressure depends on the peak overpressure, the rate of rise and the duration of the positive pressure phase. Human organs which are particularly susceptible to direct blast effects are those where a large difference in density exists between adjacent organs such as the ears and lungs.

Much of the data have been derived from nuclear experience and may slightly overestimate the fatalities from non-nuclear explosions. Information on injury to people from explosions has been given by Glasstone[25] (1962), White[32] (1968), White[33] (1971), Fugelso[29] et al., (1972), and Eisenberg[34] et al., (1975).

A probit equation has been derived for lethality from direct blast effects (Eisenberg[34] et al., 1975).

$$Y = -77.1 + 6.91 \ln p^0 \tag{3.18}$$

where p^o is the peak overpressure (N/m^2).

A similar probit equation has been derived which relates eardrum rupture from direct blast effects to peak overpressure:

$$Y = -15.6 + 1.93 \ln p^0 \tag{3.19}$$

For other type of injury use is made of the blast impulse which is define as:

$$J = \int_0^{t_d} p(t)\,\mathrm{d}t \tag{3.20}$$

where

J = Impulse (N s/m^2)
p = Overpressure (N/m^2)
t_d = Duration time (s)

A probit equation relating serious injury from missiles, particularly glass, has then been derived by using the impulse value J:

$$Y = -27.1 + 4.26 \ln J \tag{3.21}$$

It is assumed in above equation that all persons not inside buildings who are in a region traversed by a blast wave of sufficient strength will suffer injury from glass missiles. These are defined *(Eisenberg et al.)* as flying fragment of glass of 10 g (density 2.65 g/cm^3). This probit equation represents an upper bound and overestimates the extent of injury from flying fragments.

Another probit equation is relating lethality for whole body translation to blast impulse:

$$Y = -46.1 + 4.82 \ln J \tag{3.22}$$

Glasstone and Dolan[35] (1980) propose a scaling law based on experimental work using animals and dummies (70 kg man equivalent) for calculating the distance corresponding to 50% casualties:

$$d = d_r W^{0.4} \tag{3.23}$$

where

d = Distance for 50% casualties
W = Mass of explosive in kilotons (kt)
d_r = Distance corresponding to a one kt explosion

The following formula can be regarded as the 50% lethality contour for primary blast deaths, giving the distance in terms of the primary independent variable, explosive mass:

$$L_{50} = \frac{2\ 10^{-2}\ T^{1/3}}{\left[1 + (20/T)^2\right]^{1/6}} \tag{3.24}$$

where

T = Explosive mass (tons)
L_{50} = Distance (km)

Short-duration time pressure waves associated with small explosions require a much higher pressure for lethal effect than do the longer duration time (associated with larger explosions).

The probit relating serious injury from whole body translation to blast impulse is given below:

$$Y = -39.1 + 4.45 \ln J \tag{3.25}$$

The aforementioned probit relations for injury from explosion are applicable to exposed populations in general.

For the purpose of hazard assessment, the ratio of deaths to inhabitable houses has been estimated to be of the order of 1 to 10-15, and the ratio of death to injuries is of the order of 1 to 10 for industrial chemical and gas explosions. Taking an average figure for home occupancy of 2.5 (corresponding to the UK conditions), the number of casualties in a 45° sector is given by the formulae (3.23):

$$n = D(q + s/25) \tag{3.26}$$

with $q = (x - v)/8$, and $s = (z - v)/8$.
where

x and z = Primary and secondary damage zone around the hazard source
D = Population density.

TABLE 3.9. Effects of explosion overpressure on people and buildings

Kind of Explosion Overpressure Effects	Overpressure (bar)
(A) Effects on Buildings	
Building almost completely destroyed	0.70
Heavy building damage	0.35
Repairable building damage	0.10
Widespread glass damage	0.05
10 % broken glass	0.02
(B) Effects on People	
100 % lethality	5 - 8
50 % lethality	3.5 - 5
Threshold lethality	2 - 3
Severe lung damage	1.33 - 2
50 % eardrum rupture	2 - 2.33 (over 20 years of age)
50 % eardrum rupture	1 - 1.33 (under 20 years of age)

TABLE 3.10. Primary and secondary deaths resulting from an explosion in a populated area (assuming 4,000 persons per km^2)

Explosive Mass	Primary Deaths		Secondary Deaths		Total Deaths (corrected)
(t)	Radius (m)	Number	Radius (m)	Number	
1,000	170	363	1310	862	1210
100	80	80	607	186	263
10	34	14	256	33	47
1	13	2	88	4	6
0.1	5	1	19	0	1

The clear space, in above equation, is given by Eq. (3.27):

$$v = \pi R^2 \tag{3.27}$$

where R is the distance (m) from the hazard source to the population boundary, and ν is the clear space (m^2).

Because of the much larger 50% damage radius there will be more fatalities arising from secondary effects than from primary effects, even though only one fatality is to be expected for every ten homes destroyed.

TABLE 3.9 summarizes the effects of explosion overpressure on people and buildings, whereas TABLE 3.10 illustrates the effect of an explosion in a populated area.

Discussion
The strength of explosion effect models is their solid base of experimental data and the general simplicity of approach. Attention has been drawn to the fact that people may be killed indoors due to building collapse at a much lower overpressure than outdoors due to overpressure alone. Explosions in built-up areas are rarely uniform in effects. UVCEs are often directional and this effect is not accounted for in many current effect models.

3.5. Mitigating Effects

The objective of this section is to draw attention to some of the factors which may mitigate against the consequences of incident involving hazardous materials.

It has been observed in many accidents that the consequences to people and property were less severe than would have been predicted using the approaches described earlier. Obviously there are uncertainties in all the various stages of analysis and there are also modeling limitations which may lead to conservative assumptions and hence results. However, in addition to these factors, the results may be less serious then predicted to topographical factors, physical obstructions and to evasive action taken by people. Such evasive action can include evacuation, sheltering and medical treatment. These are briefly described thereafter.

Evacuation. This is a mitigating factor which can only be usefully employed if there is sufficient time for it to be effectively carried out. Evacuation is not without its own risks - useful references include ***Prugh***[36] ***(1985)*** and ***Aumonier and Morrey***[37] ***(1990)***.

Sheltering. It has been observed that, following an incident, the effects on people who take shelter differ markedly from those for people in the open. This has been discussed by *Davies and Purdy*[38] *(1986)* in relation to building types and human behavior. The effects of sheltering depend on:

■ The nature of the hazard - shelters can have a beneficial effect for thermal and toxic effects but can be of limited benefit for flash fires due to the possibility of vapor ingress. In the case of explosion overpressure the hazards may be increased due to the increased risk of collapse of the structure providing shelter.

■ The time available - escape to a shelter can be very beneficial in the case of pool and jet fires. There may well be insufficient time to shelter from a fireball and there may be no time to escape from explosion overpressure or missiles. There may be benefit in sheltering from releases of toxic materials, particular if time allows to reach shelter before there has been a significant exposure. However where the shelter has been exposed to a cloud of toxic material for some time it should be recognized that, once the outside concentrations decrease, an indoor concentration, albeit lower than the peak values experienced outdoors, may persist for some time and the total exposure could be reduced by leaving the shelter once the cloud outside has passed.

Medical Treatment. The effectiveness of training and the availability of equipment for emergency response and medical treatment can greatly improve the chance of survival for those seriously injured as a consequence of an incident involving hazardous materials. Of particular interest to those treating persons exposed to toxic materials will be the name, and the basic hazards of the material(s) involved. Modern methods of treating those who have experienced severe burn injuries have greatly increased the chances of survival. It should however be recognized that whereas facilities may exist for treating a few seriously burnt people at the same time there may be problems in treating tens or even hundreds of such people.

Discussion
The omission of mitigating effects in a risk assessment will nearly always lead to an overestimation of the numbers of casualties. There are considerable uncertainties in estimating the factors that account for evasive actions. For these reasons many risk assessments do not consider the effects of mitigating effects such as sheltering and evacuation - in such cases the possibility of having overestimated the number of casualties should be acknowledged.

Summary (Chapter 3)

This chapter considers the effect of the dispersion of airborne flammable or toxic materials, the creation of high levels of thermal radiation from fires, the production of over pressure from explosions, including missiles and the assessment of damages to people and property. The probit function concept is introduced and used to estimate the risk from various substances. Models for calculating the effects on people and buildings from fire are based on expe5rimental data. Effects of explosions calculated in this chapter include mainly the blast damages. Some recommendations (e.g. evacuation, sheltering, medical treatment) on mitigating effects in case of fire and explosions are included.

References (Chapter 3)

[1] Nicolet-Monnier, M.; Gheorghe, A.V., Integrated Regional Risk Assessment - Continuous and Non-point Source Emissions, Vol. I, Kluwer Academic Publishers, Dordrecht, The Netherlands, (1995).

[2] Eisenberg, N.A.; Lynch, C.J.; Breeding, R.J., Vulnerability model, Nat Tech. Inf. Service report AF-A105-245, Springfield, Va., (1975).

[3] Withers, J., Major Industrial Hazards, Their Appraisal and Control, Gower technical Press, (1988).

[4] Withers, J., Toxic Effects of Chlorine, First Report of the Toxicity working Party, in Ellis Horwood (ed.), Modern Chlor-Alkali Technology, Vol. 3. (1985).

[5] Major Hazard Assessment Panel. The Chlorine Toxicity Monograph., UK I. Chem. Eng., Rugby, England, (1987).

[6] Withers, J., The lethal toxicity of ammonia, Refinement of estimates of the consequences of heavy toxic vapour releases, I. Chem. Eng., New Branch, Paper No. 1, (1986).

[7] Fisher and Yates, Statistical Tables for Biological, Agricultural and Medical Research, Oliver and Boyd Ltd, Edinburgh.

[8] Withers, R.M.J.; Lees, F.P., The Assessment of Major Hazards: - the lethal toxicity of chlorine, Part 2, Model of toxicity to man, J. Hazard. Mater. 12 (1985) 283.

[9] Withers, R.M.J., Second Report of MHAP Toxicity Panel, I. Chem. E. NW Branch, Symposium Papers No. 1 (6.1) (1986).

[10] NIOSH / OSHA (1978). Pocket Guide to Chemical Hazards, US Dept. of Health and Human Services, and US Dept. of Labor. Available from: US Govt. Printing Office, Washington, USA., (1978).

[11] Haber, The Poisonous Cloud-chemical Warfare in the First World War, Clarendon Press, Oxford, (1986).

[12] Bridges, J.W., The problems with toxic chemicals, Symposium on European Major Hazards, Sept. 1984. OYEZ, London, (1984).

[13] AIChE / CCPS (1988). Guidelines for Safe Storage and Handling of High Toxic Hazard Chemicals, Center for Chemical Process safety, American Institute of Chemical Engineers, New York, (1988).

[14] RTECS-NIOSH (1987). Registry of Toxic Effects of Chemical substances, RTECS, US Government Printing Office, (1987).

[15] TOXLINE (1990). Data from the public domain files of TOXLINE 1981-1987, CD-ROM, Silver Platter International NV., (1990).

[16] Bützer, P., Toxic-Effect-Model (TEM), ein Modell zur Berechnung der Effekte von toxischen Wolken, Kanton St. Gallen, Switzerland, (1985).

[17] Technischer Behelf für den Schutz bei C-Ereignissen, "MET" (Modell für Effekte mit toxischen Gasen), Zentralstelle für Gesamtverteidigung, Bern, (1991). *Note*: The Toxic Effect Model (MET) can be ordered from: "Eidgenössische Drucksachen- und Materialzentrale (EDMZ)", Fellerstrasse 21, 3000 Bern, Switzerland.

[18] Bowen, J.R., Graphical calculation of toxic ranges for a release of 1000 t ammonia, - Canvey, an investigation, 1st report, Appendix 15 (1978) 170.

[19] Poblete, B.R.; Lees, F.P.; Simpson, G.B., The Assessment of major hazards, J. Hazard. Mater. 9 (1984) 355-371.

[20] Ariens, et al.,

[21] Lupke, N. P., et al., Toxikologie, Ullmann Encyclopädie der technischen Chemie, 6 (1981) 65.

[22] Cox , R.A.; Comer, P.J., Development of low-cost risk assessment methods for process plant dispersion, The assessment of major hazards, The Institution of Chemical Engineers, (1982) 353.

[23] Lawson, D.I.; Simms, D.I., The Ignition of wood by radiation, Br. J. Appl. Phys., 3 (1952) 288.

[24] Robinson, C.S., Explosions - Their Anatomy and Destructiveness, McGraw-Hill, New York, (1944).

[25] Glasstone, S., The Effects of Nuclear Weapons, US Atomic Energy Commission, Washington DC, (1962).

[26] Glasstone, S., The Effects of Nuclear Weapons, US Atomic Energy Commission, Washington DC, (1964).

[27] Clancey, V.G., Diagnostic features of explosion damage, Sixth Int. Meeting of Forensic Sciences, Edinburgh, (1972).

[28] Sadée, C.; Samuels, D.E.; and O'Brien, T.P., The characteristics of the explosion of cyclohexane at the Nypro (UK) Flixborough plant on June 1st 1974. J. Occup. Accid., 1 (1976-77) 203.

[29] Fugelso, L.E.; Weiner, L.M.; Schiffman, T.H., Explosion effects computation aids, Gen. Am. Transportation Co., Gen. Am. Res. Div. GARD Prog. 1540 AD903279, Niles, Ill., (1972)

[30] Methods for the calculation of physical effects, CPR 14E, Second edition, (1992). Committee for the Prevention of Disasters, TNO (ISSN 0921-9633/2.10.014/9203).

[31] Langefors, U., Kihlstrom, B., Rock Blasting. Almqvist & Wiksell, Upsala, (1963).
[32] White, C.S., The scope of blast and shock biology and problem areas in relating physical and biological parameters, Ann. N.Y. Acad. Sci., 152(art. 1), (1968) 89.
[33] White, C.S., The nature of the problems involved in estimating the immediate casualties from nuclear explosions, Lovelace Foundation for Med. Educ. and Res., Albuquerque, New Mexico, (1971).
[34] Eisenberg, N.A. et al., Vulnerability Model, A Simulation System for Assessing Damage Resulting from Marine Spills, Nat. Tech. Inf. Service, Report AD-A015-245, Springfield, Va., (1975).
[35] Glasstone, S.; Dolan, P.J., The effects of Nuclear Weapons, London: Castle House Publications, (1980).
[36] Prugh. Mitigation of Vapour Cloud Hazards - Plant/Operations Progress, 4 (2 April), (1985).
[37] Aumonier and Morrey, Non-radiological risks of evacuation, J. of Radiological Protection, (UK), 10(4) (1990).
[38] Davies and Purdy, Toxic Gas Risk Assessment - The effects of being indoors, I. Chem. E., NW Branch, Rugby, UK, (1986).

CHAPTER 4

PROBABILISTIC SAFETY CRITERIA FOR ACCIDENTAL SITUATIONS

4.1. Overview

In going about their daily-life individuals continuously assess situations and make decisions on whether the risk associated to a particular action is justified. Such decisions are mostly made under conditions of uncertainty and involve value judgments that normally cannot be explicitly expressed in terms of quantitative criteria. This is often the case when the risk is of a *voluntary nature*, i.e., it is taken as a free choice (i.e., smoking, down hill skiing). On the other hand, when the individual cannot fully chose to avoid exposure to risk, it is termed as *involuntary risk* (e.g., natural disasters, large industrial accidents) and the decision making process needs to be more explicit using quantitative data. Moreover, people are generally willing to expose themselves to quite different levels of risk depending on whether it is of a voluntary or non-voluntary nature.

Risk is defined as the likelihood of any adverse outcome. The suggested risk criteria are probabilistic in nature. That is, they account for both the involuntarily consequences (effects) and likelihood probability of hazardous events.

All activities associated with the storage, transportation and handling of dangerous goods have an associated level of risk. Risks can be assessed and managed, but it is not possible to eliminate totally a risk unless the activity itself is eliminated. In many cases this simply leads to risk transfer (for example from train to road, or vice versa) which is an important concept in risk assessment and management. The criteria are therefore based on the concept of a residual risk, the acceptability of which should be established in relation to various land uses.

The increased societal awareness on the need to protect the environment, the complexity of modern industries and their potential to cause accidents of large consequences are related to involuntary risks. Decisions involving these issues are often dominated by emotional arguments. *Therefore, a rational decision making process requires the establishment of a consistent framework with standards to express the desired level of safety. Probabilistic Safety Criteria (PSC), which are quantitative expressions for the probability of occurrence of an undesirable event within a given period of time, can play the role of such standards.*

The purpose of the next sections is to provide a general guidance concerning the setting and applications of such criteria.

Risk Categories

In addition to the voluntary versus involuntary nature of risks, a broader categorization is needed to put risks in proper perspective and to develop risk management strategies. Firstly, ***public health risk*** should be assessed separately from ***environmental risk.***

TABLE 4.1 and TABLE 4.2 outline the broad categories of risks usually adopted to assess and compare the health and environmental impacts of different hazardous activities. In all cases, risks to the environment should be assessed and compared separately from risks to human health.

TABLE 4.1. Health Risk

Source	People at Risk	Exposure	Effects
Routine, or Accidents	Workers and Public	Short, or Medium and Long Term	- Fatal and Non-fatal - Immediate / Delayed

TABLE 4.2. Environmental risk

Source	Effects	
	Duration	Extent
Routine or Accidents	Short, or Medium and Long Term	Local, Regional and Global

In terms of health impacts, *occupational* and *public risks* should be treated separately. Two categories of risk apply as a result of direct or indirect impacts, namely consequences with fatal effects and with non-fatal effects:

- *Fatal effects*, either immediate (resulting from direct exposure or accidental situation) or delayed (resulting from chronicle exposure to hazardous substances);
- *Non-fatal effects*, (injuries, diseases) of either an immediate or delayed nature.

In relation to environmental risk, categorization of risks can be made on the basis of extent: local, regional and global; and on the duration of the effect: short or medium term and long term. Some environmental effects are of such a long term nature that they are virtually, or actually irreversible. The complete destruction of vegetation and soil cover in certain mining operations is one such example; widespread loss of species in an area is another. Planning is the only way in which such irreversible environmental effects can be avoided.

Risks arising from normal processing operations, transport, transfer, or handling (loading/unloading, or pumping operation) should also be differentiated from those resulting from major accidents. The criteria proposed in this chapter refer to this latter type.

To date, emphasis has been place on the development of risk criteria in terms of acute (immediate) health effects, mostly fatalities, and in some cases immediate injuries to people. Few examples exist (The Netherlands being an exception) where an encompassing overall quantified risk criteria have been established for long-term chronic exposure to chemicals from one-off or repeated accidental exposures. For the long-term effects of chemicals, the assessments have until now relied mostly on translating animal tests results to people. Recommendations established by National Health Councils are relied upon in that regard.

There are also very few cases of probabilistic safety criteria that apply to accidental releases of chemicals into the natural environment. The diversity of response mechanisms (in type and nature) to the multitude of species within the different ecosystems, including the issue of irreversibility and/or recoverability of damage make it difficult to establish a uniform criteria in this area. Such criteria will largely depend on local circumstances and may need to be developed on a case by case basis (see Vol. I).

Definitions

Some useful definitions[1] are given below:

- *Accident consequence analysis:* An analysis of the expected effects of an accident, independent of frequency and probability.
- *Check-list analysis:* A method for identifying hazards by comparison with experience in the form of a list of failure modes and hazardous situations.
- *Code of practice:* A document offering practical guidance on the policy, standard-setting and practice in occupational and general public safety and health for use by governments, employers, and workers in order to promote safety and health at the national level and at the level of the installation. A code of practice is not necessarily a substitute for existing national legislation, regulations and safety standards.
- *Competent authority:* A minister, government department or other public authority with the power to issue regulations, orders or other instructions having the force of law.
- *Cumulative risk or population risk:* It is the number of cases of a specific effect expected in a given population.
- *Dose:* It is the quantity or concentration of chemical intake in either an experimental or real-life situation.
- *Dose rate:* For chronic exposure, the dose rate is the rate of intake per time unit; for irradiation exposure it is the rate of irradiation per time unit.
- *Emergency plan:* A formal written plan which, on the basis of identified potential accidents at the installation together with their consequences, describes how such accidents and their consequences should be handled either on site or off site.
- *Emergency services:* External bodies which are available to handle major accidents and their consequences both on site and off site, e.g., fire authorities, police, health services.
- *Excess risk:* It is the increase in probability of an effect associated with a specific cause (e.g., exposure to a toxic substance).

- *Exposure:* Human inhalation, ingestion, or dermal adsorption of a chemical substance in an environmental, occupational, dietary, or other context. The *exposure route* or *pathway* is the specific mechanism whereby the exposure occurs (e.g., drinking of contaminated groundwater). *Intake:* The quantity of a chemical inhaled, ingested, or absorbed is called intake. There is an important distinction between *acute (i.e., short-term)* and *chronic (i.e., long-term) exposure,* which usually have different effects.
- *Event tree analysis:* A method for illustrating the intermediate and final outcomes which may arise after the occurrence of a selected initial event.
- *Failure mode and effect analysis:* A process of hazard identification where all known failure modes of components or features of a system are considered in turn and undesired outcomes are noted.
- *Fault tree analysis:* A method for representing the logical combination of various system states which lead to a particular outcome (top event).
- *Hazard:* A physical situation with a potential for human injury, damage to property, damage to environment, or some combination of these.
- *Hazard analysis:* The identification of undesired events that lead to the materialization of the hazard, the analysis of the mechanisms by which those undesired events could occur and usually the estimation of extent, magnitude and relative likelihood of any harmful effects.
- *Hazard assessment:* An evaluation of the results of a hazard analysis including judgments as to their acceptability and, as a guide, comparison with relevant codes, standards, laws and policies.
- *Hazard and operability study (HAZOP):* A study carried out by application of guide words to identify all deviations from design intent having undesirable effects on safety or operability, with the aim of identifying potential hazards.
- *Hazardous substance (or dangerous good):* A substance which by virtue of its chemical, physical, or toxicological properties constitutes a hazard.
- *Hot work:* An activity involving a source of ignition such as welding, brazing or spark-producing operations.
- *Major accident:* An unexpected sudden occurrence including in particular, a major emission, fire, or explosion, resulting from abnormal developments in the course of an industrial activity, leading to a serious danger to workers, the public or the environment, whether immediate or delayed, inside or outside the installation and involving one or more hazardous substances.
- *Major hazard installation:* An industrial installation which stores, processes or produces hazardous substances in such a form and such a quantity that they possess the potential to cause a major accident. The term is also used for an installation which has on its premises, either permanently or temporarily, a quantity of hazard-

ous substance which exceeds the amount prescribed in national or state major legislation.

- *Operational safety concept:* Strategy for process control, incorporating a hierarchy of monitoring and controlling process parameters and of protective action to be taken.
- *Preliminary hazard analysis (PHA):* A procedure for identifying hazards early in the design phase of a project before the final design has been established. Its purpose is to identify opportunities for design modifications which would reduce or eliminate hazards, mitigate the consequences of accidents, or both.
- *Rapid ranking method:* A means of classifying the hazards of separate elements of plant within industrial complex, to enable areas for priority attention to be quickly established.
- *Risk:* The likelihood of an undesired event with specified consequences occurring within a specified period or in specified circumstances. It may be expressed either as a *frequency* (the number of specified events in a unit time), or as a *probability* (the probability of a specified event following a prior event), depending on the circumstances.
- *Risk management:* The whole of actions taken to achieve, maintain or improve the safety of an installation and its operation.
- *Safety audit:* A methodological in-depth examination of all or part of a total operating system with relevance to safety.
- *Safety report:* The written presentation of the technical management and operational information covering the hazards of a major hazard installation and their control in support of a justification for the safety of the installation.
- *Safety team:* A group which may be established by the works management for specific safety purposes, e.g., inspections, or emergency planing. The team should include workers, their representatives where appropriate, and other persons with expertise relevant to the task.
- *Susceptibility:* It is the vulnerability of humans (biota) to toxic effects when exposed to chemical agents. Susceptibility is usually variable in human populations, so that the effective potency of a chemical will differ among exposed individuals.
- *Threshold quantity:* That quantity of a listed hazardous substance present or liable to be present in an installation which, if exceeded, results in the classification of the installation as a major hazard installation.
- *Toxicity:* It is the capability of a chemical substance to induce adverse effects in exposed human beings. *Carcinogenicity* is the capability of a chemical to induce cancer. More specific classes of toxicity include *teratogenicity, mutagenicity, leukemogenicity, etc.* The degree of a chemical's toxicity relative to other substances is called *potency.* The lower the dose required for a particular toxic effect, the higher the potency.

- □ *Whole-body exposure:* It is often distinguished from *target-organ exposure*; the latter depends on *pharmacokinetic factors such as metabolism and excretion.*
- □ *Workers:* All employed persons.
- □ *Works management:* Employers and persons at works level having the responsibility and the authority delegated by the employer for taking decisions relevant to the safety of major hazard installations. When appropriate, the definition also includes persons at corporate level having such authority.

4.2. Risk Criteria

There are two dimensions of risk which should be considered separately, individual and societal risks. On the one hand, the individual's concern about their own life or safety is mostly independent of whether the risk is from an isolated incident or a large-scale disaster. Society's risk perception, however, is mostly influenced by multiple fatality or injury disasters.

4.2.1. SOCIETAL AND INDIVIDUAL RISK CRITERIA

- ■ *Societal Risk* is usually defined as the relationship between the number of people killed in a single accident and the chance or likelihood that this number will be exceeded. It is usually presented in the form of an "F-N curve", which is a graphic indicating the cumulative frequency (F) of killing N or more people.
- ■ *Individual Risk* is usually defined as the probability per year that anyone person will suffer a detrimental effect as the result of exposure to an activity.

Societal Risk

There is a general agreement that societal or group risks should be considered when assessing the acceptability of any hazardous activity or industrial facility. There are two components to the societal risk concept. First, the number of people exposed to levels of risk is important. Second, society is more averse to incidents that involve multiple fatalities or injuries than to the same number of deaths or injuries occurring through a large number of smaller incidents. For example, society reacts differently (and generally with greater concern) to an aircraft crash involving a number of injuries or fatalities than to repeated motor vehicle accidents involving smaller numbers of injuries or fatalities at a time. To deal with this aspect of risk, the intensity of use and the density of people need to be considered. The nature and scale of the incidents contributing to the particular risk levels at particular points and the outcomes of those incidents in terms of fatality and injury, also need to be considered.

Societal risk analysis combines the consequences and likelihood information with population information. It is usually presented in the form of an *F-N curve,* which is a graph indicating the cumulative frequency *(F)* of killing *(N)*, or more people.

Societal risk is calculated from the same basic consequence calculations used to estimate the individual risk. However, instead of presenting the risk summed for all incidents, independent of the population density (which is the method used for individual risk), each possible incident outcome is considered in turn and its frequency *(F)* and the numbers of people *(N)* that could be affected by it, are recorded as an F-N pair. The calculation of the number of people affected includes an allowance for the average geographical distribution of the population.

It is worth noting that *group risk* does not involve the calculation of the "individual risk of death" but rather the "risk of a number of deaths".

There are many ways of expressing the societal impact of serious accidents, such as the number of predicted, prompt, or latent fatalities; agricultural restrictions; large scale evacuation and economic loss. There is no international consensus on which of these or other measures should be chosen to develop societal risk criteria. Individual countries will need to choose the impacts of greatest concern to them.

A number of factors should be borne in mind when developing Probabilistic Safety Criteria (PSC) based on societal risk, including public aversion to accidents with high consequences. The risk level chosen should decrease as the consequence increases. The criteria should be relatively simple to understand, and should recognize the imprecision of Quantitative Risk Analysis (QRA) estimates that predict societal effects (either health or otherwise).

Individual Risk

Probabilistic safety criteria for individual risk are proposed under the consideration that risks arising from accidents in hazardous installations/activities should present only a small increment to the risk to which individuals are already exposed.

The criteria is intended for application to an individual risk calculated using the following assumptions:

- the individual should be considered to be resident at the location off-site, or close to the transportation road, yielding the largest risk for a representative period of time,
- the individual should be considered to be an average individual with respect to dose susceptibility,
- atmospheric dispersion calculations should be realistic, i.e., making allowance for the variability in weather and wind direction.

Whilst individual fatality risk levels include all components of risk (i.e., fire, explosion, and toxicity), there may be uncertainties in correlating toxic concentrations to fatality risk levels. The interpretation of *"fatal"* should not rely on any one dose-effect relationship, but involve a review of available data.

Some Definitions Concerning Risk

- ***Annual fatality rates*** *are measures of societal risk, which is equal to the expected long-term average number of fatalities per year.*
- ***Mean fatalities per event*** *express the average number of fatalities in given accidents. This is equal to the fatality rate divided byte event frequency.*

Individual and societal risk criteria *are used for ranking risk sources in order of priority and to indicate whether the estimated risks are tolerable or whether remedial measures are to be taken.*

- ***Individual risks*** *are expressed as the likelihood of deaths per year for an individual at risk.*
- ***Societal risk*** *is usually defined as the relationship between the number of people killed in a single accident and the chance or likelihood that this number will be exceeded. It is usually presented in the form of an "F-N curve".*
- ***F-N curves*** *are presented as a graphic indicating the cumulative frequency (F) of killing N or more people. They show the relationship between the number of fatalities (N) and frequency (F) of 'N' or more fatalities per year.*

4.2.2. ENVIRONMENTAL RISK CRITERIA

In addition to the risk to people and property, the impact assessment process for potentially hazardous storage, or loading/unloading installations, or transport systems, must also consider the risk from accidental releases to the biophysical environment. Fire and explosion hazards are of less relevance to the environment, in comparison to the effect these hazards may have on people. Acute and chronic toxicity impacts are those which must be chiefly addressed. Generally, there is less concern over the effects on individual plants or animals. The main concern of fire and explosions hazards is instead with whole systems or populations.

The assessment of the ultimate effects from toxic releases into the natural ecosystem is difficult, particularly in the case of atypical accidental releases. Data are limited and factors influencing the outcome variable and complex. There may be no immediate loss of plants or animals or other observable effects from single releases, but there may be cumulative and synergistic effects. It is therefore appropriate to ensure that a thorough review of available data are undertaken and best available information used in the assessment process.

In many cases, it may not be possible or practicable to establish the final impact of any particular release. It may be appropriate in such circumstances to assess the likelihood of identified concentrations of concern occurring in the air, water or soil. Where such criteria are used the assessment should remain on the conservative side.

Because of the complexities of such assessment and case-to-case differences, it is inappropriate to specify general criteria. The acceptability of the risk will ultimately depend on the value of the potentially affected area or system to the local community and wider society. For example, where a rare or endangered ecosystem or species is involved, a much lower risk level is necessary than where the potentially exposed area or system is degraded and/or common.

4.3. Quantification of Risk

4.3.1. PROCEDURE FOR ESTIMATING INDIVIDUAL FATALITY RISKS

Following the definition of individual fatality risk given in the previous section, the individual risk for a single event and consequence type, can be determined from the following basic expression:

$$R_c = P_{haz} \, P_i \, P_{occ} \qquad (4.1)$$

where

- P_{haz} = probability of a hazardous event, such as a pool fire, torch, flash fire, fire ball or explosion
- P_c = likelihood that an individual at a defined location will be subject to a casualty with a specified level of injury from such a hazardous event
- P_{occ} = likelihood that the individual will be at such a location when the hazardous event occur.

Procedure for Estimating Individual Fatality Risks

__Step 1:__ Define the location of the individual relative to the point of release

__Step 2:__ Define the type of injury for which the risk is to be evaluated

__Step 3:__ Select a release scenario, with a probability P_{rel}

__Step 4:__ Establish the probability of the hazardous event occurring, given the release: This is equal to P_{haz}*, (with* $P_{haz} = P_{rel}$ *x Probability of escape for an event becoming a hazardous event).*

__Step 5:__ Use an appropriate consequence model to determine P_c *for the location selected -* P_c *being the casualty probability due to the event consequence at this location. For hazardous events with immediate ignition at or near the source,* P_c *is directly related to* P_{haz}*. For dispersing toxic clouds, the probability of wind direction and other meteorological conditions (e.g., stability class) would have to be taken into account.*

__Step 6:__ Postulate the probability that an individual will be present at that location - P_{occ} *(assume* $P_{occ} = 1$ *for the person most at risk approach).*

__Step 7:__ Calculate the individual risk $R_c = P_{haz} \, P_c \, P_{occ}$.

When evaluating individual risks, assuming the *"person at most risk approach"* is adopted for a given location, then it is usually assumed that $P_{occ} = 1$. Expanding P_{haz} in terms of the likelihood of a release P_{rel} and the probability that it will give rise to the hazardous events P_{esc} leads to the following expression for the individual most at risk:

$$R_c = P_{rel} \, P_{esc} \, P_c \qquad (4.2)$$

where

- P_{rel} = probability of gas/liquid release

P_{esc} = probability of the hazardous event occurring as the result of the release

For a number of releases and hazardous events, individual risks are summed, over all possible accident scenarios and types of accidents, i.e.,

$$R_c = \Sigma P_{rel} (\Sigma P_{esc} P_c) \tag{4.3}$$

The following steps may be give guidance in the computation of individual risk of fatality:

4.3.2. HOW TO COMBINE CONSEQUENCES AND FREQUENCY ESTIMATES TO PRODUCE RISK MATRICES

The basis for a quantification of the risk from a hazardous or industrial (*Figure 4.1*) is a list of hazardous events, or groupings of like events which can be considered to produce similar consequences. *The frequencies of these events may be estimated.* There may be a range of possible outcomes from each event, depending on the different circumstances which may prevail: for example wind direction, weather category, location of people etc. *Each of these circumstances must be defined and a probability assigned to them.* The aggregation of frequency and consequence analysis can therefore be complex, although it is conceptually simple, because all analyses follow essentially the same procedure.

Damage-causing events must be related to the undesired initiating events: for example, the various possible outcomes arising from a release of flammable material may be modeled using an event tree. The conditional probability for factors such as wind direction towards ignition sources and chance of ignition at each source can then be used to produce a frequency for the damage-causing event from the frequency of the initiating event.

The consequences of each damage-causing event must be assessed: The usual approach is to define ranges to selected casualty probabilities from a combination of effect and vulnerability models. These causality probabilities may be selected and limiting ranges to each value estimated - for example, the probability of casualties occurring at various over pressures could be used in conjunction with an explosion overpressure model to produce radii to selected casualty probabilities. The selection of probabilities will usually depend on the available data underlying the vulnerability model used - the analyst should be worry of using probit type relationships to produce a large number of casualty probability bands as this not only complicates the analysis, but the degree of detail would not be supported by the basic data.

Having obtained the frequency and casualty probabilities for each damage-causing event under consideration, *the risk relationships are derived* in the following manner. Taking each event in turn, the number of people present in the area covered by each casualty probability band are multiplied by the appropriate casualty probability, producing, by summation, the total number of people predicted to be affected by each event. The overall frequency-consequence relationship can then be drawn up from the number (e.g., people, property, land, or water resources) affected and the frequency for each event. Expressing the risk in terms of the frequency distribution of multiple casualty events (F-N-curves) is known as *"calculating the societal risk"*.

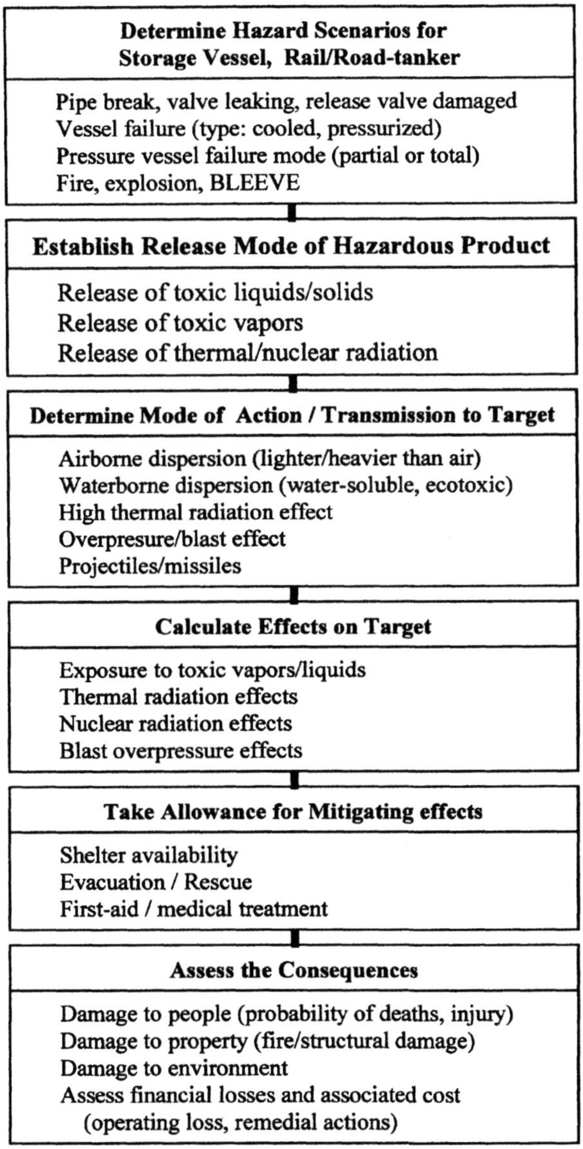

Figure 4.1. Method for assessing the consequences of hazardous product release

The *individual risk* at a location is obtained by taking the casualty probability at that location for each damage-causing event and multiplying it by the frequency of that event. The individual risk from all such events, and therefore from the activity as a whole, is obtained by summation over all the events.

The final expression of individual and societal risk then incorporates the likelihood and severity of all the outcomes of the scenarios that have been considered for particular location.

Quantification of Risks

Step 1: *Estimate the frequencies of these events.*

Step 2: *Define the circumstances concerning a given scenario and assign a conditional probability to each of them to produce a frequency for the damage-causing event from the frequency of the initiating event.*

Step 3: *Define, from a combination of effect and vulnerability models, several consequence ranges which are related to the selected casualty probabilities.*

Step 4: *Derive the risk relationships for each event, and multiply the number of people present in the area (corresponding to each casualty probability band) by the appropriate casualty probability. Finally the total number of people affected by each potential event is obtained by summation.*

Step 5: *Calculate the societal risk by estimate the overall frequency-consequence relationship from the number (e.g., people, property, land, or water resources) affected, and the frequency for each event. Express the resulting risk in terms of the frequency distribution of multiple casualty events.*

Step 6: *Calculate the individual risk at a given location by taking the casualty probability at that location for each damage-causing event, and multiply it by the frequency of that event. Obtain the individual risk from all such events, (and therefore from the activity as a whole) by summation over all the events.*

The following sections illustrate the way in which the calculations can be carried out.

Procedure for Estimating Societal Fatality Risk Levels

For estimating the societal risk R_s from a single event, one must evaluate the probability s of a given number of people n being subjected to a specified level of injury in a specified time interval.

$$R_s = P_{haz}, \; for \;\; n = \Sigma P_c \, P_{occ} \tag{4.4}$$

where

P_{haz} = probability of a hazardous event, such as a pool fire, torch, flash fire, fire ball or explosion

P_c = likelihood that an individual at a defined location will be subject to a casualty with a specified level of injury from such a hazardous event

P_{occ} = likelihood that the individual will be at such a location when the hazardous event occur.

Note that in above formula the sum is computed over all people within the area affected by the hazardous event.

When evaluating the number of people affected by a given hazardous event, the following expressions are often used:

$$R_s = P_{rel}\, P_{esc} \tag{4.5}$$
$$n = \sum n(k)\, P_c(k)$$

where

P_{rel} = probability of a gas/liquid release
P_{esc} = probability of the hazardous event occurring as the result of the release
$n(k)$ = k-time-averaged number of people subject to an average casualty probability $P_c(k)$.

Societal risk from a number of events is normally expressed in a cumulative form, i.e., as a probability of arriving at N or more casualties. This is usually expressed for a number of values of N and is obtained by summing all values of R_s for events where $n \geq N$.

The following procedural steps may be of relevance in the *estimation of societal risk levels*.

Procedure for Estimating Societal Fatality Risk Levels

Step 1: *Define the type of injury for which risk is to be evaluated.*

Step 2: *Define the casualty numbers, N, which are to be used in expressing societal risks.*

Step 3: *Select a release, j, with associated probability* P_{rel}.

Step 4: *Define the hazardous event from that release with a probability* $P_{haz} = s$.

Step 5: *Use damage models to evaluate the number of casualties, n. This can be done by allocating ranges of exposure to discrete values of* P_c *and evaluating the number of people n who fall within each exposure range.*

Step 6: *Repeat steps 3-5 for all releases.*

Step 7: *Select a value for N.*

Step 8: *Sum all values of* P_s *for which* $n > N$.

Step 9: *Repeat all steps 7 and 8 for all values of N.*

Risk Matrices

A *risk matrix* is necessary to rank the risk objects in order to allocate resources, to decide where preventive measures should be taken first, to develop emergency plans, etc. When attempting to rank risk objects systematically, by means of risk matrix, it is necessary to weigh up different kinds of hazards within the risk object. This will be a matter of judgment for the coordinating group involved in the overall process of integrated risk assessment at regional level. Both probability and consequences must be considered. It is common to concentrate on the risks with the greatest consequences.

When attempting to reduce risk levels systematically, however, it may be necessary to weigh an event with low probability but serious consequences against one which is more likely but causes less damage. The result of a ranking process should influence the development of a concrete program of action necessary to protect and save lives, the environment and property on-site and off-site.

4.4. Definition and Acceptance of Risk Levels

4.4.1. FRAMEWORK FOR PROBABILISTIC SAFETY CRITERIA

The basis adopted in many cases in setting probabilistic safety criteria (PSC) is that the criteria ought to be set below (and in many cases well below) known voluntary and non-voluntary risks associated with the different daily activities to which any one person or the society as a whole is exposed. Although it has been argued that by setting assessment criteria in this way, such criteria should provide an "tolerable" level of risk, the notion of risk "tolerability" has been and still is the subject of significant debate. Attention is now being given to the setting of a "tolerable" level of risk. The tolerability of such risks may be suggested by both reference to other levels of risks experienced by the society, and that it may be tolerated in relation to both the costs and benefits associated with the activities under consideration. Social and economic considerations become therefore integral aspects of the setting of such tolerable risk levels.

Based on the approach promulgated in a United Kingdom Policy paper on the "Tolerability of Risks from Nuclear Power Stations", it has been suggested that in setting PSC, *one has to consider three "regions of risk"*:

- An *upper region* (I) in which the risk is judged to be so high as to make the practice or activity intolerable, whatever its benefits are;
- An *intermediate region* (II), where the risk is acceptable subject to the overriding requirement that all reasonable practical measures have been taken to reduce the risk; and
- A *lower region* (III), in which the risk is judged sufficiently low as to be broadly acceptable with no additional effort required to further reduce it. *Figure 4.2* depicts the three regions described above.

It is recognized that it is difficult to define the boundaries between these three regions (I-III) as single precise values. In addition, the practical application of QRA inevitable involves uncertainty and imprecision in the estimation of risks. These factors need to be taken into account in assessing QRA results within this framework and the criteria must not be used as absolute "go/non-go" rules, hence they are shown as hatched zones, rather than single values in *Figure 4.2*. Within such a framework it is unnecessary to define separate levels for old and new plants. However, it is recognized that it will generally not be reasonably practicable to reduce the risks from plants in operation to the levels achievable on new plants.

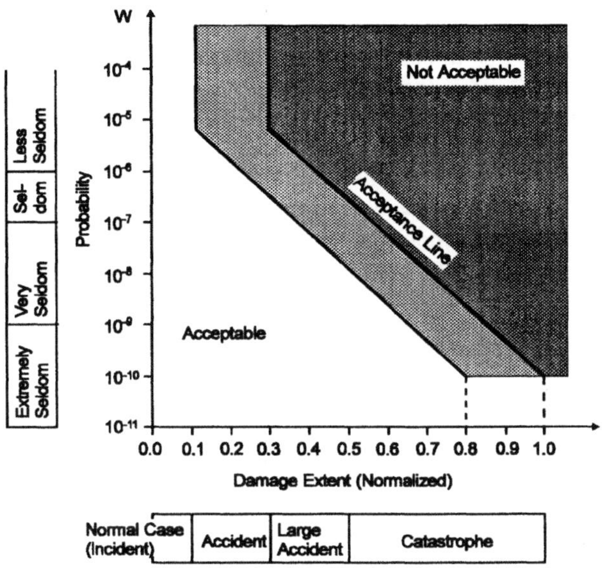

Figure 4.2. Definition of risk levels

The establishment of specific upper and lower risk criteria may be influenced by many considerations which will vary with the type of risk addressed. These considerations include public health, social and economic factors. The basic choice of the appropriate levels of the public health and societal impact related criteria is essentially a socio-political decision and can only be made in a national context. The translation of this decision into a technical definition is, however, a process in which judgment will inevitably be involved.

Principles and procedures used in establishing compliance with existing PSC in the presence of the quantified uncertainties are still evolving. *It is recommended that where distribution of frequencies has been calculated in QRA, the mean value rather than an upper or lower bound should be used. Where only point values have been used they should be representative of a central value. In all cases it should be recognized that the criteria developed should be guidelines only and not treated as absolute rules.*

Acceptability of a given level of risk involves many considerations of which safety is only one, although safety is playing an increasingly important role in planning considerations. Attitudes towards risk acceptability can vary widely depending on local situations. In some cases, certain risks may only be acceptable when they are outweighed by certain advantages which people associate with the considered activity. However, unacceptable risks can be shown to exist in a region, whatever the advantages may be.

The basis for risk criteria is that, generally, various levels of risks are tolerated on a daily basis, both to individuals and to society as a whole. Where risk is taken with the choice and full knowledge, that risk can be described as voluntary risk. Examples of voluntary risk include smoking, driving, and rock climbing, provided that the individual knows and understands the risks. Where the individual does not have knowledge of the risks, or is not entirely free to choose to avoid the risk exposure, then the risk can be termed non-voluntary. Examples of non-voluntary risks include meteorite strike, some illnesses and some natural disasters. In reality, most types of risk exposure have degrees of both the voluntary and non-voluntary. People in general are willing to expose themselves to quite high levels of individual risk by undertaking certain activities. On the other hand, society offers growing resistance to risks perceived as being imposed on one group of people for the benefits of others, or where the risk exposure of one group does not fit with their share of benefits is usually perceived as a non-voluntary risk.

4.4.2. SOCIETAL RISK LEVELS

When a risk is to be imposed on an individual or a group of people (e.g., by locating hazardous activities - e.g., transportation, storage, industry - in an area), the concept of acceptability of that risk for the decision-making process is that it should be low relative to other known and tolerated risks. As societal risk acceptability is specific to each society, it is very important that allowance be made to reflect differences between societies and cultures. The F-N approach is a complex one which does not lend itself to easy implementation. It may not be possible in practice, for example, to account for small but incremental increases in population over a period of time which may eventually have significant implications on a cumulative basis.

In view of these considerations we suggest that:

- Judgments on societal risk shall be made on the basis of a qualitative approach on the merit of each case rather than on specifically set numerical values.
- Individual fatality and injury risk contours at the individual risk criteria levels (applicable to the various land use categories) should be established. People density within each risk band should then be established and the suitability of a particular regional development judged on the basis of the location, density and vulnerability of the exposed population.

In practice, the frequency of each potential accident and the number of people that may be affected by each accident should be estimated wherever practicable. F-N curves should be generated and used as an input to the risk analysis. An example of societal risk curves for some human-caused events, that was adopted by the USA, is shown in *Figure 4.3.*

In general, the individual risk criteria for particular land uses are set with regard to the lower end of the range of typical population densities for that use. Where individual developments (industrial or residential), or a number of developments taken together result in substantially higher densities, their location relative to the suggested criteria contour should be carefully considered. Generally the closer to the limit for that type of land use, the lower the acceptable population density should be.

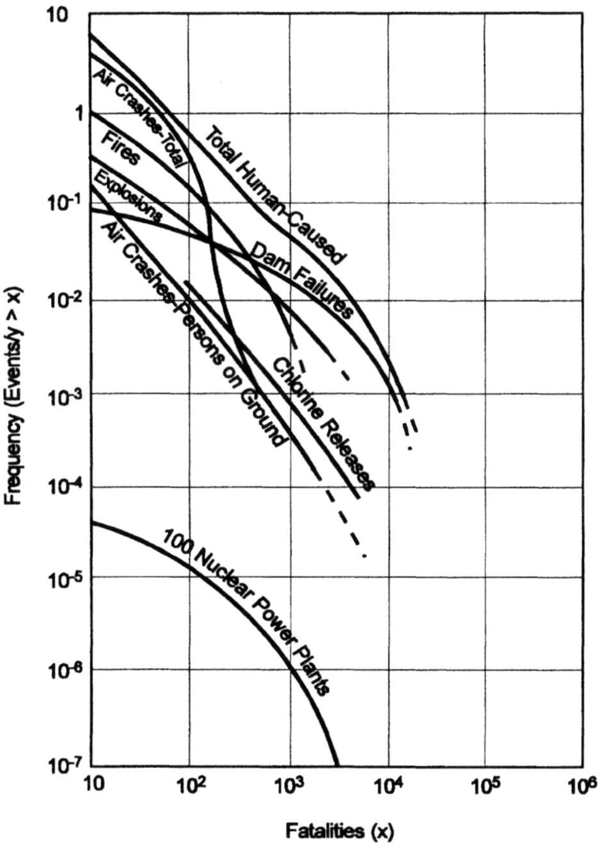

Figure 4.3. Societal risk curves

The analysis should also consider the specific vulnerability of such development and populations, taking into account the specific vulnerability of the people (aged, young, disabled, etc.); the topography and access to and egress from the development and locality for emergency response and evacuation; other emergency infrastructure and design of developments (e.g., glass facades towards an explosion hazard, people in the open air where the hazard is toxic gas).

Individual Fatality Risk Levels

Individual fatality risk is the risk of death to a person at a particular location. The International Community has recommended a fatality risk level of one in a million per year (1×10^{-6} per year) as the limit for risk acceptability for residential area exposure. Experience with implementation indicates that the criteria is practical and appropriate, and as such should be maintained. It is necessary, however, to account for variations in the duration of exposure to that risk at any particular location by any one individual. It is also necessary to account for variations in people's vulnerability to the hazard and their ability to take evasive action when exposed to the hazard.

The one in a million criterion assumes that residents will be at their place of residence and exposed to the risk 24 hours a day and continuously day after day for the whole year. In practice this is not the case and this criterion is therefore conservative. People in hospitals, children at school or old-aged people are more vulnerable to hazards and less able to take evasive action, as compared to the average residential population. A lower risk than the one in a million criterion (applicable for residential land use areas) may be more appropriate for such cases. On the other hand, land uses such as commercial and open space do not involve continuous occupancy by the same people. The individual's occupancy of these areas is on an intermittent basis and the people present are generally mobile. As such, a higher level of risk (relative to the permanent housing occupancy exposure) may be tolerated. A higher level of risk still is generally considered acceptable in industrial areas.

Accordingly, the following risk assessment criteria are suggested for the assessment of the safety of location of a proposed development of a potentially hazardous nature of the land use planing in the vicinity of existing hazardous installations:

■ Hospitals, schools, child-care facilities and old age housing development should not be exposed to individual fatality risk levels in excess of half in one million per year (0.5×10^{-6} per year).

■ Residential developments and places of continuous occupancy, such as hotels and tourist resorts, should not be exposed to individual fatality risk levels in excess of one in a million year (1×10^{-6} per year).

■ Commercial developments, including offices, retail centers, warehouses with showrooms. restaurants and entertainment centers, should not be exposed to individual fatality risk levels in excess of five in a million per year (5×10^{-6} per year).

■ Sporting complexes and active open space areas should not be exposed to individual fatality risk levels in excess of ten in a million per year (10×10^{-6} per year).

■ Individual fatality risk levels for industrial sites at levels of 50 in a million per year (50×10^{-6} per year) should be contained within the boundaries of the site where applicable. Whilst individual fatality risk levels should include all components of risk (i.e., fires, explosions and toxicity), there may be uncertainties in correlating toxic concentrations to fatality risk levels. The interpretation of "fatal" should not rely on any one dose-effect relationship, but involve a review of available data.

Injury Risk Levels

Relying entirely upon fatality risk criteria may not account for the following factors:

- Society is concerned about risk of injury as well as risk of death.
- Fatality risk levels may not entirely reflect variations in people's vulnerability to risk. Some people may be affected at a lower level of hazard exposure than others. It is therefore appropriate that risk criteria also be set in terms of injury,

i.e., in terms of levels of effects that may cause injury to people, but will not necessarily cause fatality.

Toxic Exposure Criteria
Depending on the concentration, the nature of the material, the duration and mode of exposure (i.e., via the respiratory tract, lungs, skin or ingestion), the effects of toxicants range from fatality, injury (e.g., damage to lungs and respiratory system, damage to nervous system, emphysema, etc.) to irritation of eyes, throat, or skin through to a nuisance effect. Effects can be classified as acute, chronic or delayed.

There are a number of assessment criteria and dose-effect relationships that vary from one chemical to another. Toxic criteria applicable to one chemical may not necessarily be appropriate for others. International experience conclusively shows that the formulation of a uniform specific criteria to cover all toxic effects is not appropriate or valid. Instead, each case should be justified on its merits using a thorough search of available and known dose-effect relationships as the basis for assessment. Incidents with injurious impact on people should be kept to low frequencies.

The suggested injury risk criteria for toxic gas/smoke/dust exposure are as follows:

- Toxic concentrations in residential areas should not exceed a level which would be seriously injurious to sensitive members of the community following a relatively short period of exposure at a maximum frequency of 10 in a million per year.
- Toxic concentrations in residential areas should not cause irritation to eyes or throat, coughing or other acute physiological responses in sensitive members of the community over a maximum frequency of 50 in a million per year.

Risk Levels for Property Damage and Accident Propagation
The siting of a new hazardous storage/installation or transport system (rail/road-tanker) must account for the potential of an accident at the installation/system causing damage to the building, and also propagating to a neighboring building (storage), or a neighboring rail/road-tanker and hence initiating further hazardous incidents (the so-called "domino effect").

The principle of setting risk criteria to reflect the potential for accident propagation is that the risk of an accident at one plant triggering another accident in the neighboring should be low and that adequate safety separation distances should be provided as part of siting and layout of plant/storage areas. A similar approach should be followed for railway shunting yards and loading/unloading stations in the vicinity of residential or industrial areas.

Heat radiation levels of 23 kW/m^2 as the result of fire incidents may affect a neighboring installation to the extent that unprotected steel can suffer thermal stress that may cause structural failure. This may trigger a hazardous event unless protection measures are adopted.

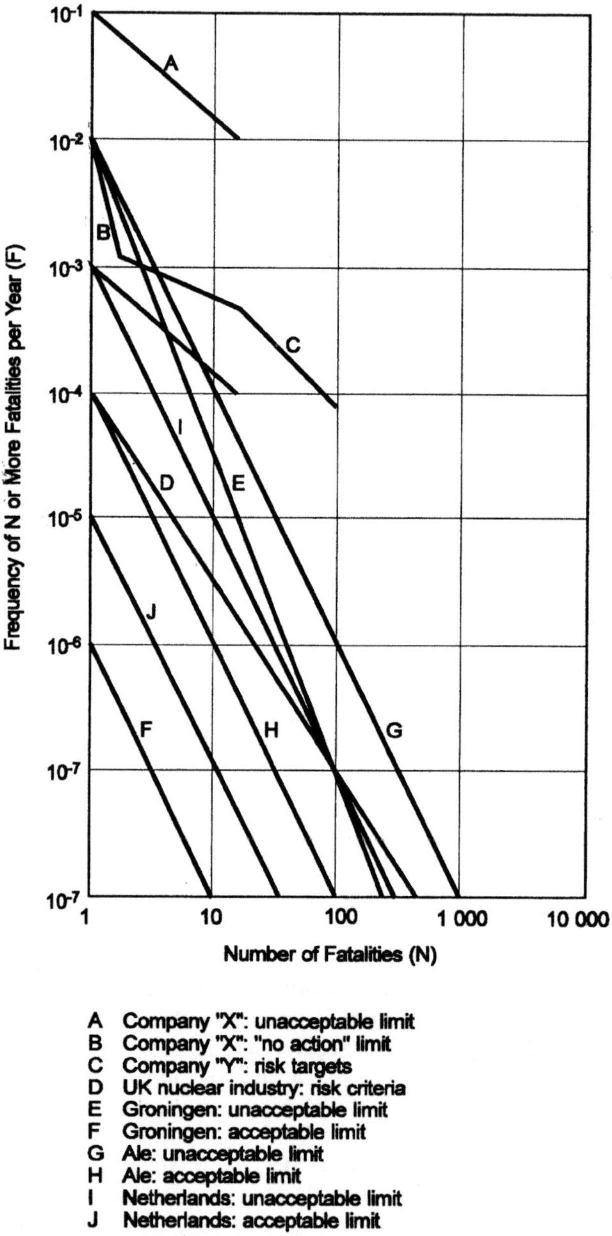

Figure 4.4. Comparison of risk criteria and targets from various sources

Explosion overpressure levels of 14 kPa may damage piping and (low-pressure) equipment at a neighboring plant/rail-tanker. It is therefore appropriate that the probability of these levels of impacts at neighboring installations be relatively low. A probability of 50 in a million per year is suggested for assessment purposes. The explosion overpressure level of 14 kPa is also sufficient to cause significant damage to buildings and as such is appropriate as the basis for a risk of damage criteria. A probability of 50 in a million per year is also suggested for assessment purposes.

These criteria do not remove the need to consider higher consequence levels at lower frequencies. The hazard analysis should consider the whole picture not just the nominated quantitative criteria.

4.4.3. FINAL ASSESSMENT OF RISK LEVELS

The qualitative and quantitative results of the risk analysis can be applied in the assessment process as follows:

- Risk impacts at various distances from a planed or existing storage, respectively loading/unloading installation, pipeline routes, or selected rail/road routes may be compared against safety targets or criteria (*Figure 4.4*). A judgment can be made about the hazard impact. A general principle of assessment is that the risk impacts from the new development should be well below the levels of risk which people and the environment are regularly exposed to from similar development and other sources.
- The analysis should particularly highlight the major contributors to the risk and their nature and extent and, secondly, areas where risk could be eliminated, or cost-effectively reduced. These results can be used to develop prevention and protection measures including priority allocation of resources for hazard control.

4.4.4. QUALITATIVE RISK ASSESSMENT CRITERIA AND SAFETY ASSURANCE

Irrespective of the numerical value of any risk criteria level for risk assessment purposes, it is essential that certain qualitative principles be adopted as yardstick for safety assessment and management. The following qualitative criteria are appropriate when assessing the risk implications of a development project of a potentially hazardous nature or the safety suitability of a development in the vicinity of a potentially hazardous installation:

■ All "avoidable risks" should be avoided. This necessitates the investigation of alternative locations and alternative technologies, wherever applicable, to ensure that risks are not introduced in an area where feasible alternatives are possible and justified.

■ The risk from a major hazard should be reduced wherever practicable. irrespective of the numerical value of the cumulative risk level from the whole installation. In all cases, if the consequences (effects) of an identified hazardous incident are significant to people and the environment, then all feasible measures (including alternative locations) should be adopted so that the likelihood of such an incident occurring is made very low. This necessitates the identification of all contributors to the resultant risk and the consequences of each potentially hazardous incident. The assessment process should addresses the adequacy and relevancy of safeguards (both technical and locational) as they relate to each risk contributor.

■ The effects/consequences of more likely hazardous events (i.e., those of high probability of occurrence) should, where ever possible, be contained within the boundaries of the installation.

■ Where there is an existing high risk from a hazardous installation, additional hazardous developments should not be allowed if they add significantly to that existing risk.

Further to proposing criteria to express the desired level of safety, it should be discussed to which extent risk estimates and their compliance with risk criteria can assure safety.

First, it should be kept in mind that severe accidents are rare events and as such their estimated probability of occurrence is the result of an engineering model representation of the reality and not the result of observable repetitive events. Therefore, when we refer to the probability of a certain undesirable outcome, we are expressing, according to the subjective concept of probability, our degree of belief that such events may happen.

Second, any model includes assumptions which have to be respected for the results to be credible. They also form the basis for the "safety assurance" which is both a fundamental safety concept and a requirement for QRA results to be a realistic qualitative and quantitative measure of plant safety.

In this context, similarly to the financial concept of "rolling forecasting" the concept of a *"living" QRA* has emerged, and is increasingly being used as a tool for operational safety management and risk monitoring. Its purpose is to keep the safety assurance plan constantly updated, should any changes in the conditions used in the base case calculations be introduced.

It goes without saying that low risk estimates are not surrogates to sound plant design and sound operational practices and to constant operators' safety awareness required for safe plant operation.

4.4.5. RECOMMENDATIONS AND GUIDANCE NOTES ON IMPLEMENTATION

The following guidance notes are provided to assist in the formulation and implementation of appropriate risk assessment criteria:

■ The individual fatality and societal risk criteria should include all components of risk, i.e., fire, explosion and toxicity.

■ The implementation of the criteria must acknowledge the limitations and in some cases the theoretical uncertainties associated with risk quantification. Two approaches are usually adopted to account for such uncertainties:

- *"Pessimistic approach":* i.e., the assumptions are on the conservative side. This result in an overestimation of the actual risk; or
- *"Best estimates' approach"* using realistic assumptions with an estimated risk that could either be an overestimate or an underestimate of the actual risk. The criteria suggested in this section are set at a realistic level.

■ In the context of defining a suitable approach, a degree of flexibility in the implementation and interpretation of the absolute values of the risk criteria may be justified in some cases. There may also be variations in local conditions. Consideration of vulnerability of people and situations is necessary. *The criteria are best implemented when used as targets rather than absolute levels.* Nevertheless, any substantial deviations from such targets should be fully justified. It is advisable that in all cases the assessment process emphasize the hazard identification and risk quantification process and procedures rather than entirely relying on absolute risk levels.

■ Given the probabilistic nature of the assessment process, care must be exercised in interpreting/assessing compliance with a risk criteria when classifying plants which exceed the suggesting criteria as "unsafe". Nevertheless, a higher resultant risk level relative to the suggested criteria indicates land use safety incompatibility and locational safety constraints.

■ The implementation of the risk criteria should differentiate between existing land use situations and new situations in terms of applicability to reflect a tighter locational and technological standard applying now than it was the case at earlier times. In the case of existing industry, compliance with a risk criteria is part of an overall strategy to mitigate existing risk levels by reducing both the risks and the number of people exposed to those risks. As such, risk criteria designed for new plants can only be used in targets for existing plants as part of an overall safety strategy.

■ The risk to an individual and/or to the public in the vicinity of an industrial site, arises from all industrial activities in the area. The basic risk criteria (to various land uses) need to be related to the site. It may also be appropriate to plan for sub-criteria for each individual site to account for cumulative impact of developments.

■ In a large industrial complex, risk criteria should also provide for the potential for accident propagation. The risk of an accident in one plant triggering another accident at another neighboring plant location should be kept low. Adequate safety separation distances should be maintained between plants, liquid/gas storage, warehouses, and loading/unloading terminals for dangerous products. Consideration should also be given to the vicinity of railways and heavy traffic roads that represent a potential danger in case of rail/road tanker accident.

■ The application of the risk criteria should also apply to planing and development or new development of residential and other sensitive land uses in the vicinity of hazardous installations.

ALARP Principle

The principle known as *ALARP (As Low As Reasonable Practicable)* is the most commonly used although there are others. It states that risks to individuals and society should be As Low As Reasonably Practicable. ALARP fits into an overall approach to risk control. Risks above a certain high level are intolerable and their causes are often prohibited by legislation. At the other end of the scale, there is a very low level of risk which is clearly negligible. Between these two levels, risks are tolerable providing that it can be demonstrated that the cost (in time, trouble and expenses) of reducing the risk further would be "disproportionate" to any improvements achieved.

Summary (Chapter 4)

Probabilistic Safety Criteria (PSC) for risk assessment and evaluation assists a rational decision making process. This chapter defines various risk categories and introduces the societal and individual risk criteria as well as the environmental risk criteria. Procedures for estimating various types of risks are given. The risk matrices integrate frequency and consequences of various types of undesired events and portrays, together with PSC instrument a framework for analysis, estimation and management of risk. Qualitative risk assessment criteria are also presented and discussed in the economy of the present chapter.

References (Chapter 4)

[1]Prevention of Major Industrial Accidents, (1991). ILO, International Labor Office, Geneva. (ISBN 92-2-107101-4).

CHAPTER 5

MODELING OF DENSE GAS DISPERSION

5.1. Introduction

As mentioned before, the principal hazard scenarios associated with accidental chemical releases which may lead to detrimental consequences to people and property include fires, explosions, and toxic and flammable vapor dispersion. The formation of toxic or flammable gas clouds at release point and their dispersion into the atmosphere has the potential for involving a large area and consequently exposing a large population to the danger resulting from toxic vapors, fire or explosion. In Volume I, we have presented a review concerning atmospheric dispersion models for continuous and non-point sources. In the present chapter we shall consider the dispersion of gases heavier than air.

A chronological literature review has been prepared to help and assist risk analysts interested in dense gas calculations and appropriate risk assessment. This work is whether part of a bench mark exercise, nor is it presenting the results of a validation effort, for which one needs the original codes and corresponding documentation. On the basis of the references presented at the end of this chapter, the interested reader may collect the original papers and deepen the subject according to his needs. It was also not possible to discuss in details all the mathematical particularities of each physical model, which would have been outside the scope of this book. It must be stated that a lot of development is under way at several universities, but it is not always possible to obtain the latest code version of a given model still under development. On the other side commercial computer codes may be purchased, that do include recent features and improvements to well-known physical models. Most, but not all the models have been checked against field test experiments and a confidence range is usually indicated.

The reviews and citations mentioned in this chapter are covering mainly the last 10 years. They were extracted from different literature sources (scientific journals, textbooks), or from databases such as CHEMICAL ABSTRACTS, INIS, NTIS, and ENERGY, and from reports from governmental agencies. Some of these databases are also available on compact discs (Silver Plater, etc.). The ETHICS system of the ETH Zürich was also used for retrieving the information. Among the journals having the most sources of information on the subject of dispersion models there are, among others, "Atmospheric Environment" and "Journal of Hazardous Materials".

A review of the international literature on atmospheric diffusion models, especially of those which are used by governmental regulatory agencies, has been prepared by Hanna[1] (1982). While the Gaussian plume model (Turner[2], 1979) is often used in western countries and in the U.S., the gradient-transport K-model is most often used in eastern countries. The most important parameters predicted by the models are the magnitude and location of maximum ground-level concentrations.

Most of the reports mentioned here are available from the aforementioned data bases, from the U.S. Government Printing Office Department, or from the DOE (Department of Energy), and also from a national library.

5.2. Dense Gas Dispersion Models

Accidental releases of hazardous gases are a serious problem. This is especially the case when the release leads to the quasi-instantaneous formation of a large cloud. The density of the cloud may be higher than that of ambient air, due to a high molecular weight of the gas or to a low temperature (under-cooled gas in liquid form). The modeling of large dense clouds heavier than air is complicated and difficult, due to gravity-driven flow phenomena, which is an unsteady process. A review of this subject has been prepared by Simpson[3] (1982). The dispersion of a heavy gas occurs in several phases. These include the source controlled dilution phase, the gravitational slumping, an interaction phase involving ambient flow and gravitational slumping, and finally the passive dispersion phase.

Three major release mechanisms may result in the quasi-instantaneous formation of a dense cloud depending on the mode of storage:

a) Pressurized Gas Stored in Liquefied Form at Ambient Temperature.
- Type of release: rapidly expanding hemispheric cloud of dense gas due to flashing
- Strong initial mixing with ambient air
- The heat source for the evaporation is the heat stored in the liquefied gas.

b) Supercooled Gas Stored in Liquefied Form
- Type of release: after a tank failure, the liquid will spread over the ground or over a water surface
- Due to heat transfer from the ground, or water surface, to the liquefied gas a dense gas cloud is formed
- The roughness and type of soil is determining the heat transfer coefficient.

c) Dense Gas Stored in Gaseous Form and Released As Such
- Type of release: gaseous release without boiling
- In the special case where the gas is stored at ambient pressure, expansion and enhanced mixing will not occur during the release.

These three types of gas release mechanisms have been simulated in field and laboratory experiments. From the observation of field experiments the following phases can be deduced[4]:

■ *Formation phase:* gas release (eventually followed by expansion and evaporation). This expansion phase is relatively rapid. A substantial volume of air is mixed with the cloud. The mass of air entrained during this phase is estimated to be between 1 and 2 times the mass of vapor released.

■ *Quasi-uncoupled slumping phase:* when the effects of turbulence and momentum of the ambient flow can be neglected. The cloud being denser than air will spread laterally with a characteristic toroidal head wave at the boundary of the cloud (Picknett and Carpenter[5], 1978; Fannelop[6], 1980, Fannelop & Jacobson[7], 1984; Havens & Spicer[8], 1984). Entrainment of air occurs during the initial stages of slumping, primarily at the boundary. There is very little mixing at the top surface of the cloud due to stable stratification. In this phase the cloud moves en masse downwind due to momentum transfer from the ambient flow field.

■ *Coupled spreading phase:* no more slumping due to atmospheric turbulence and vertical mixing. Advection by the mean wind and surface friction may affect the gravity spreading.

■ *Mixed phase:* cloud density effects can now be neglected. Formation of a diffuse cloud.

■ *Passive phase:* the cloud density effects can be neglected. In this final phase dispersion by atmospheric turbulence and advection by the mean wind are similar to those for a passive puff (neutral density cloud).

A large number of heavy gas research studies have focused on understanding the physics of dispersion of passively released, non-reacting, pure vapor clouds. Other studies are dealing with chemically reactive clouds, i.e., clouds reacting principally with air moisture (chlorine, ammonia, phosgene, nitrogen tetroxide).

Several special phenomena can also occur during gas dispersion (depending on the chemical involved and the release mode) that have significant effects as air entrainment. For example the presence of liquid aerosols lead to higher negative buoyancy, much lower cloud temperatures, and more pronounced and persistent heavy gas effects depending on the volatility of the liquid involved (Kaiser & Walker[9], 1978). In the case of vapors generated from cryogenic spills, the heat transferred from the ground or from a water surface (sea, lake) is likely to have a different effect on heating the cloud.

The dense gas problems are discussed in Van Ulden[4] (1988). It has been observed, that a dense cloud behaves differently from a puff of passive contaminant. In passive puff dispersion the vertical and horizontal scales are of the same order. The dense cloud shows a pronounced slumping effect due to a combination of high cloud density and gravity. This make the cloud negatively buoyant. Dense gases are often called heavy gases. A large dense cloud may spread over a very large area and becomes very shallow. In the case of heavy vapor injection into the atmosphere, caused by evaporation of a liquid pool, the first rapid dilution phase is absent.

Dense gas dispersion models have to reckon with three distinct phases of gas behavior:

- initial mixing (source models)
- gravity slumping (dense gas models)
- turbulent spread (neutral gas models)

There are three main categories of mathematical models used to represent dense gas dispersion: *box models, conservation models, and intermediate models.*

Box Models
These models make predictions only for overall properties of the cloud (mean radius, mean height, and mean cloud temperature) and they assume simple shapes for spatial distributions of physical properties (e.g., concentration). Included in this category are the Gaussian tracer models, and modifications of them, from which predictions are made of the standard deviations of the cloud concentration distribution which can be related to such properties as the mean cloud radius.

Box models which represent the initial development of the cloud (in the case of an instantaneous release), or a cross-sectional slice of the cloud (in the case of a steady continuous release) as a uniformly mixed volume. The shape, thermodynamic properties of the cloud are modeled using correlation's derived for the velocity of density intrusions and fluid entrainment across density interfaces. This model type often incorporate a transition, usually to a Gaussian model, to describe passive dispersion (controlled by atmospheric turbulence) of the gas in the far field. *Box models are not inferior to 3-D models; they have certain advantages. These models, in recent years, have progressively become more sophisticated in the number of physical processes considered, and in the way in which they are incorporated.*

It is essential to distinguish between models for steady state flow (i.e., steady state plume models) and models for transient flow (i.e., transient puff models).

Advanced Box Models
In advanced box models, improvement can be obtained, when the analytical solution, which is the basis for most box models, is replaced by a simple one-dimensional time integration (or distance integration, for a continuous release). Whilst box models are generally adequate for dispersion over flat unobstructed terrain, more complex mathematical models are required for a description of the effects of obstructions. The integration of the differential equations which constitute the model can be undertaken either analytically, or numerically. The latter case has been preferred by the modelers. Certain simplifications are necessary in order that analytical solutions may be obtained. The analytical integration enables the following features to be incorporated into an improved box model:

- Use of more realistic form of spreading velocity
- Incorporation of cloud inertia effects
- Gradual transition from the slumping phase to passive dispersion.

K-theory (eddy diffusivity) Models. This class of models (SIGMET model by Havens) assume constitutive relations between turbulent fluxes and the gradients in mean variables velocity, temperature, and concentration, coupled with the equations of change for mass, momentum and energy for turbulent fluid flow to predict the time and spatial variation of thermodynamic properties of the cloud.

K-ε-theory Models. They include a supplementary term for the kinetics of the energy dissipation, coupled with the parameter K.

Intermediate Models

Recent *advanced similarity models* overlap these categories, assuming self-similar concentration profiles (models described by Te Riele; by Colenbrander; or by Flothman et al.; Ooms), others have the assumption of a self-similar concentration profile coupled with a K-theory representation of turbulent mass transfer within the cloud (model of Colenbrander). In 1980, there have appeared several intermediate models (between box and 3-D models) which involve greater simplification of the equations of motion, energy and mass than is found in the K-theory models, but which still require solution of partial differential equations to predict cloud state variables. These are the models proposed by Zeman, by Rosenzweig, and by Fannelop, which may present significant advantages over the simple box models since they allow a more realistic representation of spatial variation in the simulation of cloud dispersion, and incorporate more general turbulent mixing sub-models.

3-D Time Dependent Conservation Models

This type of models are based on the full 3-D solution of the Navier-Stokes equations, derived from the conservation of contaminant, conservation of total mass, of total heat, and of momentum. Such models incorporate the features of either K-theory or K-ε-theory assumptions. The 3-D models make a *turbulence closure assumption*, where as box models make a *rate of entrainment assumption.*

Both types of models (3-D and box models) calculate average gas concentrations. The modeling of dispersion over obstacles is only possible with 3-D models, but difficult and time consuming. Both model types can fairly well model effects of atmospheric stratification and of surface heat transfer

5.2.1. BOX MODELS

The box models are based upon the concept of initial gas slumping as a whole. Some models do not consider air entrainment or edge mixing at the end of the slumping phase, while other do take this into account. The various models attempt to quantify the *heat transfer effects* in different ways. The heat absorption of dense clouds colder than air may vary considerably depending on the roughness and constitution of the ground. This will alter the entrainment rate and modify the transition time period to neutral density conditions.

A review of box models has been made by Havens[10] (1980). Box models present many disadvantages:

- the results differ quite a lot for large spills
- they do not allow for the effects of buildings and topography
- the time varying nature of the source cannot be taken into account

– it is not possible to estimate the relationship between peak and mean concentrations.

Box Models Applicable to Pure Vapors
All of the box models are essentially similar in their consideration of gravity sampling, but differ considerably in accounting for air entrainment. They assume homogeneity of properties over the volume of the cloud (or a transverse slice in the case of a plume). They cannot be applied to problems involving varying time releases or complex terrain.

It appears that some of the older models reviewed by Havens (i.e., earlier models up to the year 1977) have been recognized as inadequately describing essential features of the heavy gas dispersion process (Havens[11], 1980). The reader must apply some caution when choosing a dense gas model. The recent models (as from 1980) are in the process of being validated by comparing the model predictions with selected field experiment data (LPG-, ammonia-, chlorine releases).

Webber[12] (1983) reviewed the variety of box models and derived analytical solutions for two basic classes of models, subject to the restriction that buoyancy was conserved. He found that all the simple box models fall into one or other of the two classes, and was able to highlight the differences in scaling properties incorporated in the models in an explicit fashion. The merits of the box models have been discussed by McQuaid[13] (1984). The earliest models were based on Gaussian models for passive dispersion. They were followed by so called "box" models (also called "slab" or "top hat" profile models), the first of which was due to Van Ulden[14] (1974), the main improvement of this box models being to account for gravity spreading. Differences between models occur in the formulation of entrainment parameters, cloud advection velocity, etc.

Box Models for Heavy Gas Plumes
There are not as many box models describing plume dispersion as there are for puff dispersion. Plumes of heavy gas are generated when the gas is released over a long duration of time from a continuous source (for instance pipeline breaks, or liquid leaks from large storage tanks). The box models used for plumes are similar in principle to those developed for puff dispersion (Cox & Carpenter, 1979)[15].

Steady Plume. Both top entrainment and edge entrainment are used. The plume cross-section is assumed to be rectangular, with its axis along the wind direction. The properties such as concentration, temperature, and velocity are assumed to be uniform over a given cross-section of the plume. They vary with downwind position. The plume width and height change as a result of gravity spreading (assumed to occur laterally only) and air entrainment across the outer surfaces.

Instantaneously Released Puff. The principal difference between the plume box models and the puff models is in the concentration resulting from the effect of wind speed on the downwind translational velocity of the plume. The initial volume is assumed to be known, and is generally represented as a vertically oriented cylinder whose radial and height dimensions change as a result of gravity spreading and air entrainment across the outer surfaces of the cloud. The cylindrical cloud is usually assumed to be translated with the wind.

In either case the model requires analytical expressions for the spreading velocity (i.e., the velocity of the cloud edge) and the entrainment of air at the cloud boundaries. The velocity of the cloud edge as a density intrusion has usually been modeled using the Boussinesq approximation, or neglect of inertial effects of density variations. It must be noted that the density of the cloud (which is treated as spatially uniform in the box model) is affected by energy transfer from the cloud surroundings as well as the entrainment of air into the cloud. Some of the models provide for heat transfer from the earth's surface to the cloud. The vertical density stratification of the flow is measured by a form of the overall Richardson number.

The *entrainment of air into a jet of anhydrous ammonia*, resulting from the release of pressurized liquid ammonia and the subsequent motion of the plume, have been modeled using the box model approach by Raj[16] (1980), and Raj and Aranamuden[17] (1980). Jagger[18] (1983) has developed a plume model for non-reactive clouds which includes the side and top entrainment of air. A variation of the *plume box model* has been proposed by Colenbrander[19] (1980). The latter model is formulated in terms of eddy diffusivity rather than entrainment velocities. The model has been extensively reviewed by Havens[20] (1982), and Wheatley and Webber[21] (1984)

Box Models for Clouds with Aerosol and/or Reactive Chemicals

When the gas cloud released contains significant amounts of volatile aerosols, or when the gas cloud reacts with the ambient air, the cloud buoyancy is not conserved during its dilution. The density of the cloud has to be calculated by taking into consideration the evaporation of liquids, their reaction, or both. Also, when there is a chemical reaction (as is the case with anhydrous ammonia, chlorine, phosgene, nitrogen tetroxide, etc.) the total mass of the chemical of interest will change. Raj[22] (1983) has developed a modified box model in which air entrainment is considered using the eddy diffusivity approach. The model assumes a cylindrical cloud expanding radially. The cloud depth increase is modeled, taking into account the vertical and horizontal turbulent eddy diffusivities of the atmosphere as a function of the atmospheric stability and cloud depth. The correlation's given by Havens[23] (1982) are used.

The key element of this model is the determination of the mean density of the cloud at every instant. Equilibrium thermodynamic approach (with or without external heat supply) is used. The mass of original chemical remaining after mixing, and the mass faction of new chemicals formed are calculated. The equations for radial expansion rate and cloud depth increase are solved numerically by a simple first order difference scheme. The box model transits to a Gaussian dispersion model when the rate of expansion of the cloud radius by gravitational slumping is less than the rate of increase of the standard deviation of the Gaussian profile, consistent with the atmospheric stability. The source at the transition point is assumed to be a cylindrical volume source of chemical with the concentration prevalent in the cloud at transition time. The concentration at any point downwind is calculated using the Green's functions (Carslaw and Jaeger[24], 1947). Eventual mass loss is also accounted for. This simple box model seems to provide reasonably accurate predictions.

5.2.2. INTERMEDIATE AND NUMERICAL MODELS

Such *intermediate (or slab) models*, which are more complex than the box models, solve for the spatial and temporal variations of properties within the dispersing cloud. They retain most of the advantages of 3-D codes but largely avoid the possible numerical solution problems. Analytical solutions are possible for some cases. Development in this area has been reported by Rottman[25] (1985). The increased complexity range from so-called *"shallow layer" models* (in which the vertical gradients are neglected or simplified distribution in the vertical direction are assumed) to fully 3-D solutions of the Navier-Stokes equations with turbulence. The shallow layer models (of which there are not too many) represent intermediate complexities incorporating some of the features of the box models. The true representation of the gravity head, within which most of the dispersing mass appears to be located during the initial phase of dispersion of an instantaneously released dense gas, is one of the important features of this type or models. Shallow layer models have been analyzed by Wheatley and Webber[21] (1984). Intermediate models can account for non-uniform terrain, but cannot cope with all aspects of flow field (i.e., flow around a building in urban areas). The effect of time-varying releases and other, similar, transient effects can be accounted for.

Generalized Box Models
These are box models in which greater structure is given to spatial variations of field than merely assuming either a Gaussian or a rectangular profile for each independent direction. This enables a greater range of physical processes to be modeled more realistically (HEGADAS model of Colenbrander[19], 1980).

5.2.3. 3-D MODELS

In literature there are more than ten 3-D time-dependent models for the description of heavy gas dispersion. Detailed review of 3-D models have been given by Havens[26] (1979), Havens[27] (1982) and Wheatley and Webber[21] (1984).

More recent mathematical models are based on fundamental laws of conservation of mass, momentum and energy (Jagger[28], 1982). A typical conservation model will comprises six basic non-linear partial differential equations, which are to be solved using finite element techniques. Analytical and numerical model development continues along two distinct lines. The simpler approach is based on the solution of integral equations. Such models, though limited, have only a small number of adjustable constants whose effect may be easily interpreted physically. The model should be checked from experiments and the constants adjusted accordingly. The second approach is based on solution of the relevant differential equations simplified by the introduction of eddy diffusivities. These models are far more flexible than the precedent one, however they suffer from considerable uncertainty in the specification of the eddy diffusivities. In many problems of turbulent diffusion the use of eddy diffusivities is regarded as out-dated and appeal to higher-order closure models is made.

Three-dimensional models have fields f(x,y,z,t) for quantities such as concentration, density, temperature and velocity. These quantities are determined from equations derived from conservation of contaminant, conservation of total mass, of total heat, and of momentum (Navier-Stokes equations). An equation of state and, because the equations for the fields are averaged, a turbulence closure are also required. These equations are solved numerically.

All the 3-D codes use the gradient transfer hypothesis for turbulence closure, whose validity for heavy gas dispersion is highly questionable. The main differences in the various models arise in the consideration of certain phenomena and the simplifications made, for instance, hydrostatic approximation, or Boussinesq, or anelastic approximations. Progress in this area, however, has been made. The use of the hydrostatic approximation is equivalent to neglecting vertical inertia and vertical shear forces. Incompressibility was also assumed in some of the models. The turbulent diffusion of mass and momentum are represented using the eddy diffusivity approach. Different correlation's were used in the models in representing the diffusivity tensor. he basic equations are reasonable approximations and are in principle capable of accommodating non-uniform terrain and time varying releases. Another drawback is that a turbulence closure has to be made, which is necessarily phenomenological. The spectrum of eddy sizes that are present in any turbulent flow and their effect on turbulent diffusion are not properly taken into account (assumption of gradient transport).

The range of physical processes which are taken into account depend strictly on what is assumed for the turbulence closure. The numerical solutions involve coarse discretization schemes. Other numerical problems not directly related to the grid size, such as instability, may also be present.

Some of these models were developed specifically to determine the distances corresponding to lower flammability limit (LFL) of Liquid natural gas (LNG) releases. In general, the coupled equations of mass, momentum, and energy are solved, together with equations of state and equations describing turbulence. The solutions are obtained using numerical schemes expressing the equations in difference form or in finite element form. The ambient flow field is assumed to be steady and 2-dimensional in character.

Robins[29] (1986) has published an introduction to the numerical modeling of dispersion in the atmospheric boundary layer. The numerical treatment of passive contaminant dispersion is considered, starting from diffusivity theory and leading to second order closure techniques, and including the prediction of both mean and fluctuating concentration levels. Lagrangian statistical models are also discussed. Building-affected dispersion and buoyant plume behavior are treated. Mesoscale and dense gas dispersion modeling are considered. The restrictions imposed by natural variability are commented and a preference for simple models is sustained.

5.2.4. CONCLUSIONS ON THE PRACTICAL USE OF HEAVY GAS DISPERSION MODELS

The following assumptions are valid for most of the older models up to 1982, as reviewed by Blackmore[30] (1982):

– the dispersing cloud moves over flat terrain or water
– the ground has constant roughness and thermal properties
– there are no obstacles, such as trees, fences or buildings, to the wind or to the moving cloud
– the gas cloud undergoes, unless stated differently, no chemical or physical reaction during dispersion
– local concentration fluctuations are not predicted.

Special assumptions apply to the following model types:

- The *K-theory models* can be used in the case of terrain with obstacles
- In *box (slab, or top-hat) models* the cloud is assumed to have a flat top:
  - the concentration of pollutants is uniform across the cloud, which means that similarity profiles of velocity or concentration are imposed on the gas cloud
  - the cloud slumping velocity is described by the gravity intrusion models.

Some of the difficulties with 3-D codes are inherent to large numerical calculations, i.e., on of stability and convergence. Significant error can result if the cell sizes for numerical computation are not properly chosen. Coarse cell sizes result in improper representation of density interface leading to erroneous diffusion rate predictions. The error is particularly severe during the initial slumping phase, when the cloud depth is small. The conclusions reached by Wheatley and Webber[21] (1984) are that existing 3-D models have serious deficiencies (due for one part to the use of coarse grids and inadequate turbulence models) which cause predictions of these models to be unreliable. Nowadays with the availability of large mainframe computers, like a Cray computer, this problem could be partly overcome. All the 3-D codes reviewed by these authors have unsolved problems in turbulence modeling, density fluctuations, averaging, and elimination of sound wave solutions.

Conclusions
Current 3-D codes are not very useful for determining hazard distances for toxic gas releases where toxic concentrations of interest are in the 10 ppm levels. They however can be used to describe cloud dispersion up to the point where the concentrations are of the order of 1% (such as LFL). It is the opinion of many researchers in the heavy gas field that for most calculation purposes, the simple box model approaches may be more than adequate.

5.3. Modeling Problems

The estimation of the dispersion of dense gases in the atmosphere is a major factor in the assessment of the hazards resulting from the loss of containment of flammable and toxic gases. The models generally agree with the experimental observations, but they disagree with each other when used to forecast what might happen in situations that have not been subject to experimentation.

The predictions of the models described above are still strongly affected by weather dependent diffusion coefficients. Dense gas clouds can be very flat and a width to height ratio of 100 is easily attained. The Reynolds number based on the height of the cloud in a wind tunnel is often lower than under real field conditions. Some box models relate entrainment coefficients to turbulent length and velocity scales which in turn are related to broad ranges of Pasquill stability categories. Others relates distributions of eddy diffusivities to vertical temperature gradients and these in turns are related to Pasquill stability categories.

Fannelop and Waldman[31] (1971), Van Ulden[32] (1979), Byggstoyl and Saetran[33] (1983) have proposed models that describe the spreading process by momentum equations with suitable boundary conditions for the leading edge of gravity currents. In these models however, vertical acceleration in the cloud and the reaction of ambient fluid to accelerations of the front edge and of the top of the cloud have been neglected. Such models are not capable of describing the early spreading stage, when the cloud is not shallow.

The problem of entrainment and initial acceleration of the cloud has been handled by Van Ulden[32] (1979), assuming a rectangular shape with a linear velocity distribution.

Improvement to these equations was done at a later period (Van Ulden,[34] 1984): a new bulk model for the spreading of a dense fluid (fixed volume release) in an infinitely deep channel was developed using dimensions density differences. The model is non-hydrostatic and makes use of three rate equations (lengths, momentum-integral, and volume) and four diagnostic equations (describing the shape of the current and the velocity distribution). Numerical integration with respect to time is applied. The Boussinesq approximation has not been used. A new parametrization of entrainment is proposed, which does not violate the conservation law for potential and kinetic energy. This model has been used to analyze laboratory experiments by Havens and Spicer[35] (1985) and Thorney Island trials with low atmospheric turbulence. The model describes satisfactorily the radial gravity spreading observed in laboratory and the field data for low wind speed conditions. The mixing process (area-averaged concentrations) have been simulated as well. The model gives predictions for the cloud height and the volume-averaged concentration. It was found that the concentration in dense clouds decrease rapidly with height near the ground surface and more slowly in the upper part of the cloud. The Gaussian profile does not seem to be a proper approximation to vertical concentration profiles.

The modeling of the intermediate phase includes the following points: advection by the mean wind, vertical mixing, gravity spreading, stretching by wind shear, relative horizontal diffusion and meandering motion.

When vertical mixing is enhanced by atmospheric turbulence, the reduction of gravity spreading will also be enhanced. No box model seem to include this feature.

Jagger and Kaiser[36] (1981) have attempted to link a model of the cloud formation phase with that for the dispersion phase, but the lack of experimental evidence is presently a considerable inhibiting factor.

The activities in the field of research concerning the formation and spreading of flammable and toxic gas clouds has been reviewed by Thaning[37] (1988). The aim of the study of the literature was to survey the state of the art to get a basis for future national research efforts in Sweden. The current situation as regards general knowledge and modeling appeared to be fairly good within the following areas: Source strength associated with a leak form the liquid space in pressure vessels; Gas escape from boiling liquid pools; and Heavy gas dispersion in open and flat terrain. Modeling appeared to be absent or uncertain within the following areas: Evaporation from subcooled ground deposits; Formation of liquid pools due to impaction and rain out; Initial entrainment and spreading associated with large instantaneous releases, and Spreading of heavy gases in landscapes with complicated topography or large roughness (forest, urban areas, etc.).

5.3.1. MODELING THE DISPERSION OF AMMONIA ACCIDENTALLY RELEASED INTO ATMOSPHERE

The *consequences of a catastrophic failure of a vessel containing ammonia liquefied under pressure can be predicted quite successfully*[9, 38, 39, 40] under the following conditions:

- If ambient air, wet or dry, is mixed with pure ammonia vapor, the mixture is always less dense than air
- When the initial liquid fraction, *F*, is above 16-20% at 20°C, the cloud will become denser than air
- The dilution with air and the humidity play also a role.

It is important to know how much air is entrained and what fraction of the airborne ammonia is entrained as droplets that are fine enough not to fall to the ground.

The modes of release of ammonia are of two kinds:

Pressurized Tank With a Small Hole in the Vapor Phase. The definition of a small hole is that $R = area_{(hole)} / A_s < 1$, where $area_{(hole)}$ is the area of the hole and A_s is the area of the liquid surface. For ammonia stored at ambient temperature, pure vapor will be released and the plume will be buoyant and its dispersion can be modeled by using standard plume rise modeling.

Pressurized Tank With a Large Hole in the Vapor Space. In such a catastrophic case, the pressure above the liquid is released instantaneously and bulk boiling occurs. Most, if not all, of the contents of the vessel can be flung into the air. About 20% of the content of the vessel will be vaporized. The remaining 80% stays as a liquid at the boiling point of ammonia (-33°C), and much of this liquid may become airborne as a cloud of finely fragmented liquid droplets. During the bulk boiling considerable turbulence is generated and as much air as 10 kg of air for every kg of ammonia released will be entrained. The resulting mixture is always denser-then-air.

Other modes of release have been explained previously for dense gases.

5.4. A Chronological Description of Dense Gas Dispersion Models

5.4.1. A REVIEW BY HAVENS

A mathematical review prepared by Havens[41] (1982) and summarized here shows that the models can be arbitrarily classified in two categories.

Box or Top-hat Models

Box or "top hat" which represent the initial development of the cloud (instantaneous release) or a cross-sectional slice of the cloud (steady continuous release). The shape, thermodynamic properties and position of the cloud are modeled using correlation's derived for the velocity of the intrusions and fluid entrainment across density interfaces. This model type often incorporates a transition, usually to a Gaussian model, to describe passive dispersion controlled by atmospheric turbulence of the gas in the far field.

A number of models have been proposed in this category, but it is possible to consider a general form in which different approaches have been undertaken to the modeling of gravity spreading and air entrainment into the well mixed volume or cloud. The use of Box type model is restricted to the representation of two types of heavy gas clouds.

a) an instantaneously released quantity of gas with a known initial volume, usually represented as a vertical orientated cylinder whose radial and height dimensions change as a result of gravity spreading and air entrainment across the outer surfaces of the cloud. The cylindrical cloud is usually assumed to be translated with the wind.

b) a steady plume with rectangular cross section (axis along the wind direction) whose thermodynamic properties vary with downwind position and whose width and height change as a result of gravity spreading (assumed to occur laterally only) and air entrainment across the outer surfaces.

In either case the model requires an analytical expression for the spreading velocity (i.e., the velocity of the cloud edge) and the entrainment of air at the cloud boundaries. Usually the Boussinesq approximation is applied. The density of the cloud, which is treated as spatially uniform in the box models, is affected by energy transfer from the cloud surroundings as well as the entrainment of air into the cloud. Some of the box models provide for heat transfer from the soil surface to the cloud.

Differences between the models arise from the different approaches to modeling the surface-to-cloud heat transfer, the spreading phase and the air entrainment. Several field experiments were analyzed by the different authors and parameters and correlation derived for the respective models. An important parameter is the Richardson number, which is used as a measure for the vertical density stratification of the cloud. The widely different entrainment sub-model parameters used reflect reliance on different data sets which are difficult to compare.

Van Ulden's model (Van Ulden[42], 1974) of the gravity spread indicates minimal cloud volume growth and dilution. Based on the analysis of heavy freon gas spreading experiments, it is assumed that cloud frontal entrainment during the gravity-dominated spread phase is negligible. Germeles and Drake[43] (1975) prefer modeling top air entrainment and neglect cloud frontal entrainment. Picknett and Carpenter[5] (1978) correlated the concentration and visible cloud dimension data from a series of instantaneous releases on land of freon-air mixtures. Picknett terminates the gravity spreading phase, and transition to a Gaussian passive dispersion model, when the Richardson number of the spreading layer decreases to Ri = 7. The top hat model of Eidsvik[44] (1980) uses air top entrainment defined in a different way and does not transition to a Gaussian passive dispersion model. Eidsvik's model neglects entrainment due to shear at the interface whereas Germeles and Drake suggested that entrainment is associated with the shear at the interface. Cox and Carpenter[45] (1979) are modeling frontal air entrainment and make a transition to Gaussian model.

Advanced Similarity Models
Some of the models which assume Gaussian forms for the concentration profiles of heavy gas cloud or plume evolution have been proposed by Te Riele[46] (1977), Flothmann and Nicodem[47] (1980), Ooms[48] (1974), and Colenbrander[19] (1980).

K-Theory Models
K-Theory models which assume constitutive relations between turbulent fluxes and the gradients in mean variable velocity, temperature and concentration, coupled with the equations of change for mass, momentum and energy of turbulent fluid flow to predict the time and spatial variation of thermodynamic properties of the cloud. Representative models are: the SIGMET model (Havens[49], 1979), and its successor models ZEPHIR[50] and MARIAH.[51] The computational methods have been improved but the specification of turbulent mixing and of other physical models have not been much improved.

Further Development
Recent models overlap these categories, which makes a classification more difficult. Several investigators have assumed self-similar concentration profiles (Te Riele[52], 1977; Colenbrander[19], 1980; Flothmann[47], 1980) and some have coupled the assumption of a self similar concentration profile with a K-theory representation of turbulent mass transfer within the cloud (Colenbrander[19], 1980). In the period 1980 there have appeared several models which involve greater simplification of the equations of motion, energy and mass than is found in the K-theory models, but which still require solution of partial differential equations to predict cloud state variables. Models proposed by Zeman[53] (1980); Rosenzweig[54] (1980), and Fannelop[6] (1980) may present significant advantages over the box models since they allow a more realistic representation of spatial variation in the simulation of cloud dispersion, and incorporate more general turbulent mixing sub-models.

It appears now that some of the models reviewed by Havens[55] (1977), have been in the mean time recognized as inadequately describing essential features of the heavy gas dispersion process, which is now better understood. There is however still large disagreement among the models currently in use.

Models Reckoning with Terrain and Buildings Effects

Deaves[56] (1985) has prepared a review of the current status of 3-D modeling, with special reference to the inclusion of turbulence effects. The code HEAVYGAS is also described and recommendations for further improvement of the computer code are given.

A three-dimensional numerical program has been developed by Jacobson[57] (1987) for simulation of heavy gas dispersion. The K-ε-model with standard constants has been used for calculation of the diffusion coefficients where the effect of density gradients on the mixing process is accounted for by applying correction factors given as functions of local Richardson number. It was found that the results of a numerical simulation of the Thorney Island Trial 008 are in good agreement with experimental data.

5.4.2. TOPICAL ISSUES ON THE DEVELOPMENT OF DENSE GAS DISPERSION MODELS

One of the first review of models for dense gas dispersion was prepared by *Havens*[58] *(1978)*, who made a *review of the predictability of catastrophic LNG spills into water* and showed that a number of models produced widely varying estimates for the distance to which a flammable cloud would extent under a wind of 5 mph. Some of the models presented in Havens report in 1978 are now considered obsolete and superseded by revised versions.

Schnatz et al.[59] (1980) presented *formulas for the modification of the wind and turbulence profiles,* which are based on the experimentally gained knowledge from hydrodynamics, meteorology, and oceanography, and are transferable to the processes of heavy-gas dispersion. The knowledge in the experimental field of heavy-gas dispersion is outlined, and a micro-turbulent system equation is introduced which is suitable for simulation of heavy-gas dispersion. The formulas for the modification of the wind and turbulence profile are applied to a simplified K-dispersion model, and the effect of this modification is discussed by an example.

A review of fifteen heavy gas dispersion models (TABLE 5.1 - TABLE 5.4) have been prepared by Blackmore et al.[30] (1982). It contains 60 references of *K-theory and slab mathematical models* for the dispersion of accidental releases of heavy gases until they dilute with air to nonflammable or nontoxic concentrations. The models are broadly classified into K-theory models (5 models) and slab models (10 models). The description of four jet release models is also given. Denser-than-air vapors usually form low, flat clouds which spread because of their own density even in the absence of wind. It was recognized that attempting to describe such systems by adapting Gaussian models suitable for neutrally or positively buoyant clouds was inadequate.

Deaves[60] *(1983) has developed a K-theory turbulence flow model for the 2-dimensional dispersion of heavy gases in complex situations,* including transient convection and diffusion from a low-momentum (i.e., non-jet) source, with either an instantaneous or continuous release (e.g., chlorine within in a building); steady-state near-field predictions with low-, or high-momentum sources (e.g., spray curtain dispersion); and the treatment of more general cases of heavy gas dispersion, which could be described as turbulent buoyant natural convection. However, in the last case, such treatment is most satisfactory from the point of view of the stability of the numerical methods employed in the code, when there is a non-zero ambient wind. A review with 15 references concerning the *application of advanced turbulence models* in determining the structure and dispersion of heavy gas clouds is also presented by *Deaves*[61] *(1984)*.

Knox[62] *(1984) made a review,* with 17 references, of models for the dispersion of heavy gases in air, focusing primarily on *liquefied natural gas and* H_2S.

Tasker[63, 64] *(1984)* has made a *review of the basic concepts of dense gas dispersion* with special regard to *modeling of heat transfer*. A basic physical picture of dense gas dispersion is provided in his paper. Mathematical and wind-tunnel models of dense-gas flow are presented and discussed, including the constraints and disadvantages of the different modeling techniques. Special emphasis is given to heat transfer in dense gas dispersion concerning such dense gases as Cl_2, SO_2, liquefied natural gas, and liquefied C_3H_8.

Farmer[65] *(1982)* has prepared a *survey of turbulence models* with particular reference to dense gas dispersion. *Havens*[41] *(1982)* has prepared another review of mathematical models for prediction of heavy gas atmospheric dispersion, with 46 references.

A detailed *review of experimental results and some of the models* was presented in a "Workshop on Heavy Gas Dispersal", held at the Von Karman Institute (Raj[66], 1982).

A *comparison of model predictions with experimental data* made by Woodward et al.[67] (1982) showed that the predictions of four of the most recent models (at that time) had improved.

Another *extensive review of dense gas dispersion modeling* is that of *Webber*[12] *(1983)* and *Wheatley and Webber*[68,69] *(1984)*. The latter was prepared for the Commission of the European Communities.

TABLE 5.1. Comparison of K-theory models (source: Blackmore, J. Hazard. Mater., 6 (1982))

	Model Name	Zephyr	Transloc	Sigmet-N	Mariah	Disco
1.	**General description**					
	<> K-Theory models	x	x	x	x	x
	<> Numerical, 3-Dimensional transient heavy gas models	x		x	x	
	<> K-models modified for describing heavy gas clouds		x			
	<> Numerical, 2-dimensional steady- state heavy gas model: uses stream function					x
2.	**Conditions treated**					
2.1	Sources and types of spills	All	All	All	All	Multiple runs
2.2	Type of release					
	- jet momentum	With fine grid	No	With fine grid	With fine grid	No
	- buoyant plume	Yes	No	Yes	Yes	Yes
	- aerosol	No	Yes	No	No	No
2.3	Dispersing regimes					
	- open	Yes	Yes	Yes	Yes	Yes
	- obstructed (wakes)	Yes	No	Yes	Yes	No
	- topography (variable terrain)	Yes	No	Yes	Yes	No
	- deposition from cloud	No	Yes	No	No	No
2.4	Meteorology					
	- atmospheric stability	Yes	Yes	Yes	Yes	Yes
	- very low wind speed (<1 m/s)	Yes	No	Yes	Yes	No
	- humidity and heat of condensation	Yes	No	Yes	Yes	Yes

TABLE 5.1. (continued)

Model Name:	Zephyr	Transloc	Sigmet-N	Mariah	Disco
3. Mechanistic features					
3.1 Mathematical treatment of air entrainment	3-D K-theory, Particle-in-cell, Lagrangian diffusion	2-D K-theory, modified from nuclear fallout model; with variable terrain	3-D K-theory	3-D K-theory; uses implicit form of equations	2-D K-theory; Steady-state
3.2 Transition from gravity spreading / entrainment to non-buoyancy passive dispersion	Not needed Use reasoning	Not needed	Not needed Use reasoning	Not needed Automatic reasoning	Not needed
3.3 Heat transfer from surface to gas cloud	Yes	No	Yes	Yes	Yes
3.4 Other significant features		Momentum balances not solved	Allows water vapor entrainment		

TABLE 5.2. Comparison of slab models (source: Blackmore, J. Hazard. Mater., 6 (1982))

Model Name	Denz	Germeles & Drake	Picknett	van Buijtenen	van Ulden
1. General Description					
<> Slab models	x		x	x	x
<> Transient behavior of heavy gas clouds from area sources		x		x	x
<> Heavy gas clouds spreading behavior	x		x		
2. Conditions Treated					
2.1 Sources and types of spills	Instantaneous (cylindrical source)	Instantaneous (cylindrical) and constant (vertical rectangular) source	Instantaneous (cylindrical) sources	Instantaneous, constant and time varying source	Instantaneous (cylindrical) sources
2.2 Type of release					
- jet momentum	No	No	No	No	No
- buoyant plume	No	No	No	No	No
- aerosol	No	No	No	No	No
2.3 Dispersing regimes					
- open	Yes	Yes	Yes	Yes	Yes
- obstructed (wakes)	No	No	No	No	No
- topography (variable terrain)	No	No	No	No	No
- deposition from cloud	No	No	No	No	No

TABLE 5.2. (continued)

	Model Name	Denz	Germeles & Drake	Picknett	van Buijtenen	van Ulden
2.4	Meteorology					
	- atmospheric stability	Yes	Yes, Gaussian port.	Yes	Yes	Yes
	-very low wind speed	No	Yes	Yes	Yes	Yes
	- humidity and heat of condensation	No	Yes	Yes	Yes	No
3.	**Mechanistic features**					
3.1	Mathematical treatment of air entrainment	Entrainment velocity for top-surface mixing is a function of atmospheric turbulence and Richardson number; no edge-mixing; well mixed cloud assumed	Entrainment velocity for top-surface mixing is a function of gravity spreading velocity and Richardson number; no edge-mixing; well mixed cloud assumed	Entrainment velocity for top-surface mixing is a function of atmospheric turbulence and Richardson number; entrainment velocity for edge mixing is dependent on gravity spreading velocity; well-mixed cloud assumed	Take-up rates from an assumed stationary gas-pool above the liquid pool; Gaussian plume treatment with empirically modified vertical dispersion coefficients equivalent to Pasquill F behavior	Only cloud radius calculations according to publication; extension expected in future publication to air entrainment with edge and top-surface mixing as a function of Richardson number
3.2	Transition from gravity spreading / entrainment to non-buoyancy passive dispersion	When density difference between cloud and air becomes less than specified value	When cloud edge speed equals wind speed	When an eddy can reach the ground *3-sigma = height of cloud*, or *Richardson number = 1*	Not needed: gas entrained from the gas-pool is considered to be the source	When Richardson number equals 0.5
3.3	Heat transfer from surface to gas cloud	Yes	Yes	No	No	No

TABLE 5.2. (continued)

	Model Name	Denz	Germeles & Drake	Picknett	van Buijtenen	van Ulden
3.4	Other significant features	Alternative transition criterion: if the radial increase due to gravitational slumping does not exceed that due to turbulence alone, and the top entrainment velocity is greater than the longitudinal velocity				

TABLE 5.3. Comparison of slab models (source: Blackmore, J. Hazard. Mater., 6 (1982))

Model Name	Hegadas-II	Cox & Carpenter	Eidsvik	Fay	Flothmann & Nicodem
1. General description					
<> Transient behavior of heavy gas clouds from area sources	Yes	No	No	No	No
<> Heavy gas cloud behavior	No	Yes	Yes	Yes	Yes
2. Conditions treated					
2.1 Sources and types of spills	Instantaneous, constant and time varying (horizontal rectangular) sources	Instantaneous (cylindrical) and constant (vertical rectangular) sources	Instantaneous (cylindrical, time-varying gas input) and constant (vertical rectangular) sources	Instantaneous (cylindrical source)	Instantaneous (cylindrical) and constant (vertical rectangular) sources
2.2 Type of release					
- jet momentum	No	No	No	No	No
- buoyant plume	No	No	No	No	No
- aerosol	No	No	No	No	Yes, thermodynamic equilibrium
2.3 Dispersing regimes					
- open	Yes	Yes	Yes	Yes	Yes
- obstructed (wakes)	No	Yes; approx. method	No	No	No
- variable terrain)	No	No	No	No	No
- deposition from cloud	No	No	No	No	No

TABLE 5.3. (continued)

Model Name	Hegadas-II	Cox & Carpenter	Eidsvik	Fay	Flothmann & Nicodem
2.4 Meteorology					
- atmospheric stability	Yes	Yes	Yes	Yes	Yes
- very low wind speed (<1 m/s)	No	Yes	Yes	Yes	No
- heat (humidity+condens.)	Yes	Yes	Yes	Yes	Yes
3. Mechanistic features					
3.1 Mathematical treatment of air entrainment	Dispersion in vertical and horizontal directions dependent on turbulence; vertical dispersion is also a function of Richardson number; similarity concentration profiles assumed	Entrainment velocity for top-surface mixing is a function of atmospheric turbulence and Richardson number; entrainment velocity for edge mixing is dependent on gravity spreading velocity; well mixed cloud assumed	Entrainment velocity for top-surface mixing is a function of atmospheric and convective turbulence and Richardson number; entrainment velocity for edge mixing is dependent on gravity spreading velocity; well mixed cloud assumed	Entrainment velocity for top-surface mixing is a function of atmospheric turbulence and Richardson number; entrainment velocity for edge mixing is not important in the later stage; well mixed cloud assumed	Entrainment velocity for top-surface mixing is a function of atmospheric turbulence and Richardson number; entrainment velocity for edge mixing is dependent on gravity spreading velocity; Gaussian concentration profiles assumed

TABLE 5.3. (continued)

	Model Name	Hegadas-II	Cox & Carpenter	Eidsvik	Fay	Flothmann & Nicodem
3.2	Transition from gravity spreading / entrainment to non-buoyancy passive dispersion	Continuous and smooth	When lateral turbulence velocity > lateral spread velocity	Continuous and smooth	No transition	Continuous and smooth
3.3	Heat transfer from surface to gas cloud					
3.4	Other significant features	The "observer concept" accounts for the time-varying sources	Thermodynamics of 2-phase mixtures included		Not a model that describes the growth of the cloud; it only gives	

TABLE 5.4. Comparison of jet release models for elevated sources (source: Blackmore, J. Hazard. Mater., 6 (1982))

	Model Name:	Astleford et al.	Bloom	Cox et al.	Ooms et al.
1.	**General Description**				
	<> Jet-release models	Yes	Yes	Yes	Yes
	<> Descending plume and ground level heavy gas models	Yes	Yes		
	<> Heavy gas plume and cloud behavior; plume converting to ground level cloud			Yes	
	<> Plume path of heavy gases				Yes
2.	**Conditions treated**				
2.1	Sources and types of spills	Continuous jet relea-se from point source; Ooms model and Riele model are combined	Continuous elevated release from point or finite area source	Continuous elevated point source; Ooms model, and Cox & Carpenter model are combined	Continuous jet relea-se from point source
2.1	Type of release				
	-jet momentum	Yes	Yes	Yes	Yes
	- buoyant plume	Yes	Yes	Yes	Yes
	- aerosol	No	No	No	No
2.3	Dispersing regimes				
	- open	Yes	Yes	Yes	Yes
	- obstructed (wakes)	No	No	No	No
	- topography (variable terrain)	No	No	No	No
	- deposition from cloud	No	reactive gases	No	No
2.4	Meteorology				
	- atmospheric stability	Yes	Yes	Yes	Yes
	- very low wind speed (<1 m/s)	Yes	Yes	Yes	Yes
	- humidity and heat of condensation	No	Yes	No (but ground level model part does)	No

TABLE 5.4. (continued)

	Model Name:	Astleford, et al.	Bloom	Cox, et al.	Ooms, et al.
3.	**Mechanistic features**				
3.1	Mathematical treatment of air entrainment	Heavy gas behavior is followed once the plume centerline hits the ground; entrainment rates of air into the plume depend on jet action, buoyancy and ambient turbulence	Entrainment rates of air into the plume depend on jet action, buoyancy close to the source and ambient turbulence further away	Heavy gas behavior is followed once the plume centerline hits the ground; entrainment rates of air into the plume dependent on jet action, buoyancy and ambient turbulence	Wind drag forces across the plume is allowed for; entrainment rates of air into the plume dependent on jet action, buoyancy and ambient turbulence
3.2	Transition from gravity spreading / entrainment to non-buoyancy passive dispersion	Continuous and smooth	Not needed	As for Cox & Carpenter	Not needed
3.3	Heat transfer from surface to gas cloud	No	Not in referenced version of model	As for Cox & Carpenter	Not relevant
3.4	Other significant features		Not designed to describe events after the plume reaches the ground.; energy equation includes phase changes, chemical reactions, and gain or loss of material; including rainout of particles		Not designed to describe events after the plume reaches the ground

A three dimensional Lagrangian pollutant dispersion model for passive contaminants was modified by Gaffen[70] *(1985) to allow simulation of dispersal of heavy gases, and moist and buoyant plumes.* Outputs from a two dimensional Navier-Stokes atmospheric boundary layer solver are used. Model runs were made to simulate experimental tests of gas dispersion in atmospheric surface layers. Predicted concentration fields are in accord with measurement. Model simulation of water vapor plume dispersal in the planetary boundary layer were made using field data from cooling towers. Predicted plume rise and visible plume outline are in good agreement with observations for various meteorological and source conditions.

As part of an effort to replace the *Air Force's Ocean Breeze/Dry Gulch dispersion model* with a more state-of-the-art model, a *three-dimensional Gaussian puff atmospheric dispersion model has been developed for predicting the dispersion of toxic gases resulting from a chemical release (Kunkel*[71]*, 1985).* The model is similar to the SPILLS model (Shell Development Co.), but includes several improvements. A Continuous stability parameter is used instead of the discrete Pasquill stability categories. The surface heat flux, friction velocity, Monin-Obukhov length, and surface roughness are used to determine the stability parameter. There is also an option to use the standard deviations of the horizontal wind direction for computing the stability parameter. The computer code was written in BASIC language for the Z-100 microcomputer. Future improvements in the model will include incorporating the heavy gas effect and better methods of defining the source strength.

The experiences of Trans Canada Pipelines in heavy gas dispersion research and its application to the hazard assessment process have been summarized by Frayne[72]*, (1985).* The company was one of the co-sponsors of large-scale field tests conducted at Thorney Island, England, which released large volumes of a heavy gas in simulations of gas spills. Results from some of these tests were compared with results obtained using 4 state-of-the-art computer models: the Cox/Carpenter and Eidsvik box models, the 3-dimensional hydrodynamic model MARIAH II, and the advanced box model HEDAGAS II. For high wind speeds, the latter 2 models were found to give better predictions of downwind travel and lateral extent of the gas cloud, maximum cloud gas concentration, and maximum distance to 1% concentration level. Gas spill incidents at Bhopal and Mexico City are used to illustrate the importance of hazard assessment techniques to real-world applications.

An overview of current state of knowledge on heavy gas dispersion is presented by McQuaid[73] *(1985).* The relative merits of different types of mathematical models (box, 3-dimensional, and intermediate types) are compared. Ways to validate dispersion models are evaluated, and some short-cut methods of dispersion estimation are outlined. Special topics of current interest include physical modeling, investigations on the effects of obstructions on gas dispersion, and the influence of turbulence.

A summary of heavy gas spills modeling research was prepared by Raj[74] *(1985).* The physical processes occurring during the dispersion of a heavy gas in the atmosphere are outlined, including source-controlled dilution, gravitational slumping, combined slumping and ambient flow interaction, passive dispersion, and ancillary phenomena such as aerosol effects, chemical reactions with ambient air or water vapor, and heat transfer during cryogenic liquid spills. Some results from field tests and laboratory experiments on heavy gas spill dispersion are described. Different modeling approaches indicated in the literature are reviewed, including box models applicable to pure vapors and plumes, box models for clouds with aerosols or reactive chemicals, and intermediate and numerical models which are more complex and which solve for the spatial and temporal variations of properties within the dispersing cloud. The utilization of spill modeling research results by ultimate users is also discussed.

The effect of the ground on entrainment rates at the top of a dense gas cloud due to atmospheric turbulence has been modeled in the laboratory. Entrainment rates obtained in the laboratory are often used to estimate the top mixing calculated by computer models which try to predict the dispersion of a dense gas cloud. Due to the large horizontal extension of the cloud, small variations in assumed entrainment rates or mixing coefficients can affect substantially the dilution of the cloud. From experimental results it is seen that a solid boundary near a density interface can affect the turbulent mixing taking place through the interface. Several authors have shown the reduction in vertical turbulent velocity near a wall, and furthermore that the larger scales of turbulence are the most effected by the reduction. Since the larger vertical scales are more effective at mixing in a stratified fluid, the present observations indicate, that the blocking of these scales by the bottom boundary will reduce mixing rates for low Richardson numbers. *With the dispersion of a dense gas cloud in mind, it is possible to give a simple relation to determine whether the proximity of the ground will reduce mixing at the top of the cloud (Redondo[75], 1987).* If the vertical thickness of the cloud, d, is less than or comparable with λ, the proximity of the ground will reduce the mixing at the top of the cloud. The expression $Ri \le a(1/d)^{1/n}$ gives the condition for the ground to affect mixing at the top of the dense gas cloud.

Webber and Wheatley[76] (1987) have presented an integral model of the behavior of an instantaneously released heavy gas cloud in calm conditions, or sufficiently close to the source, that gravity effects dominating the ambient turbulence generated from the initial potential energy of the cloud may effect the subsequent dilution. The model treats the turbulent energy in the cloud as a dynamic variable which determines the entrainment rate. It is constructed such that overall dissipation of mechanical energy is guaranteed. The turbulent energy of the cloud released from rest is thus generated explicitly from the initial potential energy, and the entrainment rate may depend on the initial aspect (height to radius), ratio, and the initial density of the cloud. An investigation of the properties of the model indicates that these effects, whilst present, are small. Consequently, as a result of this more detailed study of the energy budget of the cloud, *considerable support in favor of simple models, which treat the early dilution of the cloud as "edge entrainment" with an entrainment velocity proportional to the spreading rate, can be given.*

Dunst et al.[77] (1987) have developed a new energy-conserving Box model for the heavy gas diffusion dominated by gravity. Comparison with other models have shown that especially for large volumes of heavy gases, being emitted during real accidents very often, the new Box model yields reliable distributions of the concentration. Furthermore it could be clarified, in what cases the thermal effects should be accounted for. *Finally the combination of the new Box model and the numerical diffusion model is an efficient tool, to simulate the whole dispersion process approximately.*

The thesis of Vanulden[78] (1987) relates a *new model of dense gas clouds, such as chlorine, for the phase where gravity induced spreading and mixing dominates dispersion.* From the fundamental equations of motions, integral equations for the spreading, slumping, and mixing of the cloud are derived. These equations account for radial and vertical accelerations in and around the cloud and for the effect of large density differences between the cloud and the environment. Turbulent mixing is described with an entrainment equation, which uses the turbulent kinetic energy of the cloud. This turbulent energy is described with a time-dependent turbulent-energy equation. The model is compared with observations on axis-symmetric dense clouds from the literature. For radial gravity spreading agreement between model simulations and experimental data is satisfactory. Analysis shows that near-surface concentrations are likely to be much higher than indicated by previous studies of the same experimental data. The reason for this is that usually a Gaussian or uniform profile is assumed for the vertical distribution of area-averaged cloud concentrations. However, such profiles are poor approximations to the observed profiles, especially close to the ground where strong vertical gradients are observed. A new ad-hoc similarity profile is proposed, and gives a satisfactory simulation of observed concentrations.

A top hat model, based on energy conservation, was developed at Hamburg University, Germany, that is used for the dispersion of instantaneously released heavy gases at ambient air temperature (Fischer[79], 1987). The energy balance equation used involves only the kinetic energy of the mean motion and the available potential energy. In the energetically consistent model, cloud slumping after release starts realistically from rest, assuming a three-dimensional, radially symmetric velocity field. The free parameters of the model are determined using data from field experiments. The edge entrainment coefficient thus obtained is consistent with the appropriate parameter fitted to wind tunnel data. *The model results were compared with those of Picknett's model and with data not used in fitting the parameters. The concentration predictions are only satisfactory for low densimetric Froude numbers.* The ratio of the modeled concentration to the measured peak concentration in the front vortex has been estimated.

Schreurs and Mewis[80] (1987) express a justified *critique concerning the use of transport phenomena models for heavy gas dispersion simulation,* since such models entail several difficulties. These can be related into two categories: (a) the difficulties related to the adequate description of the relevant turbulent transport process; and (b) numerical complications. The numerical problems of a Lagrangian particle model are of the following type:

a) In convection-turbulent diffusion problems, the objective is to predict the distribution of a scalar quantity in a fluid flow. The scalar quantity is simultaneously convected with the mean flow and diffused by the turbulence. In the case of heavy gas release into the atmosphere, the flow may be affected or even dominated by the presence of the contaminant.
b) Most distributed parameter models of such transfer processes have used finite-difference methods for approximating the model (partial differential equations), and the central difference approximation is commonly used for the diffusion terms. However when the central difference approximation is used for the

convection terms, numerical instability may occur when the convective flux dominates the diffusive flux. The relative importance of convection and diffusion is indicated by the grid Peclet number. Unrealistic results are obtained when the Peclet number is greater than 2. In complex flow fields, the streamlines may not be aligned with the numerical grid lines. If there is a significant gradient of the transported quantity in the direction normal to the velocity vector, the associated numerical diffusion error may even obscure the physical diffusion being modeled.

The physical model, which has been described by Schreurs and Mewis, belongs to the *generic category of primitive equation models,* since it attempts to solve the dynamical equations for the wind field and the temperature and contaminant field. It describes the transient dispersion of accidentally released contaminant in the near field, and takes into account the primary heavy gas effects. The model also makes it possible to incorporate the effect of obstacles in the ambient flow field. It uses a Lagrangian particle method to solve the convection diffusion equations governing atmospheric dispersion of heavy gases. The proposed algorithm uses pseudo discrete fluid particles (representing the contaminant gas) which can be advected and diffused by flow field. The prescription for particle displacement must be consistent with the governing transport equation. The influence of the contaminant on the flow is accounted by integrating the influence of individual particles over the Eulerian mesh cells. The Eulerian computational cells are used to convert particle positions to concentrations at times and places of interest. The false diffusion problem is thus effectively eliminated. Among the numerical problems, numerical diffusion is one of the major ones. The discretization requirements to control numerical errors were examined for typical applications.

The model is demonstrated to alleviate numerical diffusion errors which result from application of low-order finite difference methods, while allowing affordable discretization for heavy gas dispersion predictions of practical interest. Comparison with a large scale experiment on heavy gas dispersion shows a good agreement.

Tasker[81] *(1987) has studied the scaling requirements for modeling heat transfer effects on the dispersion of cold dense gas clouds in wind tunnel tests.* Mathematical simulations using a box type model show that convection induced mixing can significantly dilute a cold dense gas cloud assuming an entrainment coefficient of 2.5. A rig was built to measure convective entrainment rates into a 1 m deep cloud of nitrogen, having a bulk temperature of 170K, and heat fluxes of up to 1100 W/m were applied. At low heat fluxes, no convective entrainment occurs. It is concluded that convective entrainment may only affect the dispersion of a cold dense gas cloud if the gas is buoyant at ambient temperature, i.e., for gases such as methane and ammonia.

The mathematical modeling of the accidental spills of liquefied hydrocarbons into the environment has been presented in a thesis by Sherbrooke[82] *(1987).* For the spreading and vaporization of the liquids, a model resolving the conservation equations has been compared to a simple model based on the intrusion equation; Three distinct models have been developed, which apply respectively to continuous, instantaneous and transient spilling. The spreading model had to be coupled with the transient spilling model so as to describe the time evolution of the dimensions and of the vaporization rates. By comparing the predictions with experiments realized by Shell Ltd. in Britain, the models could be adequately calibrated.

Dispersion evaluations in the framework of the implementation of the *German Nuclear Accident Ordinance* have taken place and have been reported at a meeting[83] hold in 1987. Some topics of the lectures are: features of the Lagrange particle dispersion model and comparative compilation of different models of this type; applications of the simulation model LASAT in level and complex areas; application of a Lagrange dispersion model; application of a K-theory model in the practice of emission simulation; dispersion of pollutants in the recirculation area of buildings; application of the DEGADIS model; presentation of the TUeV (Technical Control Association) heavy gas model SINDIM and first comparison with recent dispersion experiments; modeling of the dispersion of cryogenic gases; possible application of 3D-heavy gas models within the Nuclear Accident Ordinance (1989) by FIZ.

A methodology for evaluating the effectiveness of mathematical models in predicting the atmospheric dispersion of heavier-than-air vapor releases is reported by Ermak[84] (1988). The methodology is based upon *ratio comparisons of the model-predicted value to the observed value* in field-scale experiments involving continuous releases of denser-than-air gases. Plume characteristics used in the ratio comparisons include maximum concentration, center-line concentration, plume half-width, and plume height, all as a function of downwind distance from the source. Also included in the report is a review of the scientific efforts in the field of atmospheric dispersion model evaluation during the past two decades.

> *A mathematical model for the release and dispersion of hazardous materials in the atmosphere should include a submodel for the modeling of phase change in combination with chemical reactions.*

Such models are presented by Rodean[85] *(1989) for mixtures of UF_6 and N_2O_4 with dry and moist air,* which are based on thermodynamic equilibrium, the ideal gas law, temperature-dependent saturation-vapor pressure equations, a temperature-dependent equilibrium constant for the $N_2O_4 = 2\,NO_2$ reaction, and the reactions of UF_6 and N_2O_4/NO_2 with H_2O. The material model equations are written in terms of pressure ratios and dimensions parameters. These equations are used to construct equilibrium diagrams with temperature and the mass fraction of the material in the mixture as the coordinates.

In his paper Ermak[86] *(1991) presents a limited review of atmospheric dispersion models.* The models were separated into the two classes: three-dimensional, conservation equation models, and one-dimensional (quasi-three-dimensional) similarity models. These two classes of models differ considerably in physical completeness, numerical complexity, computer capability required to run the models, and ease of use. Each class of models is discussed a brief description of their distinguishing features, examples of models developed in the past decade, their advantages and disadvantages as a class, and some preliminary results from independent model evaluations. Special attention is given to the *FEM3* and *SLAB models* developed at Lawrence Livermore National Laboratory.

Guidance on the application of refined dispersion models for air toxics' releases can be found in the report of Touma[87] *(1991).* The purpose of the document is to provide general guidance considerations for applying dispersion models to such releases and to show the thought process required by the non-expert user to develop all model input parameters. Two example applications for each model are provided with a step-by-step explanation of all model input parameters and model output. Four specific models are currently included in the document. These are the *DEGADIS, HEGADAS,* and *SLAB models* appropriate for denser-than-air releases and the *AFTOX model* for neutrally buoyant releases of toxic air pollutants.

A method is presented by Ermak[88] *(1990), and Ermak*[89] *(1991) for including dense-gas effects in an existing advection-diffusion (particle-in-cell) type model,* capable of dispersion simulations over terrain, and with time-varying synoptic winds. The physical processes associated with dense-gas dispersion affect both the windfield and the turbulent diffusivity. These effects are included by perturbing the ambient windfield and diffusivity within the local region of the dense-gas cloud. The perturbed local windfield is calculated by using a vertical or layer averaging approach and is nested between the windfield and dispersion calculations. The ambient diffusivity is replaced by an adapted form of a dense-gas, K-theory diffusivity that has the property that it approaches the ambient diffusivity level as the cloud density approaches the ambient value. For numerical models, the low-lying nature of dense-gas clouds often presents a problem with resolution in the vertical direction. This is largely overcome in the proposed approach by:

(1) the use of vertical averaging and assumed vertical cloud profiles to calculate the dense-gas perturbations on the flow field and diffusivity, and
(2) the use of the particle-in-cell technique with a Monte Carlo (stochastic) displacement equation that does not rely on the concentration gradient to calculate the trajectories of the concentration marker particles.

These dense-gas dispersion modifications attempt to preserve the main features of the advection-diffusion, particle-in-cell model and, thereby, minimize the impact on the existing code. He reports also on the mathematical description and the thermodynamics of the thermal transport within a cold, heavy gas cloud for the case of adiabatic mixing, and for the more complex situation with ground heating being considered.

A simple analytical semi-empirical model has been presented by Matthias[90] *(1990), which describes the concentration field in a collapsing gas cloud of cylindrical shape.* The model examines the process of top and side entrainment, the occurrence of a leading torus and a trailing disk, and Gaussian distribution in the entrapment zones. These processes are described for cases in which there are *no atmospheric effects, i.e., no wind or ambient turbulence.* Turbulence within the cloud is self-generated, due to the sudden collapse of the cloud. The model was compared to wind tunnel and field experiments' results.

A mathematical model describing the motion of a dense gas, released continuously into an environment is presented by Bidokhti[91] *(1991).* The rate of gas release is mainly constant, in a case in which it varies with time will be considered. The model predicts the concentration and size of the cloud formed by the released gas as function of downstream distance. It also correctly reproduces the experimental results obtained. Such a model can be used in the analyses of accidental releases of hazardous materials in nuclear industry and other environments.

Sloping terrain is known to have an important effect on gas clouds heavier than air. *Models of the behavior of a heavy cloud released instantaneously on a slope have been presented by Deaves and Hall*[92] *(1990)*, and *Nikmo and Kukkonen*[93] *(1990)*. Each of these models is an intuitively appealing generalization of the flat ground integral model approach to include the effect of slopes. However the effect of slope is found with a numerically computed solution to a set of differential equations. Webber et al.[94] (1993) have presented a model of the motion of a heavy gas cloud down a uniform slope in calm ambient conditions. The objective was to examine the very simple case of a heavy cloud released instantaneously onto a uniform slope and dispersing isothermally in a way which is known to conserve buoyancy. The study focuses on the effect of the slope on the overall motion of the cloud. The model is, at present, restricted to zero wind and no mixing is assumed. The model is derived from solutions of the shallow water equations with appropriate boundary conditions. Its predictions are shown to agree adequately with experimental results in calm conditions, and possible generalization to allow for the presence of wind is discussed.

Heavy gas models are also being surveyed by a working group under the auspices of the UK's CIAA.

5.5. Description of Computer Codes for Dense Gas Modeling

Most of the models in use are of two categories: *K-theory (eddy diffusivity) models* and *slab (box or top-hat) models* (TABLE 5.5 - TABLE 5.6). Recently 3D-models have been presented, which are used for atmospheric dispersion, but these are still under development and not yet fully tested for the case of dense gas dispersion.

TABLE 5.5. List of computer codes for dense gas modeling

Model Type	Model Characteristics
Box or Slab or Top-Hat Models	
Germeles and Drake	Well mixed cylindrical source, slumping transition to Gaussian plume, no edge mixing
DENZ	Well mixed cylindrical source, transition to Gaussian plume, no edge mixing.
Cox and Carpenter	Well mixed cylindrical source, transition with heat transfer and edge mixing.
Shell HEDAGAZ	Needs a wind but no transition needed, no heat transfer but edge mixing.
EIDSVIK	Well mixed source, no transition needed, with edge mixing.
Conservation Models	
MARIAH ZEPHYR SLAB FEM	Conservation models, fully three-dimensional, no transition needed. All these models use virtually the same basic equations but have differing numerical solution methods with differing approximation in matters of detail.
3D-Hydrodynamic Models	
SIGMET	A three-dimensional time-dependent hydrodynamic computer model

TABLE 5.6. Alphabetical list of dense gas models

Model Name	Model Type	Application	Author	Development Year
ADAM	Advanced thermodynamic code, with different atmospheric, dense gas modules	Several source types, chemical reactions, 2-phase-jet release, explosive release of high pressure liquids, cryogenic liquids and gases. Supplement to OB / DG and AFTOX codes	Raj	1987
ADPIC	Particle-in-cell, advection-diffusion model, conservation of energy and momentum	Needs a mass-conserving wind flow model to run. Cold dense gas release at ground level	Ermak	1992
AFTOX -3.1	Evaporative emission source for liquid spills, and Gaussian plume dispersion model	Similar to Shell SPILLS model	U.S. Air Force, Hanscom	1988
ALOHA	Gaussian plume model	Transport and dispersion model (no dense gas model)	U.S. National Oceanic & Atmospheric Administration	1988
ARCHIE	Integrated methods	Vapor dispersion fire, explosion	U.S. Department of Transportation; U.S. Environmental Protection Agency; Federal Emergency Management Agency	1986
B+M	Set of simple equations; not a code	Dense gas dispersion	Britter & McQuaid	1988
CHARM	Gaussian puff model	Complex Hazardous Air Release Model	Radian Corp., Austin, TX	1986
Cigale2	Box model (improved)	Based on DENZ	Crabol et al.	1986
Cobra III	LNG Gas model	Heavy gas / liquid spill dispersion	Alp	1985-1991
CRUNCH	Plume box model		Jagger	1983

TABLE 5.6. (continued)

Model Name	Model Type	Application	Author	Development Year
Germeles & Drake	Top-hat model (first generation)			1975
Hegadas-II (Shell)	Box model (integral, advanced), K-Theory	Puff or constant source release rate	Colenbrander, Witlox, Puttock	1988
HT-Heavy Gas model 1 and 2		Heavy gas dispersion	Sing	1991
MARIAH	Three-dimensional numerical solution, finite difference, K-theory		Taft	1981
SIGMET-LNG	Finite difference, K-theory	Catastrophic LNG spills on water	Havens	1982
OB / DG	Purely empirical correlation's	Limited to source scenarios (continuous release from ground level source); Inert gas with no initial buoyancy	Ocean Breeze / Dry Gulch, Hanscom	1963
Ooms / DEGADIS		Elevated dense gas dipsersion	Havens, Guinnup	1988
LLNL	Atmospheric dispersion model and Predictive mesoscale model	Used for emergency response	Gudiksen	1991
MERCURE-GL	3-D model	Complex terrain		1985
Nikmo & Kukkonen	Advection of dense gas	Complex terrain	Nikmo & Kukkonen	1990-1991
PHAST	similar to SAFETI		Technica, Inc., London	1989
RISKAT	Complete code for atmospheric and dense gas dispersion		Pape & Nussey, HSE (GB)	1985
SAFER	Atmospheric and topographic gas dispersion	Emergency response	DuPONT Co.	

TABLE 5.6. (continued)

Model Name	Model Type	Application	Author	Development Year
SAFETI		Risk assessment system	Technica, Inc., London	1988
SLAB	1-D model, time-dependent, conservation of mass, momentum, energy, and species	Ground-level evaporating pool; Elevated horizontal jet; Stack or elevated jet; Instantaneous source	Ermak	1990
TRACE	Similar to SAFER	Two-phase dense gas releases; Elevated releases, Flashing jets; Aerosols; Instantaneous, steady or transient; Neutral or buoyant	Chickhliwala et al.	1987
TRAUMA	Two-phase jet model, conservation of mass, momentum, energy, and species		Kukkonen	1990
TRIO-VF	Multi-purpose thermal-hydraulic code	Local scale (1 km) atmospheric / dense gas dispersion	Grand	1991
WHAZAN	Similar to SAFETI	Risk assessment	Technica, Inc., London	1988
ZEPHIR	Three-dimensional numerical solution, particle-in-cell, finite difference, K-theory		McBride	1981

5.5.1. CHRONOLOGICAL LIST OF DENSE GAS MODELS

For the sake of convenience we present below (TABLE 5.7) a chronological list of the computer codes and models found in literature from 1968 up to 1993 .

TABLE 5.7. Chronological list of dense gas dispersion models

Author	Year	Model Name	Model Type	Application
Parker & Spata	1968			
Welker et al.	1969			
Burgess et al.	1970			
Murphy et al.	1970			
Parker	1970			
Wilcox	1971			
Feldbauer et al.	1972			
Humbert-Basset & Montet	1972			
May et al.	1973			
Clancy	1974			
Drake et al.	1974			
Lewis	1974			
Lutzke	1974			
Raj & Kalelkar	1974			
Simons	1974			
Van Ulden	1974		Box model	First model of this type. Accounts for gravity spreading
Fay & Lewis	1975		Box model with 1-D spatial variation of quantities	
Gerard & Picol	1975			
Germeles & Drake	1975		Box model; First generation	

TABLE 5.7. (continuation)

Author	Year	Model Name	Model Type	Application
SAI	1975			
Van Buijtenen	1976			
FPC	1976			
Harsha	1976			
Bellus et al.	1977			
Cox & Roe	1977			
Te Riele	1977		Similarity model	
England & Teuscher	1978	SIGMET	3-D model; Time-dependant; Hydrodynamic	
Eidsvik	1978		Box model	
Ermak et al.	1978			
Kaiser & Walker	1978			
Picknett	1978		Box model	
Nikodem et al.	1978			
Fryer & Kaiser	1979	DENZ	Box model	
Cox & Carpenter	1979		Box model	
Doo	1979			
Fiedler et al.	1979			
Flothmann and Nicodem	1979		Advanced similarity model	
Schnatz & Flothmann	1979	TRANSLOC	3-D model	
Taft & McBride	1979	SIGMET2	3-D model	
Havens	1979	SIGMET-LNG	K-theory, 3-D model	
Chan et al.	1980	FEM3	3-D model	
Colenbrander	1980		Generalized plume box model; Similarity and K-theory model	A greater range of physical processes can be modelled

TABLE 5.7. (continued)

Author	Year	Model Name	Model Type	Application
Cox & Carpenter	1980		Box model	
Eidsvik	1980		Box model second generation	
Fay	1980		Box model	
Lee & Weinstein	1980	GRID	3-D model	
Mc Bride & Teuscher	1980	ZEPHIR	3-D model	
Rosenzweig	1980		Advanced K-theory model	
Zeman	1980		Advanced K-theory model	
Raj	1980		Plume box model	Air entrainment into a jet of anhydrous ammonia
Raj & Aranamuden	1980		Plume box model	Air entrainment into a jet of anhydrous ammonia
Su & Patak	1981	SIGMETN	3-D model	
Taft	1981	MARIAH	3-D model	
Foussat	1981	Unamed	3-D model	
Fay & Rank	1981		Box model	
Fannelop et al.	1982		Advanced K-theory model	
Deaves	1983	HEAVYGAS	3-D model	
Deaves	1983/8 5	SLUMP	Enhanced box model, with numeri- cal integration	
Jagger	1983	CRUNCH	Plume box model	Air entrainment into a jet of anhydrous ammonia
Meroney	1984	DENS-20	Depth integrated numerical model	
Webber	1984		Box model	

TABLE 5.7. (continued)

Author	Year	Model Name	Model Type	Application
Alp	1985	COBRA		
	1986	ARCHIE	Integrated methods	
Spicer & Havens	1986	DEGAIS		
Crabol et al	1986	CIGALE2	Based on DENZ code	
Carpenter & Cleaver	1987	Based on Cremer & Warner's model	Calibrated Box model; with momentum relationship	
Raj	1987	ADAM	Thermodynamic 3-D model	
Colenbrander et al.	1988	HEGADAS-II	Advanced box model	
Chan	1988	FEM3A	Improved FEM3 model	
Havens & Guinnup	1988	OOMS/DEGADIS		
Spicer & Havens	1989	DEGADIS-2.1		
Nikmo & Kukkonen	1990	Unamed	Complex terrain	
Kukkonen	1990	TRAUMA		
Ermak	1990	SLAB	1-D/3-D, Time dependant model	
Grand	1991	TRIO-VF	3-D model	
Sing	1991	HT-HEAVY GAS 1+2		
Gudiksen	1991	LLNL		
Niggli	1992	DISPTOOL	Different models for gas dispersion	
Ermak	1992	ADPIC	3-D model	

Other computer models/codes are shown in TABLE 5.8:

TABLE 5.8. Some more computer codes

- *Model RISKAT: The HSE computerized toxic gas risk assessment tool RISKAT has been described by Pape and Nussey.*[95] *Such a model has been refined to take account of the protection provided to people who are either indoors, or escape indoors or walk out of the cloud. <ask for paper>.*
- *Model CRUNCH: This is a dispersion model developed by Jagger*[18] *for continuous release of a denser-than-air vapor in the atmosphere*
- *Model for the vaporization of a liquid pool by Webber and Brighton*[96]
- *Model for the evaporation from spill of hazardous liquids on land and water by Shaw and Briscoe*[97]
- *Model of Chay and Reid*[98]*: Spreading boiling model for instantaneous spills of liquefied petroleum gas (LPG) on water.*
- *Model CRITS and CRITTER: These are a simple homogeneous equilibrium critical discharge model applied to multi-component, two-phase systems*[99] *developed at UKAEA.*
- *Model of Cox and Roe*[100]*: It is used for modeling the dispersion of dense vapor clouds*

5.5.2. COMMERCIAL SOFTWARE ON DENSE GAS MODELING

Beside the above mentioned computer models, which were for a great part developed at universities and are not always made available to the public, integrated computer codes which are incorporating many features of the different mathematical models can be bought nowadays for prices ranging from a few hundred dollars to some $ 100,000.-, depending on complexity and flexibility. Among models listed below (TABLE 5.9), we find the proprietary models AIRTOX, CHARM, EHAP, PHAST, SAFETI, TRACE and WHAZAN which are offered for sales. The other models are publicly available.

Some of the models are intended for application to certain types of scenarios and should not be applied outside of these areas. The BM model, for instance, is intended for dense gas clouds only, and does not contain algorithms for proper transition to situations with neutrally-buoyant gas clouds. Also the HEGADAS model is not foreseen for use with instantaneous sources. Other models such as GPM (Gaussian Plume Model), INPUFF, OB/DG, and AFTOX models do not contain algorithms to treat dense gas slumping. The thermodynamic effects of the dense gases are important only close to the source, and such models will deliver wrong results in that range.

Hanna[185] *(1991) has tested the above mentioned models and found out that the performance of a given model is uncorrelated with either its complexity or its cost.* As long as a model properly accounts for simple physical relations such as air entrainment, it can produce good agreement with observations. More complex models tend to account for a wider range of physical phenomena and therefore may be more useful for studies of a different number of emission scenarios. A few of the models exhibit relatively good performance in their predictions of maximum concentrations on monitoring arcs, with relative mean biases of $\pm 30\%$ or less and root mean square error values that are about 40% to 60% of the mean. Better performance is found for all of the models when observed and predicted plume widths are compared. The SLAB and CHARM models are the only models that demonstrate consistently good behavior for all of the data sets tested.

TABLE 5.9. Commercial software

ADAM	
AFTOX	Kunkel (1988)which are listed below
AIRTOX	Heinold et al. (1986), Mills (1988)
ALOHA	
ARCHIE	U.S. Environmental Protection Agency
BM	Britter and McQuaid (1988)
CHARM	Radian (1988)
DEGADIS	Havens (1988), Spicer and Havens (1988)
DISPTOOL	Niggli (1992)
GPM	Hanna et al. (1982)
HEGADAS	Witlox (1988)
INPUFF	Peterson and Lavdas (1986)
OB/DG	Nou (1963)
SLAB	Ermak (1989)
TRACE	DuPont (1989)
WHAZAN	

SAFER Code

The Swiss company SANDOZ Pharma AG in Basle, has started a pilot project for local emergency management, jointly with other chemical industries and the Authorities of the Basle region. In the case of incidents with gas release, it is very important to make an appreciation of the situation as soon as possible in order to gain time for alarming and eventually evacuating the urban population.

Sandoz is using a computer program developed by the DuPONT company, which is entitled SAFER (Systematic Approach for Emergency Response). *This code calculates and models the atmospheric and topographic gas dispersion*, using data pertaining to the physical properties of the soil, such as roughness, and actual data on the weather conditions. The calculation results are presented on a graphical display, including a map with isodose lines' projections.

DispTool Code

DispTool[101] provides the risk engineers with quick, transparent and *easy-to-use gas dispersion models, in connection with a program to estimate possible effects caused by release of acute toxic gases.*

The following four model types are implemented in DispTool to model instantaneous and continuous release of neutrally or positively buoyant and dense gases, respectively:

a) Neutral Gases, released Instantaneously (NGI)
b) Neutral Gases, released Continuously (NGC)
c) Heavy Gases, released Instantaneously (HGI)
d) Heavy Gases, released Continuously (HGC)

This set of programs should enable the evaluation of potential hazards related to a broad range of conceivable scenarios. The models behind DispTool have undergone simplifications and approximations in order to achieve broad applicability. The limits of DispTool have been investigated.

The program DispTool is running on IBM-PC or compatible having an 80286 or 80386 processor unit, running under MS-DOS Version 3.30 or higher. The use of a math co-processor is optional.

The dense gas dispersion is treated by the use of box models (i.e., assuming a uniform concentration distribution within the cloud or plume), whereas the models for neutrally buoyant gases are based on Gaussian dispersion. These models can be applied to low momentum releases over flat terrain. The releases are assumed to be isothermal. The effect model integrated into DispTool considers acute toxic releases. This model is based on temporal and spatial concentration distributions obtained from the dispersion calculations. Combined with toxicological data specific to the substances released, the probabilities of certain effects caused by that substances (such as lethality for example) may be estimated. The Pasquill stability classes are used by this model.

The dispersion submodels used by DispTool are described below:

Dispersion Model Used for Describing Neutrally and Positively Buoyant Gases Released Instantaneously. The initially formed cloud is assumed to be half sphere. Within the cloud, the gas is distributed according to Gaussian concentration profiles along the x, y, and z directions. The concentration profiles are centered at the origin of the half sphere. The cloud is transported in the direction of the mean wind by advection, whereas the direction is assumed to remain constant over the time of dispersion. Cloud radius and height will increase with increasing downwind distances, whereas the peak values of the concentration distributions within the cloud will decrease due to turbulence and dilution effects.

The concentration distribution is obtained as analytical solution to the Fokker-Planck equation, which describes diffusion and advection of particles as a stochastic process (Schiegl and Schorling[102], 1986). The concentration distribution depends on the dispersion parameters and the relative coordinates. The dispersion parameters are assumed to be identical in lateral direction but different from that in vertical direction. They depend on empirical coefficients determined on the basis of the Pasquill stability classes and on the roughness of the ground.

The initial dimensions of the cloud are given by the mass released and the initial cloud density.

Dispersion Model Used for Describing Neutrally and Positively Buoyant Gases Released Continuously. The model is similar to the preceding one. The gas is assumed to be emitted at a constant rate (steady state conditions). Its downwind transport and dispersion takes place in a constant wind field. The concentration profiles within the plume are assumed to be Gaussian. The evolution of the plume is not modeled, but rather the resulting plume in the atmosphere is considered. The model equations are time-independent.

Dispersion Model Used for Describing Heavy Gases Released Instantaneously. The modeling of heavy gases has received a lot of consideration in the literature (Havens[20], 1982; and Schnatz[103], 1986). The model used is based in many aspects on the work of Fryer and Kaiser[104] (DENZ model, 1979). The cloud is assumed to be a cylinder containing the gas homogeneously distributed within its boundaries (box model). Soon after the release, the cloud slumps due to gravitation, and takes a shape similar to that of a pancake. Therefore the radius increases very rapidly whereas the height decreases until it has reached a minimum. Due to air entrainment, the cloud volume and height increases again, and also the radius grows continuously, but at a reduced rate. The rate at which the cloud is diluted will depend on its surface, since the mixing process starts at its top and edge areas. A constant wind field is assumed. When the cloud has reached a certain size, it behaves as a neutrally buoyant gas cloud. At this stage another model (a) must be applied (see above).

Dispersion Model Used for Describing Heavy Gases Released Continuously. The plume model for heavy gases is based on mass emission at a constant rate. The gas is transported downwind and dispersed in a constant wind field (steady-state conditions). As is the case for the continuous release of neutrally buoyant gases, the model for dense gases is also based on steady-state conditions. The plume is considered to consist of a series of instantaneous releases at regular time intervals. the model assumptions are closely related to that given by Jaegger[18] (CRUNCH model, 1983). The source is considered to be rectangular, which creates a plume with initial half width, L_0, and height, H_0. Due to gravity, the dense plume will slump in the vicinity of the source, which leads to a rapid increase of the plume half width. With increasing downwind distance, the plume will be increasingly diluted by entrainment of air. This leads to a continuous growth of the plume dimensions. When the dilution of the plume has become very high the gas can be treated as if it was a neutrally buoyant gas. Two criteria are used in this heavy gas plume model, to check whether a transition to neutrally buoyant gas dispersion has occurred at a given downwind distance.

The heavy gas models stemming from different authors, which were incorporated into the code, have been tested, according to the author of DispTool, through comparison with experimental data obtained from the large scale tests at Thorney Island (McQuaid[105], 1985). The cloud concentration along the downwind axis as calculated by the model was compared with the measured peak concentrations. Parts of the temporal evolution of the cloud namely its position and area, were compared with measurements (data from Brighton et al.[106], 1985). The comparison has shown for instantaneous gas release, that the agreement between measured and calculated values is in general satisfying. In the case of low wind speeds, the model tends to overestimate the concentration values (as a function of distance) in the field far from the source. For high wind speeds, the concentrations might be overestimated in the field close to the source. For low initial cloud densities, the model predictions tend to be too optimistic in the far field. The model does not consider the inertia of the cloud mass in the initial acceleration phase. The comparison between the calculated and measured cloud areas, on the other hand, agree very well giving a correct representation of the cloud evolution with time.

DENZ Code

DENZ is a computer program[107] *for the calculation of the dispersion of dense toxic or explosive gases in the atmosphere* developed for the United Kingdom Atomic Energy Authority Safety and Reliability Directorate (UKAEA-SRD) by Fryer and Kayser[104] in 1979.

It is intended to model *puff releases*, and is *a box model* that considers four basic processes: gravitational slumping, air entrainment, heat sources or sinks, and the development of a source cloud by initial dilution.

It is assumed that the vapor is at first in the form of a cylinder, a form which has been observed and is appropriate for sudden releases from pressurized containers or when refrigerated liquid boils rapidly LNG poured onto water). The puff moves with the mean wind speed at its half height. The initial height is arbitrarily taken to be equal to the radius. Once formed the source cylinder begins to slump. Cloud heating may take place by natural convection, forced convection or by the sun heating at day time. The model incorporates heating by the ground due to turbulent natural convection and neglects any other source of heat. Air entrainment is in practice taking place at the edges of the cloud and at the top. The calculation of the rate of entrainment over the top surface is not yet possible. The entrainment velocity is proportional in some way to the Richardson number.

The computer program DENZ uses two alternative test to determine whether the plume may be deemed to be passive. The concentration within the puff is based on the assumption that the material is distributed in a Gaussian fashion across the puff. The prediction for the rate of entrainment of air gives rise to the greatest uncertainties. Once slumping has terminated and the ambient atmospheric turbulence starts to work on the cloud, predictions of the height are extremely sensitive to the assumed rates of entrainment of air.

The user must provide the mass of gas and the mass of air within the cylinder together with its initial density and temperature. The ratio of cloud height to cloud radius must also be specified as well as the average wind speed. The weather conditions, defined by the wind speed and the Pasquill category, must be given. The model can predict the toxic dose, a complicated function of average concentration and exposure time, using an approximate or an exact calculation method. The user has to provide the lethal exposure time as a function of the average concentration. Upwind and downwind hazard ranges are calculated as a function of the duration of the cloud passage. *The effects of the toxic cloud on the surrounding population can be estimated as well if the user provides simplified population data and weather data.* Getting data from the Meteorological station, it is possible to compile a table of joint probability distributions in a given sector (0 - 360°). The model can also make predictions for the hazard range of explosive gases.

Recent experiments using DENZ in laboratory simulations and large scale field trials have been considered by Jagger.[108] The implications for empirical program constants, process modeling, and model predictions are discussed in Jagger's paper.

CIGALE2 Box Model

The box model DENZ has been further developed in France, at the CEA, to give the *improved box model CIGALE2.* The interpretation of the Thorney Island phase 1 trials has been tried out with both models.[109]

COBRA Code

A heavy gas/liquid spill and dispersion modeling system COBRA (Alp[110], 1985; Alp[111], 1991) has been developed for estimating the hazard zones due to accidental spills of flammable or toxic chemicals such as propane, butane, chlorine and ammonia.

It consists of three basic modules: the *spill rate module* for estimating the direct source term for an accidental release from refrigerated or pressurized storage; the *pool spread and gas generation module* for estimating the rate of spread of any liquid spill on a surface and the rate of gas generation (indirect source term) as a result of evaporation or boil-off from the pool; and the *heavy gas spread and dispersion module* for estimating time dependent downwind concentration distributions and hazard zones. The basic equations used in the dispersion module are described in relation to the physical phenomena important in the behavior of heavy gas clouds released from pressurized liquefied storage. COBRA was implemented on a computer using FORTRAN 77. Model comparisons with available field data from Maplin Sands, Thorney Island, China Lake and Frenchman Flats have been made. Results of comparative sensitivity runs are given for various scenarios such as refrigerated versus pressurized storage, diked versus unconfined spills, etc., demonstrating the capabilities of the modeling system and its usefulness in emergency planning, and also as a safety design tool.

HEGADAS-II Box Model

The HEGADAS-II model by Colenbrander[19] (1984) can be described as an *integral advanced top hat (box) model,* which assumes a particular form of concentration profile, (a quasi-steady representation of the cloud), shapes and empirical relations describing air entrainment, although it utilizes an *eddy diffusion theory approach (K-theory). This model appears less applicable for instantaneous gas releases. It also does not apply for calm wind conditions.*

The HEGADAS program models the *dispersion of releases of heavier-than-air gases to the atmosphere.* The model provides for a smooth transition from the slumping phase to passive dispersion. The source release rate may be constant or time-dependent (e.g., evaporating liquid pool).

The *steady-state model* takes an initially uniform crosswind-concentration profile, but as dispersion progresses, a Gaussian profile imposed at the edges increases in width.

Gravity-spreading contributes to the increase of width of the cloud. The effects of density gradients on vertical mixing are quantified by Richardson number expression. Such formulation is also consistent with passive dispersion in neutral, stable or unstable atmosphere. For *time-dependent gas source*, several puffs are calculated, taking into consideration the rate of an evaporating liquid pool. Allowance is made for longitudinal diffusion, and, if required, the effects of gravity spreading can be redistributed between the crosswind and the downwind directions. The entrainment relations in HEGABOX (see above) and HEGADAS have been formulated principally to predict ground level concentration and a measure of cloud height. The temperature of the entrained air is taken as the ambient temperature at half the cloud height. Descriptions of heat transfer from the ground and water vapor transfer are also included. The software is written in FORTRAN for use on an IBM-PC or compatible machines (Witlox[112], 1988; Marsden[113], 1989).

IIT Heavy Gas Dispersion Model 1 and 2
Two heavy gas dispersion models have been developed at IIT (Sing[114], 1991) in view of estimating vulnerable zones due to accidental (both instantaneous and continuous, respectively) release of dense toxic material in the atmosphere. Performance of IIT Heavy Gas dispersion model has been compared with other operational models like Degadis, Denz etc. as well as with the observed data collected during the Burro Series, Maplin Sands, and Thorney Island field trials. The IIT models include relevant features of dense gas dispersion, viz., gravity slumping, air entrainment, cloud heating, and transition to the passive phase, etc. The DEGADIS model has been considered for comparing the performance of IIT heavy gas models because it incorporates most of the physical processes of dense gas dispersion, and has also been satisfactorily tested against field observations. The predictions from IIT heavy gas models indicate a fairly similar trend to the observed values from Thorney Island, Burro Series, and Maplin experiments with a tendency toward overprediction. There is a good agreement between the prediction of IIT Heavy Gas models 1 and 2 with those from DEGADIS, except for the simulations of IIT heavy gas model-1 pertaining to very large release quantities under highly stable atmospheric conditions. There is still a scope of improvement in the IIT heavy gas models.

Singh[115] (1991) has prepared *quick look nomograms for some commonly used toxic chemicals,* like Cl_2, using the heavy gas dispersion model developed at IITD.

Nomograms for Use in Emergency Case
These nomograms can be used by non-specialists in emergencies to determine safe distance based on weather scenario and the amount of toxic chemical accidentally released. The nomogram is based on quantity released and meteorological conditions. The meteorological scenario for the whole year has been subdivided into 24 categories.

Germeles and Drake's Model
The model consists of two sub-models, a *gravity spreading* (or heavy gas submodel) and a *neutrally buoyant (Gaussian) model.* It belongs to the *first generation top-hat models* which incorporate air entrainment in the vertical via constant coefficients. It

requires that a transition be made to a neutrally buoyant Gaussian model for the far field solution. The model is restrictive in that the required transition from the former to the latter is made when the cloud edge speed falls below the wind speed.

Eidsvik's Model
The Eidsvik's model, however, typifies *second generation top hat models* which incorporate horizontal cloud edge and vertical air entrainment via non-constant coefficients which are dependent on the Richardson number. This model does not require transition to a Gaussian model.

ZEPHIR Model
It is a *three-dimensional model* which includes numerical solutions of the partial differential equations of mass, energy, and momentum transfer. The model uses a particle-in-cell technique coupled with an explicit finite difference approach.

MARIAH Model
The MARIAH model is a *three-dimensional model* which includes numerical solutions of the partial differential equations of mass, energy, and momentum transfer. An implicit finite difference approach is used for this code. The atmospheric stability class assumptions of this model are different from those assumed in the other models. The wind profile is different too. Both the ZEPHIR and the MARIAH models are similar to the SIGMET model. Significant differences between these models are found mainly in the numerical solution methods, as indicated above.

MARIAH-II Model
The MARIAH-II model incorporates simplified forms of the Navier-Stokes and energy balance equations, with initial and boundary conditions describing a specified ambient flow (which can be zero) and the placement of contaminant gas into that flow. The Boussinesq approximation is invoked in the momentum balance equation, neglecting variations in density except in the buoyancy force terms. The equations are approximated with finite differences. The advection terms are calculated using the second-order Crowley method with the FRAM (Filtering Remedy and Method) technique to damp local oscillations. The diffusion terms are calculated implicitly, with the resulting linear equation system solved using incomplete Cholesky conjugate gradient method.

At a further development stage the turbulence submodel originally used in MARIAH-II has been replaced with a local turbulence model derived from a second-order formulation incorporating simplifying approximations. A two-dimensional (Cartesian or cylindrical coordinate) version of MARIAH-II has also been developed.

SIGMET LNG Vapor Dispersion Model
The SIGMET vapor dispersion model, which has been used to predict *dispersion of vapor clouds from catastrophic LNG spills on water* (of 10 m^3 to 25,000 m^3), has been described by Havens.[116]

Turbulent transport of mass, momentum, and energy is modeled using the *first order, eddy diffusivity approach (K-theory).* The equations are written in finite difference form. Specification of vertical eddy cinematic viscosity coefficients in SIGMET is based on methodology proposed by Smith and Howard.[117, 118] One main drawback is that correlations of height-dependent vertical momentum diffusivity in the atmospheric boundary layer are differing by several hundred percent for any stability conditions, and there are no data available to check directly the applicability of the diffusivity specification methods (including SIGMET method) proposed in literature.

The *primary uncertainties in the SIGMET simulation of catastrophic LNG release* are probably associated with vertical turbulent transfer modeling and vertical advection of the cold vapors.

Ooms/DEGADIS Model

A version of the *elevated dense gas dispersion model,* Ooms/DEGADIS, is presented by Havens[119] and Guinnup[120] (1988) which represents intermediate development of a dense gas modeling package which is undergoing refinement. The computer program, Ooms/DEGADIS, described in the EPA report entitled, "A Dispersion Model for Elevated Dense Gas Jet Chemical Releases - Volumes 1 & 2", is a VAX-operational program designed to *simulate the dispersion of heavier-than-air gases which are emitted into the atmosphere with significant velocity through elevated ports.* The program incorporates the sequential execution of two models. The first one (Ooms) calculates the trajectory and dispersion of the gas plume as it falls to the ground. The second (DEGADIS) calculates the downwind dispersion of the plume after it touches the ground. DEGADIS can also be used to describe the release and dilution from a low-momentum, ground-level release. The program is written in FORTRAN to run on a VAX computer.

DEGADIS 2.1 Dense Gas Dispersion Model

DEGADIS is an adaptation of the Shell HEGADAS model by Colenbrander and Colenbrander and Puttock; DEGADIS incorporates some techniques used by van Ulden for heavy gas dispersion. The DEGADIS model (Havens and Spicer[121, 122, 123], 1985) accounts for the three regimes of heavy gas dispersion processes and can be used to simulate *instantaneous, steady state, and transient releases.* Mathematical modeling techniques use to predict atmospheric dispersion of heavy gases are reviewed and criticized by Havens and Spicer. The dispersion processes are characterized by three phases:

a) *Near field, negative buoyancy-dominated flow regime:* this near-field, buoyancy-dominated regime is modeled using a lumped parameter model of a denser-than-air gas "secondary source" cloud which incorporates air entrainment at the gravity spreading front using a frontal entrainment velocity. For the negative buoyancy-dominated dispersion phase, the model is based on laboratory-scale releases of dense gas in a calm environment.

b) *Intermediate field, stably stratified shear flow regime:* For the stably stratified shear flow phase, the model is based on the laboratory experiments of

McQuaid, of Kantha et al. and of Lofquist. It is emphasized that DEGADIS was not calibrated to any dense gas field-scale tests.

c) *Far field, passive turbulent diffusion regime:* the downwind dispersion phase assumes a power law concentration distribution in the vertical direction and a modified Gaussian profile in the horizontal direction with a power law specification for the wind profile.

The model provides predictions of downwind concentration decay which are in good agreement with the full range of field experimental data currently available.

A user guide for the dense gas dispersion model, DEGADIS 2.1, is descried by Spicer et al.[124] (1989). Version 2.1 incorporates a newly developed jet-plume model which will result in a plume which disperses in a fashion consistent with current Gaussian plume models of the plume become neutrally buoyant before it is predicted to face to the ground. Conventional air pollutant dispersion models may not be applicable to assess episodic releases of hazardous gas from chem. process pressure relief operations and DEGADIS describes the dispersion processes which occur in high velocity releases (from pressure relief valves). The jet-plume part of the model was modified to allow application to scenarios where the plume remains aloft, to incorporate ground reflection when the plume lower boundary reaches ground level, and to improve computational efficiency of the model. DEGADIS predicts the ensuing ground-level plume dispersion.

DEGADIS Model
Hofmann[125] *has applied the dense gas dispersion DEGADIS model developed by Havens and Spicer (1985) to calculate* NH_3 *air concentrations at several heights above the ground level and at several distances downwind of the release point. Linear density approximation and the TRAUMA program of C. J. Wheatley*[126] *(1987) are used for describing the air/*NH_3 *density function to obtain differing* NH_3 *air concentration distributions. The model has been adapted to run on an IBM-PC.*

The Jet-Plume model provides for: automatic adjustment of integration step-size (using the Runge-Kutta-Gill method as in DEGADIS); elliptical plume shape (cross-section), with air entrainment specified consistent with the Pasquill-Gifford plume dispersion coefficient representation of atmospheric turbulent entrainment; user specification of averaging time; ground reflection when the plume (lower) boundary reaches ground level; and application to scenarios where the plume remains aloft.
The program is written in FORTRAN for compilation and execution on a DEC VAX Computer.

A version of the DEGADIS Dense Gas Dispersion Model written in Microsoft FORTRAN for microcomputers (IBM-PC, VAX) has been developed by Gas Research Institute[127] (1990), with emphasis on LNG vapor dispersion prediction. The DEGADIS model is consistent with a wide range of laboratory and field test data for dense gas releases on a flat surface with dispersion over unobstructed flat terrain.

Comparisons[128] *indicate that the DEGADIS model, which is used for regulatory application, is superior in performance to the Gaussian line source (GLS) model prescribed in liquefied natural gas (LNG) storage facility USA standards (49 CFR 193).*

Research is continuing on vapor dispersion estimation methods which provide for effects of vapor detention systems, turbulence induced by plant structure, and plant/area topographical features.

DEGAIS Model

Developed for the Coast Guard, *DEGAIS describes the negative buoyancy-driven flows and reduced vertical mixing observed for releases of heavier-than-air gases in the atmosphere (Spicer*[129]*, 1986). The model was developed for N_2O_4*, and tested using the data of two field scale releases (Eagle 3 and 6) of nitrogen tetroxide (N_2O_4) conducted by the Lawrence Livermore National Lab. at the Nevada Test Site during 1983. An analysis of the chemical interaction of N_2O_4 with the ambient humidity and oxygen is made. The reported source mass evolution rate is adjusted to account for these reactions; Reported nitrogen dioxide (NO_2) concentrations downwind of the source are adjusted for the source mass evolution rate, and these observed conditions are compared with predicted concentrations using the Ocean Breeze / Dry Gulch model, the Pasquill-Hanna Gaussian plume model, and DEGADIS.

FEM3 and FEM3A Models

FEM3 is a numerical model[130, 131, 132, 133] developed at the Lawrence Livermore National Laboratory (LLNL) primarily to simulate *heavy gas dispersion in the atmosphere*, such as the gravitational spread and vapor dispersion that result from an accidental spill of liquefied natural gas (LNG). The model is based on solving the time-dependent, multidimensional conservation equations of mass, momentum, energy, and species. The model is time-dependent and one-dimensional in that the equations depend on only one spatial dimension, downwind distance. On the other hand, the model is quasi-three-dimensional since it contains equations to determine the crosswind dimensions of the dispersing cloud. Specific submodels are included to describe wind speed as a function of height, air entrainment into the cloud as the result of turbulent mixing, and heat and momentum flow between the cloud and the ground beneath it and the air surrounding it.

To accommodate large density changes and yet preclude sound waves, a generalized anelastic approximation is employed (i.e., the fluid is considered to be basically incompressible but to have variable density). A modified Galerkin finite element method (GFEM) with eight-node isoparametric hexahedron elements (four-node quadrilaterals in two dimensions) is used for spatial discretization, and a modified forward Euler method is employed for time integration. The discretized Poisson equation for pressure is solved implicitly by means of a skyline solver. In this report, the theoretical and numerical aspects of the model are described briefly, user's guides for using the model are given, and a numerical example is presented to illustrate the use of the model.

FEM3 solves both two and three-dimensional problems and, in addition to the anelastic formulation, has options for using either the Boussinesq approximation or an isothermal assumption, when appropriate. The FEM3 model is running on a CRAY-1 computer and is composed of three parts: a preprocessor PREFEM3, the main code FEM3, and two postprocessors TESSERA and THPLOTX. PREFEM3 and FEM3 use a limited number of LLNL computing environment system subroutines, which are not included on the magnetic tape. These are identified in the program and presumably can be replaced easily with equivalent subroutines suited to the local computing environment. Descriptions of the 'missing' subroutines are included in the reference report. The postprocessors, used for plotting velocity vectors, contours, time histories, etc., depend heavily on LLNL graphics software which is not exportable; they are not included in the package.

A phase-change submodel was added to FEM3 to account for the phase changes of atmospheric water vapor.[134] This phase-change submodel has been generalized so that the release and dispersion of hazardous liquids with boiling points that are approximately equal to or less than normal ambient temperature can be simulated. A submodel for instantaneous sources has also been included. When the phase-change submodel is used in combination with the instantaneous source model, the initial evaporative cooling of some or all of the liquid in the source should be accounted for. A submodel for this evaporative cooling is developed and numerical results have been presented for two hazardous liquids: ammonia and hydrogen cyanide.

An overview of the improved FEM3 model was made by Ermak[135] (1986). The model employs a modified finite element method (Gresho et al.[136], 1984) to solve the time-dependent conservation equations of mass, momentum, energy, and species along with the ideal gas law for the equation of state. Turbulence is treated by using a K-theory submodel. These equations provide a mathematical description of the physics of heavy gas dispersion including gravity spread, the effect of density stratification on turbulent mixing, and ground heating into the cloud and its effects on density stratification and turbulence. In addition, FEM3 can treat flow over variable terrain and around obstructions such as cylinders and cubes. Since it is fully three-dimensional, FEM3 can simulate complicated cloud structures such as the vortices that are typical of dense gas flows, cloud bifurcation that has been observed during heavy gas releases under low wind speed, stable, ambient conditions, and cloud deflection caused by sloping terrain.

The model FEM3A is a further development of FEM3 code (Chan[137], 1988). A generalized anelastic approximation is invoked to preclude sound waves and yet allow large density variations in space and time. Turbulence is parameterized via a K-theory submodel and heat transfer between the ground surface and the vapor cloud is also included. The model can solve both two-dimensional and three-dimensional problems, including treatment of variable terrain and obstructions. It will handle instantaneous sources, finite-duration, and continuous releases. A simple phase-change submodel based on local thermodynamic equilibrium is available for handling the phase transition between vapor and liquid. Over the past six years, the model has been evaluated using a wide range of data obtained from both laboratory-scale and field-scale heavy-gas dispersion experiments. In the above mentioned reports, an overview of the model is given, the theoretical and numerical aspects of the model are briefly described, user's guides for using the model are provided, and three example problems are presented to illustrate the use of the model.

However, certain recent applications revealed that, for problems involving large density variations, the model was deficient in conserving both species and total mass. To extend the applicability of FEM3/FEM3A for such problems, a new and cost-effective algorithm, based on solving a slightly modified set of equations, was recently developed (Chan[138], 1991). In this paper, this algorithm is described and numerical results are presented to demonstrate the improvements obtained.

ARCHIE Model

The key purpose of ARCHIE[139] is to provide planning personnel with *integrated methods for use in assessing the vapor dispersion, fire, and explosion impacts related to discharges of hazardous materials into the terrestrial environment.* By doing so, the program not only enhances understanding of the nature and sequence of events and hazards associated with potential accidents, but also provides a basis for emergency planning.

Chart A. *Discharges of gas or liquid from a tank or pipeline leading to flammable, or toxic vapor dispersion, or fires, or explosions involving such materials*

- Compute discharge rate and duration for vapor/gas discharge direct to atmosphere
- Compute pool area for liquid spills
- Compute pool evaporation rate
- Evaluate downwind toxic gas or vapor dispersion hazards
- Evaluate pool fire hazards
- Evaluate vapor cloud fire hazard
- Evaluate vapor cloud explosions.

Chart B. *Fireball and Blast Effects Resulting from the Explosion of an Inadequately Vented Tank Engulfed in a Fire (i.e., a BLEVE)*

- Evaluate tank overpressurization explosion hazard
- Evaluate fireball radiation hazard upon tank rupture

Note that fire/explosion models of ARCHIE only apply to flammable or combustible materials upon their ignition.

Chart C. *Detonation of a solid or liquid*

- Evaluate true explosive force of solid or liquid explosive.

Note that explosion models of ARCHIE do not consider hazard of airborne fragments !

Figure 5.1. Hazard assessment procedures used by the ARCHIE *code*

Three model selection charts display the sequence of procedures suggested for assessing the hazards of toxic vapor dispersion, fire, or explosion (*Figure 5.1*).

In each case, the user should start at the top of the chart and work down the various branches to define the desired sequence for any particular hazard. A great deal of effort was put into the program to ensure that users do not apply the evaluation procedures in an inconsistent and inappropriate fashion. Nevertheless, the complexity of the processes being evaluated did not permit the development of a fully foolproof program, so it is necessary for users to apply common sense at each stage of an analysis to ensure that input provided to the program and the scenario being considered are consistent and reasonable.

A Word of Caution
First, it must realized that the procedures provided with ARCHIE are simplified versions of more sophisticated (and more difficult to use) methods available to experts in the field. Thus, be advised that ARCHIE is intended to provide approximate answers sufficient for general planning. It will usually (when used as instructed) produce results that overestimate hazards but occasional exceptions are possible and likely. Application of safety factors by users is encouraged. A second issue involves the fact that the procedures in ARCHIE are designed for spills of relatively pure substances. Mixtures can only be considered in special cases by the knowledgeable user by provision of appropriate input data to the program. The units used by the program are US units. The program is not restricted to gases heavier than air.

The program comprises a series of options and is based on the *"Handbook of Chemical Hazard Analysis Procedures"*:

Option a: a system of nine procedures to estimate *the rate and duration of discharge* under various conditions. Three series of models respectively address non-pressurized tanks containing liquids, pressurized tanks containing gases or liquids, and long pipelines containing gases or liquids.

Option b: a series of procedures for roughly estimating the *area of evaporating or burning pools* due to spillage of a liquid. Under certain conditions, pool areas can be computed from discharge rates/durations estimated by the discharge models.

Option c: a model designed to estimate the *rate at which a liquid will evaporate from a pool of spilled liquid.* The model requires the evaporating pool area determined by the pool size model described under *option b.*

Option d: a model for analysis of *toxic gas/vapor dispersion hazards* due to release of such a substance into the atmosphere. The model provides the downwind distance over which toxic concentrations will exceed a user specified level as well as a table of concentration and suggested initial evacuation zone widths as a function of downwind distance. The model uses gas/vapor evolution rates and durations from the discharge rate or pool evaporation rate models depending on whether gas or liquid is discharged from the container.

Option e: a model used to estimate the radius of the zones in which *fatalities and second degree burns* may be expected if a pool of flammable or combustible liquid should ignite. The model uses the burning pool area estimated from Option b above.

Option f: a model to evaluate the fatality zones, injury zones, and other characteristics of a fireball resulting from a situation in which a closed container of a blamable liquid or compressed gas is engulfed in flame. The model essentially looks at the *thermal radiation hazards of Boiling Liquid Expanding Vapor Explosions (also known as BLEVEs).* The explosion model described under *option 1* looks at the blast and shock wave effects of such events. Neither model, except *option f*, addresses hazards due to fragments of the container that may become airborne at high velocity.

Option g: flammable gases venting from a container under pressure can form a lengthy *flame jet* if ignited. This model computes the length of such a jet and a safe separation distance.

Option h: ignition of a cloud or plume of flammable gas or vapor in air can result in a *flash fire or explosion.* This model estimates the downwind distance, hazard zone width, and weight of airborne flammable or explosive gas/vapor in air when such a cloud or plume is formed. It uses results from discharge rate models when gases are directly released to the atmosphere. When they evolve from evaporating spilled liquids, it uses results of the pool evaporation model.

Option i: ignition of a cloud or plume of flammable gas/vapor in air can sometimes result in an *unconfined vapor cloud explosion.* This model uses the weight of flammable/explosive gas/vapor determined by the model described in Option h to evaluate the explosion impacts of such events. Note that the center of the explosion can be anywhere within the boundaries of the flammable gas/vapor cloud or plume determined by the model described under *option h.*

Option j: any sealed container that is overpressurized due to internal reaction or overheating may rupture violently just as a balloon may pop when too much air is blown in. This model assesses the *explosion effects* of such events, but does not address impacts resulting from airborne fragments of the container.

Option k: this *explosion model* is designed to evaluate the blast effects of true high explosive materials like dynamite, TNT, nitroglycerin, and similar substances.

Hazardous Materials Information Exchange (HMIX)
Users of ARCHIE may contact agencies of the Federal Government. One method to communicate with these agencies is via use of the Hazardous Materials Information Exchange (HMIX) computerized bulletin board system which is operated by FEMA and the DOT. The system includes a conference section that is dedicated to ARCHIE. HMIX can be accessed by modem at telephone number 1-312-972-3275. A users manual for the system and/or technical assistance can be obtained by calling 1-800-PLAN-FOR (1-800-367-9592 in Illinois). Users unable to access HMIX may submit questions or comments regarding the computer program in writing to: ARCHIE Support (DHM-51 / Room 8104), U.S. DOT Office of Hazardous Materials Transportation, Research and Special Programs Administration, 400 7th Street, S.W. 2 Washington, D.C. 20590.

ADAM Model

The models used in the past by the Air Weather Service of the U.S. Air Force (USAF), namely "Ocean Breeze" and "Dry Gulch" model, *did not account for the variations in the chemical properties of the gas.* A comprehensive toxic chemical vapor dispersion analysis system was developed by Raj et al.[140] in 1987, on behalf of the USAF. Mathematical models were developed to described a variety of source types and the dispersion of vapor clouds/plumes in the atmosphere. *In the ADAM code sixteen different source types are modeled including pressurized liquid releases, flashing and aerosol formation, 2-phase jet releases, explosive releases and releases of high vapor pressure liquids, cryogenic liquids and gases.* The dispersion model takes into account the differences in source characteristics, higher-than-air density of clouds (due to aerosol presence, temperature or molecular weight). *Reactions of the chemicals,* if any, with water vapor in the air are modeled and considered in the dispersion model. Transition from heavy gas dispersion to near neutral density dispersion is modeled without abrupt changes in size or discontinuity in concentrations. The models have been coded in FORTRAN and run on an IBM PC/AT, using the HALO graphic software. The system has been titled ADAM ("Air Force Dispersion Assessment Model"). The different models used by the ADAM system have been assessed against the results of different field test experiments. The theoretical models are in very good agreement with the experimental data.

The dispersion assessment model ADAM comprises five scientific modules libraries:

Atmosphere models module: this contains the programs needed to calculate the Pasquill stability category of the atmosphere based on the input weather and output data for wind velocity at 10 meters and wind friction velocity.

Source models module: the code includes mathematical models to characterize various source types, providing source strength data for dispersion predictions i.e., dimension and composition of the cloud. These include: single source, confined and unconfined source, instantaneous and continuous source, liquid and/or gas release, cryogenic or non-cryogenic liquid release sources, etc.

The *types of storage conditions* considered include the ambient temperature pressurized liquid storage, gas storage, and cryogenic liquid storage. Criteria are used to classify the releases into cryogenic and non cryogenic releases. These are based on the comparison of the temperature of the liquid that hits the ground with the ground temperature. The calculation of the liquid temperature outside the tank is accomplished by modeling the flashing process for pressurized liquid release.

The *spread of liquid pool on the ground and its evaporation* have also been modeled. The low vapor pressure model, originally proposed by Ille and Springer[141] (1978), has been simplified and improved to provide better estimates of the evaporation rates.

The entrainment uptake of vapor produced by an evaporating liquid pool is also modeled. This model provides the strength of a two dimensional "window" source of vapor and its physical dimensions at the down wind edge of the liquid pool. This source description is in conformity with the source characterization used in the continuous vapor dispersion model.

The possibility of *gas venting from pressurized gas storage containers* has also been taken into account. The models used are basically compressible gas flow models and provide the values for such parameter as the mass flow rate, the density and temperature of the gas, the dimensions of the gas plume outside the tank and the velocity of the gas stream.

In the case of pressurized release through a relatively small hole it is expected that the flow will be in the form of a *two phase jet containing flashed vapor and liquid aerosols.* Because of the high velocity of the jet, the down stream distance up to which the jet effects are dominants is large. *The model used describes the characteristics of this jet, including the air entrainment and chemical reactions in the jet.*

Dispersion analysis module: it includes *dispersion models for predicting the hazard areas resulting from heavier-than-air toxic chemical vapor releases.* The source, atmosphere and user input data are passed to this module for calculating the dispersion of chemical in the air. The dispersion is controlled by heavy gas effects initially (if the density of the released vapor/aerosols is greater than that of the air) resulting in a rapid lateral expansion and a low ground hugging cloud. As additional air is entrained, the turbulent mixing of the atmosphere becomes the dominant dispersion force and the heavy gas effects become negligible. This is modeled by using a new modified Gaussian technique whereby "tails" are added to the heavy gas "boxes". Once the dispersion is entirely atmosphere turbulence driven, a smooth transition from the modified Gaussian models to Gaussian models occurs.

The dispersion models predict the concentration distribution, both vertically and horizontally at any point down the wind of the source, the dosage at any point over which the cloud passes, the thermodynamic condition of the cloud (temperature, density, species fractions in the vapor and liquid phases, if any). Similar parameters are calculated for the plume release. In the case of a "puff" or "cloud" dispersion in the heavy gas regime, the cloud is assumed to be cylindrical ("box") and the model calculates the down wind motion of this cylindrical cloud due to wind induced drag. The cloud is diluted due to the entrainment of air over the top and sides of the cylinder. The air entrainment rates are expressed as function of the gravity induced radial expansion velocity as well as the atmospheric turbulent velocities modified by the cloud stratification. In addition, the expansion of the edges due to atmospheric turbulent diffusion is super imposed on the box dispersion by the use of a modified volume source Gaussian dispersion phenomenon. The result of this hybrid model is that there results a central core region in the cylindrical cloud within which the concentration distribution is essentially uniform and the outer regions of the cloud in which the concentration falls off. This is a truer representation of the real phenomenon compared to the box model.

In the heavy gas dispersion regime the thermodynamic conditions of the cloud is calculated at every position of the cloud noting the amount of air entrained up to that position and the total amount of heat exchanged between the cloud and the surroundings. The heavy gas type of dispersion is terminated when the local Richardson number is of the order of unity. However, the volume source Gaussian dispersion is continued beyond the transition region. This ensures that the property value changing with distance are smooth and continuous. In addition, the concentration and other distribution profiles smoothly change from the initial "top hat" profiles to the Gaussian profiles at the far field. The same type of approach is used for modeling the plume (continuous release) dispersion. The dispersion model developed predict the concentration distributions, both vertically and horizontally at any point down wind of the source, and the dose is calculated. Further information obtained are the thermodynamic conditions of the cloud, the cloud translation velocity, the cloud size etc. Similar parameters are calculated for the plume. The results of the model are displayed in graphical and tabular form.

Property database library: it contains the subroutines that calculate various properties of the chemicals based on the data contained in the data files of the property database (HACS).

Thermodynamic models module: the thermodynamic models can reckon with the mixing of humid air and the chemical vapors and liquid aerosols resulting from the chemical release. The models predict thermodynamic conditions at equilibrium resulting from the mixing of a known mass of humid air with a specific mass of chemical vapor/aerosol cloud. They are used for calculating the thermodynamic equilibrium conditions for the "box" model.

> ***Note:** the Coast Guard's Hazard Assessment Computer System (HACS) contains a database with thermo-chemical properties of over 1000 chemicals*

If a reactive chemical is released into the environment, there results a reaction between the released chemical and the entrained air. Models describing the final thermodynamic state resulting from the mixing of humid air and the chemical vapor clouds have been developed based on the assumption that the reaction rate is much faster than the air mixing rate. These models take into account the energetic of the reaction, if any, and the heat exchange with the external surroundings (i.e., atmosphere and earth's surface). Regarding the reactivity of the chemicals with the ambient air, these can be grouped into three types:

a) *Passive chemicals* (for instance: chlorine, phosgene, sulfur dioxide, and hydrogen sulfide), where no reaction with water vapor occurs. During the initial phase of mixing with air when liquid aerosols are present in the cloud, the temperature of the cloud can be substantially below the saturation temperature of the chemical at ambient pressure. However the density decreases continuously as the mass of air in the mixture increases. The temperature starts to increase again when all the aerosols have evaporated

b) *Water soluble chemicals* (ammonia). The chemicals belonging to this group exhibits strong solubility in liquid water. The presence of liquid aerosols of either the chemical or of the condensed water will lead to the dissolution of the chemical vapor in the liquid phase. The heat of dissolution and the phase equilibrium properties of the chemical - water system will influence the distribution of the vapor and liquid phases in the vapor cloud. The aqueous aerosols will be present until the dilution is such that the cloud temperature goes above the wet bulb temperature of the air (depending on the relative humidity).

c) *Reactive chemicals* (nitrogen-tetroxide). The mixing of this type of chemicals with air results in reaction leading to the formation of a new chemical species.

The density of the chemical and air mixture is very insensitive to the relative humidity of the air, but the chemical reactivity or dissolution rate is not.

DENS-20 Model

Meroney (1984)[142] has developed a wind-tunnel validated, non-proprietary *depth-integrated numerical model, DENS-20,* which reproduces the essence of dense cloud behavior for *isothermal or cold dense clouds released from ground level suddenly, over a finite time, or continuously.* The model, which does not depend upon the Boussinesq assumption but does require the hydrostatic assumption, is time dependent, quasi-3-dimensional, and permits cloud heating from below and the entrainment of moist air. The model constants are set to fit a cross-section of data from wind-tunnel experiments on the transient and steady behavior of releases of heavy gases. Plume shape and concentration decay with distance and time are reproduced for comparative cases from the Porton Downs and China Lake field tests and the Colorado State University cold gas laboratory tests on dense gas behavior.

Part II of the paper of Meroney[143] relates numerical experiments on dense cloud physics. Numerical calculations with the model permit the examination of dense cloud characteristics observed in the field but difficult to reproduce or measure accurately. The numerical model reveals the characteristics of upwind motion of dense gases at the source, gravity waves induced on the cloud top by wind shear, and the variable hazard zones associated with gases released instantaneously, over a finite time, and continuously.

SLUMP Model

Whilst box models are generally adequate for dispersion over flat unobstructed terrain, more complex mathematical models are required for a description of the effects of obstructions. Such model make use of a full 3-D solution of the Navier-Stokes equations in which atmospheric turbulence and its interaction with the dense cloud is modeled (see Wheatley and Webber[144], 1984). For instance, the *Atkins ES code HEAVYGAS,* can be used in either its fully 3-D form, or in a 2-D plane or axisymetric form (when no wind is present). In spite of the obvious limitations of the results of 2-D modeling, these give an indication of the effects of obstructions on dense gas dispersion.

> *Full 3-D model are necessary to model the complex gas flow around roughly cubical buildings.*

The data which has been obtained from the Thorney Island trials has already been put to good use in model validation and development. The data has now been used by Davies[145] (1987) *to enhance a box type model SLUMP*. The integration of the differential equations which constitute the model can be undertaken either analytically, or numerically. Certain simplifications are necessary in order that analytical solutions may be obtained. These include the use of cloud advection velocity which is imposed after integration of the equations, rather than one which accurately reflects the momentum balance, and the requirement that any transition to passive dispersion should be sudden rather than gradual, do set serious limitations on the accuracy with which the models can cope with the initial stages of cloud acceleration, respectively with the final stages of cloud lift-off. The model is further subject to possible improvements, which are mentioned in the original paper.

Cloud Advection. A serious shortcoming of analytically integrated the box models is that the velocity of the cloud, and hence its position, has to be specified independently of the slumping process. Numerical integration of the model equations enables the local cloud velocity to be calculated on the basis of the initial momentum of the cloud (usually zero for instantaneous release), and the momentum of the entrained air.

Transition to Passive Dispersion. As the cloud becomes more diluted, gravity effects become less important, and passive dispersion takes over. The parameters which determines this change is the ratio of turbulence energy in the atmosphere to potential energy in the cloud. *Such effects have been incorporated in the computer code SLUMP (Box model).*

Dispersion Modeling Within Risk Assessment Studies

The maximum danger areas for heavy gases released in an accident, can be obtained from wind tunnel experiments, together with similarity theory analysis. A method for estimating the dispersion of the heavier-than-air gases released under unfavorable atmospheric conditions is briefly described in a paper by Schatzmann et al. [146, 147] (1990/91). The method is based on the results of a wind tunnel study on the dispersion of both instantaneous and continuous releases into a boundary layer shear flow distorted and undisturbed by surface obstacles. It can be used for heavy and frozen gases, which can be introduced spontaneously or continuously and the results can be transferred to any quantity. The effect of obstacles is included in the context of a rather coarse scale.

Wind tunnel experiments were compared with data from various field experiments (Maplin sands, Thorney Island, Desert Tortoise, TUV North Germany). Agreement is generally good, also in the case of toxic heavy gases in which the coupling of wind tunnel results with a simple fault case model *(VDI Guideline 3783 sheet 1)* suggested in the text for neutral or light gases was achieved. The results were converted into a Guideline *(VDI 37 82 sheet 2)* for estimating the spread of heavy gases in a fault in the context of safety analyses. To make it easier to apply, a computer program was produced, which can run on PC's with the MS-DOS operating system. A case study, using the example of the BASF Works at Ludwigshafen is included. *The method has been adopted by the German Engineering Society and is used as a basis for risk assessment studies of flammable and toxic heavy gases.*

A Heavy - Gas Dispersion Model With Continuous Transition From Gravity Spreading to Tracer Diffusion

Flothmann et al.[47], (1980) have developed a *dispersion model for dense gases,* that assumes a continuous transition between the gravity spreading phase and the tracer dispersion phase and is suitable for risk analysis of flammable gases.

ADPIC Particle-in-cell Model

A dense-gas version of the *ADPIC particle-in-cell, advection- diffusion model* was developed to simulate the atmospheric dispersion of denser-than-air releases (Ermak[148], 1992). In developing the model, it was assumed that the dense-gas effects could be described in terms of the vertically-averaged thermodynamic properties and the local height of the cloud. The dense-gas effects were treated as a perturbation to the ambient thermodynamic properties (density and temperature), ground level heat flux, turbulence level (diffusivity), and windfield (gravity flow) within the local region of the dense-gas cloud. These perturbations were calculated from conservation of energy and conservation of momentum principles along with the ideal gas law equation of state for a mixture of gases. ADPIC, which is generally run in conjunction *with a mass-conserving wind flow model to provide the advection field,* contains all the dense-gas modifications within it. This feature provides the versatility of coupling the new dense-gas ADPIC with alternative wind flow models. The new dense-gas ADPIC has been used to simulate the atmospheric dispersion of ground-level, colder-than-ambient, denser-than-air releases and has compared favorably with the results of field-scale experiments.

LLNL Atmospheric Dispersion Model

LLNL is an atmospheric dispersion model developed for emergency response (Gudiksen[149], 1991). Several improvements to this emergency response models are being made: dense gas effects, including the slumping and horizontal spreading of the dense cloud, the turbulence suppression, and the diversion of the ambient flow by the dense cloud, are being incorporated into the dispersion model. Model predictions are coupled with observations by using regression analysis techniques in order to derive more accurate values of poorly known model input parameters such as source term characteristics. *Additionally, a predictive mesoscale model is being developed to provide forecast wind and temperature fields to the dispersion models.*

SLAB Atmospheric Dispersion Model

SLAB is a one-dimensional (1-D), time-dependent computer model (Ermak[150], 1990) that simulates the atmospheric dispersion of denser-than-air releases (diffusion and gravity flow). The properties of the air-gas cloud are treated explicitly in their dependence on the downwind direction. When strong turbulent mixing occurs within the cloud, the properties are sufficiently constant in the crosswind directions to justify the 1-D simulation. The types of releases treated by the model include a ground-level evaporating pool, an elevated horizontal jet, a stack or elevated vertical jet, and an instantaneous volume source. Except for the evaporating pool source which is assumed to be all vapor, all of the remaining sources may be either pure vapor or a mixture of vapor and liquid droplets. Atmospheric dispersion of the release is calculated by solving the conservation equation of mass, momentum, energy, and species. The conservation equations are spatially averaged so as to treat the cloud as either a steady state plume, a transient puff, or a combination of the two depending upon the duration of the release. The mathematical description of the physics of heavy gas dispersion (gravity spread, reduced turbulent mixing, etc.) as well as the description of the normal atmospheric advection and turbulent diffusion processes, are inherently included in the conservation equations. The time-averaged concentration predicted by SLAB depends on not only the various physical phenomena associated with the dispersion equations, but also on the specified concentration averaging time.

HEGABOX Model

The box model HEGABOX[151] has been developed to *simulate gravity-dominated dispersion behavior soon after a sudden release of dense gas*. It is used as a front-end to the HEGADAS dense gas dispersion model (see below), and there is a smooth transition from the gravity-dominated phase to the region in which ambient turbulence has greater influence.

The dense gas dispersion model HEGADAS is used principally for *the simulation of dispersions from spills of liquefied gases*, and it can handle the time-dependent vapor evolution from an evaporating pool of liquefied gas. The gravity-driven spreading of the dense gas is done explicitly. In very low winds, or for a sudden release of gas, however, there is also strong gravity spreading along the wind direction, which HEGADAS cannot handle. To simulate the early stages of such spills the front-end model HEGABOX has been developed.

By the use of HEGABOX, a smooth transition is made from the gravity-dominated stage of dense gas dispersion to the HEGADAS model, which calculates the calculation into the region were ambient turbulence is important. HEGABOX is a Box model which treats the cloud as a cylinder of uniform gas concentration, and is thus similar to a number of published Box models. The cylindrical cloud is affected by gravity-driven slumping which causes the radius to increase. The head entrainment is taken to be proportional to the gravity-spread velocity. An initial time delay is needed to model the build-up to full entrainment. In addition, heat transfer and convective entrainment are included as in HEGADAS. A simple approach to modeling the cloud's velocity is obtained by assuming that its momentum is entirely due to the air entrained, and that the effective velocity of the air is a constant factor times the average, over the cloud height, of the external wind velocity.

MERCURE-GL Code

The 3-D computer program MGL (MERCURE-GAZ LOURDS) is described by Riou[152] (1987). *It is derived from the mesoscale non-hydrostatic model MERCURE*, which is based on the *Boussinesq equations* with finite difference and finite volume techniques. Indeed, the most commonly used box models are inadequate to simulate denser-than-air vapor dispersion in extreme meteorological conditions and in presence of obstacles.

The use of the formulation of Boussinesq assumption is justified by the good agreement between MGL predictions and field trials. To determine the state variables of the flow field (concentration, fluid speed, pressure, temperature), the following equations must be considered:

- Conservation of mass
- Conservation of momentum
- Conservation of energy
- Equation of state of the mixture air/heavy gas

- Equation for advection-diffusion of the concentration in pollutant.

The main assumptions that have been made are:

- use of Boussinesq approximation
- the volume force is only due to gravity
- correlations between speed and density and between temperature and density are neglected
- air and heavy gas are well mixed and follow the perfect gas law
- there are no chemical reactions
- we only consider thermal convection, without radiation or thermal exchanges.

Other features of the code are:

- modeling of the turbulence using the Smagorinski formula. It considers that the eddy viscosity tensor is diagonal and dependent on the deformation tensor, and use of a terrain following coordinate.

Particular attention has been paid to the choice of numerical method in order to avoid numerical diffusion and to reduce computational time. The results obtained with this code are in good agreement with measurements for different meteorological conditions and relative density of gases, close to the source. Further comparisons of predicted concentrations have to be done far from the source. The geometrical characteristics of the cloud are well predicted. The dispersion in presence of obstacles in the terrain has not yet been modeled at this stage.

Nikmo and Kukkonen's Model

A mathematical model for the *advection of a denser-than-air gas cloud in complex terrain* has been presented by Nikmo[153] (1990) and Kukkonen[154] (1991). The gas cloud is advected due to buoyancy forces, the drag force due to the ambient flow field, friction forces and the momentum of the entrained air. The model allows for any orientation of the slope with respect to the wind direction. The discussion includes the structure of the model and derivation of detailed model equations for the special case in which the wind is directly uphill or downhill. The model was designed as a hazard analysis tool, and its computer implementation can be used as a subprogram in heavy gas dispersion models. The model predictions are compared with results of the Thorney Island phase I field experiments. Although these trials were conducted on flat terrain, the comparison is useful for understanding the cloud transport processes. The numerical calculations of heavy gas cloud dispersion on a slope were also carried out.

TRAUMA Code

Hazard analysis of storage and transport of pressurized chemicals requires mathematical models of the behavior of two-phase jets. In the past different models have highlighted different aspects of this problem. The most important factors in modeling two-phase jets have been reviewed by Webber and Kukkonen[155] (1990). A model derived by Wheatley for a *two-phase ammonia jet* (including water aerosols) forms the basis of the TRAUMA code, *which has now been extended to also cover gases other than ammonia.* The TRAUMA model (Kukkonen[156], 1990) has been applied *to predict the evolution of ammonia and water aerosols.* This model includes the thermodynamic aspects of liquid and gaseous ammonia interacting with atmospheric water vapor, but neglects the wind and gravitational effects on the jet and its interaction with any solid surfaces. The two phase mixture is treated simply by assuming either that the liquid is all deposited in the early stages, or that it moves coherently with the gaseous components of the jet. In particular, the model computations give predictions of the mass fraction of species deposited on the ground. Various generalizations of the model, in particular the effects due to the water vapor of ambient air, have also been discussed.

The numerical results from the *two-phase jet model TRAUMA* provide information on the mass fraction of species deposited on the ground, on the effects of various factors on jet evolution. The two-phase pipe flow model is partly based on the homogeneous equilibrium flow assumptions and it includes a description of flow friction, flow resistance and gravity. Various generalizations of the model, in particular the effects due to the water vapor of ambient air, have also been discussed in the paper of Kukkonen. The numerical results show the influence on two-phase flow evolution of ambient conditions, pipework structure and the physical and chemical properties of various species. New solutions have been derived analytically, to the set of differential equations governing the dispersion of turbulent jets. These equations describe the species, mass and momemtum conservation of the jet, for two entrainment submodels. In particular, the analytic model gives explicit estimates of length scales over which gravity and wind effects will be significant. The most important factors for future modeling efforts are gravitational spreading, deposition of substance liquid, ambient wind and the transition to heavy gas dispersion. A new model describing two-phase jets is also presented in Kukkonen's paper.

Three-Dimensional Code TRIO VF

The TRIO VF code (Grand[157], 1991) is a *multi-purpose thermal-hydraulic software which can simulate local scale atmospheric dispersion (within 1 km).* Turbulence can be treated by one of the following approaches: use of the Reynolds averaged equations with an eddy viscosity calculated with the two equations model, use of a large eddy simulation and a subgrid scale model. Comparison with experimental results have been made for the dispersion of a heavy gas (Thorney Island experiment).

5.6. Field Trial Tests and Laboratory Experiments

The principal objective of the heavy gas research efforts has been to develop an ability to predict the extent of downwind dispersion hazards for a specified mass of a chemical released under given environment conditions, and possibly to delimit the potential area of hazard. Two approaches have been taken, namely the experimental and theoretical modeling.

Field experiments on dense gas dispersion are performed because there is a great need for data to confirm or contradict theoretical dispersion predictions. The objectives are:

i) to obtain reliable data, also at large scale, with which to test the predictability of mathematical and physical models. Such data comprise primarily the distribution of concentrations as a function of time and position in the terrain for a variety of weather conditions, and the meteorological parameters required to specify the weather conditions.
ii) to obtain the data required for improving physical understanding of the mechanisms of heavy gas dispersion and to test the fundamental hypothesis in mathematical models. Such data comprise measurement of turbulent fluctuating wind velocities and gas concentration distributions. The cloud behavior is usually recorded by the use of photographic records.

Usually the *Pasquill categorization scheme* is adopted for convenience for the purpose of the trials planing. The Pasquill scheme utilizes meteorological parameters which can readily be observed without the need for elaborate instrumentation. There is however considerable controversy as to the most appropriate turbulence classification scheme to use. Heavy gas dispersion models generally prescribe the physical process in terms of parameters such as the turbulent velocity and length scales in relationships for entrainment, eddy diffusivities, etc. The models then relate these primary parameters to Pasquill categories or other schemes, such as for instance the temperature difference scheme[158] of the U.S. Nuclear Regulatory Commission, *wind standard deviation*, *gradient Richardson number*, and *bulk Richardson number*, which can in turn be related to the Pasquill categorization.

Laboratory experiments are of two types and can also be useful. They involve detailed simulation of dense gas spills in a wind tunnel or water flume, or they try to isolate some process occurring during the dispersion phase and study it in details. These experiments usually do not model the thermal aspects - enhanced mixing due to the thermal motions and the transition to buoyancy as the gas warms up and mixes with air. Determining the correct Reynolds' number is a problem in wind tunnel experiments, since in practice gas clouds can be very flat and wide. For the purpose of comparison, it is convenient to have a measure of the extent to which dense gas effects, such as gravity spreading and vertical mixing, influence any spill.

5.6.1. WIND TUNNEL MODELING OF A RELEASE OF A HEAVY GAS

There exist in the literature descriptions of a variety of *wind-tunnels studies on the dispersion of neutral gas plumes in the atmosphere.*[159-164]

Experiments have confirmed that the Froude number is the parameter which governs plume spread rate, trajectory, plume size and entrainment, when gases remain negatively buoyant during their entire trajectory.

In the case of spills of liquefied inflammable gases like LNG and LPG, buoyancy of the plume will be a function of both mole fraction of the gas and temperature. Thus depending upon the relative rate of entrainment of ambient gases versus rate of thermal transport from surrounding surfaces, the state of buoyancy may vary from negative to positive.

Heat Transfer Effects
Wind tunnel concentration data were obtained for continuous area releases of isothermal, cold N_2, cold CO_2 and cold CH_4 clouds. Wind tunnel results were compared to field test results and to a computer model simulation (Andreiev[165], 1983). *Heat transfer and humidity effects* on model concentration distributions were significant for methane plumes when surface Richardson numbers, Ri, were large (i.e., low wind speed and high boil-off rates conditions). At field scales heat transfer and humidity still play a role in the dispersion of methane spill cases, but plume dilution and lift off are not as exaggerated as for the model cases.

Wind Tunnel Model
Hall[166] (1982) has attempted to assess the validity of the wind tunnel modeling technique as a dispersion prediction method. His report describes a study of wind tunnel model simulations of dense gas clouds released instantaneously in the atmosphere. The comparative data used was that obtained from the field trials carried out at the Chemical Defense Establishment, Porton Down, in 1976/77. The aim was to predict the downwind distance to the point at which peak concentrations in the cloud had fallen to 2% of the source value. *Model scaling was based in this experiment on use of the correct source gas density and Froude number scaling*. Within the limits of accuracy with which such a comparison was possible, the results were generally satisfactory. The wind tunnel method appears to produce a very realistic and reliable physical model of the field trial. The results of the experiment indicate that Richardson based scaling is acceptable.

Comparison of the wind tunnel predictions with the larger scale field trials at Thorney Island is also reported by Hall et al.[167] (1985). *The Richardson number appears to be the major governing scaling parameter of the flow*. It was found in most of the comparisons, that a number of concentration measurements had been made sufficiently closely in position between model and full scale to be worth comparing. Thus it was possible to compare cloud appearance, growth and travel rates and internal concentrations. For values of the Richardson number greater than ca. three, the cloud behaved much as a release in still air being driven along but not greatly dispersed by the wind. For lower values of Richardson number atmospheric turbulence considerably increases the dispersion rate over that from gravity driven flow in the cloud, thus reducing concentrations in the cloud and also reducing the distance from the source to the point where a 2% peak concentration was measured.

The model clouds in the wind tunnel are very similar in appearance to the full scale field trials and generally spread and travel at the correct rates. Measured concentrations within the gas cloud also compare very well, and the peak concentrations are correctly predicted within a factor of two, adequate for hazard analysis work.

Model Calculations of Thorney Island Trial No. 008

Model calculations of Thorney Island Trial No. 008 were compared with experimental results, using the TNO model (Van Heugten et al.[168], 1985). An interesting conclusion could be drawn, which is reported below:

> ***Some Conclusions from Wind Tunnel Simulations***
> *There is some evidence that bulk Richardson scaling in a wind tunnel in extreme cases (neutral stability conditions, Pasquill class D, and low wind speed of 2.4 m/s at 10 m height) leads to concentration levels which are low compared to the full-scale values. Even a simple heavy gas dispersion model, which only makes allowance for the initial slumping due to gravity forces, proved to give reliable results in simplified source and environmental conditions.*

Simulations of Near-field Dispersion of Hydrogen Sulfide Gas

Baynes[169] (1985) has reported *simulations of the near-field dispersion in the atmosphere of postulated releases of hydrogen sulfide gas (H_2S) at a heavy water plant.* The size and extent of the flammable, or detonable gas clouds, which might result are estimated. This work was undertaken to support experimental studies of the detonability of H_2S releases. Thirty-six different cases were simulated involving the catastrophic failure of a liquid H_2S storage tank or tank car of H_2S. The major variables were the size of the release, the initial mixing ratio of gas with ambient air, and the wind speed. Since the gas/air mixture is initially heavier than air, *an existing heavy gas mathematical model (DENZ) was used for these simulations.* The model was modified to provide the outputs needed to support the experimental studies. The outputs were the mass of H_2S in the cloud, the mass and volume of the cloud, its radius at ground level and its temperature, all as functions of distance and time from release. The edge of the cloud was defined by a given concentration of H_2S in air. The simulations were repeated for ten different values of this parameter, ranging between 3% and 40% H_2S by volume. *Simulations were also performed using a simple 'top-hat' mixing model to predict the length of the flammable or detonable jet formed at the break in a pipe carrying H_2S vapor under pressure.* The analysis was conducted for four postulated pipe break diameters and repeated for the same ten concentration levels used in the storage tank studies. The complete outputs from the 36 storage tank failure simulations are available on floppy disks for IBM-PC.

Dispersion of Dense Gas Clouds Near a Building

The outdoor release of material that is negatively buoyant can produce cloud or plumes that cover large surface areas with limited vertical extent. The important density difference between the released gas and the environment depends strongly upon the volume of release or the release rate and the ambient wind speed. Comparison between wind tunnel and field experiments are necessary for a proper understanding of the dispersion mechanism involved, especially in the presence of obstacles such as buildings, fences or trees.

Krogstad et al.[170] *(1986) have reported wind tunnel modeling of a release of a heavy gas near a building:* a heavy gas cloud (C_3H_8) from a hemispherical continuous release was investigated experimentally. The cloud generated was bifurcated, with strong concentration gradients towards the center line as well as the edges of the cloud. The shape of the cloud was documented by measurements of velocities and concentrations. Models of buildings were introduced into the cloud and the cloud was strongly modified by interacting with the model. The horseshoe vortex system found at the base of an obstacle in a shear flow reduces the concentration level on the walls to a very low level (<2% by vol. for most of the tests). This effect depends on the height of the cloud compared to the model height.

The basic nature of the transport and dispersion of a dense gas plume in the simulated neutral atmospheric boundary layer of a wind tunnel was investigated by Britter[171] (1988), both in *flat terrain* and *over an inclined ramp*. For this simulation a ground-level circular source was used. It was observed that in *flat terrain* the lateral profile of the dense gas plume displayed very little variation of mean concentration over the central part of the plume. The concentration distribution was non Gaussian. The buoyancy-driven lateral velocity produced near-uniform ground-level concentrations across the plume, and in some case a more diffuse edge. The vertical concentration profiles were nearly exponential, quite distinct from Gaussian or top-hat shapes. The *effect of a ramp* was a slight reduction in the ground level concentration.

Comparison of the Model HEGABOX / HEGADAS with Field Experiments

The box model HEGABOX has been developed to simulate gravity-dominated dispersion behavior soon after a *sudden release of dense gas*. It is used as a front-end to the HEGADAS dense gas dispersion model, and there is a smooth transition from the gravity-dominated phase to the region in which ambient turbulence has greater influence.

Comparisons have been made between the model predictions and data from the instantaneous gas releases performed at Thorney Island (Puttock[151], 1987). Various features of the data have been compared to give confidence in the applicability of the models; these include the time-development of the cloud-averaged concentration, peak concentration as a function of distance, and cloud speed. Agreement between model and data was generally very good. However distances were overpredicted for the one trial with very high gas density, were the release produced are very low, less homogeneous cloud, and for the trial with the highest wind speed. Far downwind, a slow rise to the peak concentration is predicted, as observed. The effective roughness length at the Thorney Island site is dependent on upwind terrain and therefore on wind direction. For low concentrations, the best predictions are obtained when this is taken into account.

The dense gas dispersion model HEGADAS has been used principally for the simulation of dispersions from spills of liquefied gases, and it can handle the time-dependent vapor evolution from an evaporating pool of liquefied gas. The gravity-driven spreading of the dense gas is done explicitly. In very low winds, or for a sudden release of gas, however, there is also strong gravity spreading along the wind direction, which HEGADAS cannot handle. To simulate the early stages of such spills the front-end model HEGABOX has been developed.

By the use of HEGABOX, a smooth transition is made from the gravity-dominated stage of dense gas dispersion to the HEGADAS model, which calculates the calculation into the region were ambient turbulence is important. HEGABOX is a Box model which treats the cloud as a cylinder of uniform gas concentration, and is thus similar to a number of published Box models. The cylindrical cloud is affected by gravity-driven slumping which causes the radius to increase. The head entrainment is taken to be proportional to the gravity-spread velocity. An initial time delay is needed to model the build-up to full entrainment. In addition, heat transfer and convective entrainment are included as in HEGADAS. A simple approach to modeling the cloud's velocity is obtained by assuming that its momentum is entirely due to the air entrained, and that the effective velocity of the air is a constant factor times the average, over the cloud height, of the external wind velocity.

The HEGADAS model exists in *steady-state* and *time-dependent versions.* The steady-state model takes an initially uniform crosswind-concentration profile, but as dispersion progresses, a Gaussian profile imposed at the edges increases in width. Gravity-spreading contributes to the increase of width of the cloud. The effects of density gradients on vertical mixing are quantified by Richardson number expression. Such formulation is also consistent with passive dispersion in neutral, stable or unstable atmosphere. For time-dependent gas source, several puffs are calculated, taking into consideration the rate of an evaporating liquid pool. Allowance is made for longitudinal diffusion, and, if required, the effects of gravity spreading can be redistributed between the crosswind and the downwind directions. The entrainment relations in HEGABOX and HEGADAS have been formulated principally to predict ground level concentration and a measure of cloud height. The temperature of the entrained air is taken as the ambient temperature at half the cloud height..

Wind Tunnel Modeling of a Continuous Heavy Gas Release

A report by Hall (1989)[172] describes an *investigation, using a wind tunnel model,* of two specific aspects of the dispersion of continuously released, ground based, heavy gas plumes. The range of releases were examined, which might be described as 'slightly heavy' gases, in which the plume passes from a neutrally buoyant state to that of a heavier-than-air plume with a high degree of stability and strongly density driven flow components. The effect of source size on plume dispersion was examined for two distinct plume types, one of relatively small scale with respect to the source and with a low degree of stability, and one of relatively large scale and with a high degree of stability. The plume is sensitive to the detailed form of the source.

Physical Modeling of Heavy Plume Dispersion

An LNG vapor plume at boil-off conditions is heavier than air. Although the plume will eventually become positively buoyant due to heat absorbed from the surroundings, much of the dispersion will occur while the plume density is greater than that of air. Neff[173] (1989) has studied in laboratory tests the *LNG dispersion* during the heavier-than-air phase, and this *was approximated by means of isothermal-model plumes produced by high-molecular-weight gases.* Methods for scaling between model tests and field scale events are defined. The concentration scaling theory used by previous investigators was found to cause large errors and a new theory was developed. An extensive data base on the structure of different laboratory heavy plumes was obtained. These experiments include a large range of conditions for source gas specific gravity, gas flow rate and wind speed. Three different procedures were formulated whereby a single model test can be predictive of a larger class of field events. These enhanced scaling procedures are analyzed with respect to the measured laboratory data base to assist in specification of their capabilities and limitations. A useful empirical description of all the continuous plume tests was developed, and its applicability to field conditions discussed. Model tests on measured field scale LNG spills were performed to validate physical modeling capabilities. When the model tests reproduced the approach flow wind characteristics properly the plume concentration field was in good agreement with the field test results.

Scale Model Studies of Heavy Gas Dispersion
Scale model testing conducted in the RWDI boundary layer wind tunnel *to determine the impact of arrays of obstacles on the path of a heavy gas cloud* has been reported by Murphy et al.[174, 175, 176] (1988-1989). The parameters of interest included terrain roughness, flow path, arrival time, dilution rate, and concentration fluctuations. The obstacle ways consisted of cube blocks arranged on a rectangular grid with uniform spacing. Flow visualization of the cloud passage during calm conditions showed that immediately after a release with uniformly-spaced obstacles, a circular cloud formation was observed while at later times the cloud tended toward a surprisingly exact diamond shape. As the wind speed was increased, the diamond shape became less distinguishable. The observed concentration profiles showed an increased dose-time relation with obstacles present. The concept of a threshold wind speed was identified, above which cloud behavior was not greatly affected by the presence of the obstacles.

In another article (Krogstad et al.[177], 1989) *a brief treatment of the dispersion of heavy gas clouds* is given. The authors give general guidelines to the numerical analysis of the dispersion process, which include the necessary equations. Frequently used approximations and model simplifications are also summarized. The dispersing cloud as a phenomenon is described, and it is pointed out that unlike the dispersion of passive gases, the heavy gas cloud has inherently a self-propagation mechanism, which causes it to spread upwind and laterally to distances far beyond where the ambient wind could bring it. Because of its density, the cloud tends to spread out as a flat disk, which therefore may cover large areas. The large density also causes an interaction with the ambient flow. The requirements for model experiments of heavy gas dispersion are discussed *and it is pointed out that due to the large scale differences, model experiments must be carefully planned and executed in order to represent the full-scale flow field properly.* If the experiment planing is correctly performed, reliable results may be obtained. The possibility of having laminar flow in a region which at full-scale is turbulent makes simulations of releases into a quiescent atmosphere questionable. The transition from heavy gas to passive dispersion may be detected as the point where the concentrations start to decay as $c \approx x^{-m}$. The exponent m is of the order of 1, and x represents the distance from the source.

Dispersion of Dense Cylindrical Cloud in Calm Air
One problem relevant to most heavy gas models is the simulation of a dense gas dispersion when no wind is present. The model presented by Matthias[90] (1990) examines the process of top and side entrainment, the occurrence of a leading torus and a trailing disk, and Gaussian distribution in the entrainment zones. A simple analytical semi-empirical model has been developed which describes the concentration field in a collapsing gas cloud of cylindrical shape, for cases in which there are no atmospheric effects (i.e., no wind, or ambient turbulence). Turbulence within the cloud is self-generated, due to the sudden collapse of the cloud. This model was compared to wind tunnel and field experiment results.

Evaluation of Four Box Models for Instantaneous Dense-Gas Releases
Four Box models (OME, DENS1, SLAB, and the BOX) for instantaneous hazardous gas releases have been evaluated with data from the Burro field scale tests and also laboratory scale experiments performed at the University of Arkansa (Gudivaka et al.[178]*, 1990).* A new box model "BOX" was developed, which incorporates the proposed equations, by *Wheatley and Prince,* for translational speed of a dense gas cloud. The models were evaluated using statistical measures similar to the interim guidelines on air quality model evaluation given by the United State Environmental Protection Agency. The *non-parametric bootstrap resampling procedure,* which is relatively new method, was used to obtain confidence limits on the various statistical measures. The following results were obtained for the prediction of:

Ground Level Concentrations

- In predicting "ground level" concentrations, the SLAB model performed well in all Pasquill atmospheric conditions, as well as calm conditions
- The BOX model showed a similar performance in unstable cases, neutral cases, and calm conditions.

Prediction of Cloud Speed

- The DENS1 model predictions of the "cloud speed" under all the conditions were in good agreement with observation
- The BOX model performed well in stable and neutral atmospheric conditions.

LNG Vapor Barrier and Obstacle Evaluation
The *final report on the Wind Tunnel Simulation of 1987 Falcon Spill Series* has been published in 1991 (Shin[179] et al.). Measurements of the behavior of simulated liquefied natural gas clouds dispersing over small-scale model placed in environmental wind tunnels permits evaluations of the fluid physics of dense cloud movement and dispersion in a controlled environment. A large data base on the interaction of simulated LNG plumes with the Falcon test configuration of vapor barrier fences and vortex generators was obtained. The purpose of the reported test program is to provide post-field-spill wind tunnel experiments to augment the "LNG Vapor Fence Field Program"

data obtained during the Falcon Test Series in 1987. The goal of the program is to determine the probable response of a dense LNG vapor cloud to vortex inducing obstacles and fences, examine the sensitivity of results to various scaling arguments which might augment limit, or extend the value of the field and wind-tunnel tests, and identify important details of the spill behavior which were not predicted during the pretest planning phase.

Effects of Obstacles on the Spread of a Heavy Gas (Wind Tunnel Simulations)
The spread of a heavy gas cloud in terrain with obstacles has been studied in a wind tunnel (Heidorn et al.[180]*, 1992).* Several different obstacle configurations were examined. The results of flow visualization experiments in still air have been used to mathematically describe the spreading rate (slumping phase only) using a box model approach.

> *A review of the approaches used show that the majority of the models separate the spread of quasi-instantaneous release of a heavy gas into three phases: the initial release phase, the slumping phase, and the dispersion phase.*

The box model approach used for this study assumes that the heavy gas is released as a cylindrical cloud of known dimensions. The gas slumping phase is controlled by gravitational effects, resulting from greater gas density over that of the ambient air, and the shape of the advancing cloud. In this phase the aerodynamic drag on the cloud is balanced by the net hydrostatic head. A constant cloud volume is assumed during this slumping phase in an unobstructed terrain. The equations used assume no entrainment of ambient air into the heavy gas cloud (constant volume assumption).

> *It could be shown that entrainment of air into the cloud is not important in determining the cloud spread rate in a calm environment.*

A comparison of wind tunnel simulations results with radial spread equation was made. Obstacles were simulated using little blocks. In the case of flow through obstructions, the shape of the cloud alters from the cylindrical form. *These and other tests show that a cylinder of gas released quasi-instantaneously spreads as a vortex ring rather than a cylinder. The addition of a term to the standard box model cloud spread equation which accounted for the effects of block coverage fits the observation very well.* Although these new equations are based upon idealized obstacle arrays, they point toward a general equation for spread which includes a term dependent on the areal coverage of the obstacles. *In emergency response situation, this equation could be used to determine the minimal time required before a heavy gas cloud would reach a given point.*

Dry Adiabatic Model (Dense Cold Gas vs Isothermal Gases)
The exact variation of the density ratio for the entire life of a plume model is difficult to simulate for plumes which simultaneously vary in molecular weight and temperature. The influence of latent heat release by moisture upon the plume buoyancy is a function of the quantity of water vapor present in the plume and the humidity of the ambient atmosphere. It is believed that using a *dry adiabatic model* should give conservative results.

5.6.2. PROCEDURES FOR ESTIMATING THE UNCERTAINTIES OF HAZARDOUS GAS MODELS

Mathematical models of gas dispersion are in constant development. The most widely used approach was the so-called Gaussian model as described in practical terms by Turner or box models which have shown relative merits. In the field of heavy gas dispersion, the use of such approaches appeared somewhat limited and therefore new models have been proposed. Some of these new generation models were making use of the latest progress in turbulence modeling as derived from laboratory work as well as numerical advances, and of three dimensional codes that were computing both flow field and gas dispersion taking into account details of the ground obstacles, heat exchange and possibly phase changes as well. The description of these new types of models makes them appear as a considerable improvement over the simpler approaches. *However, recent comparisons between many of these have led to the conclusion that the scatter between predictions attained with sophisticated models was just as large as with other ones.* It is therefore felt necessary to analyze the key features of both approaches and put in evidence their relative merits and degree of realism when being really applied.

Riethmuller[181] *(1983) has made a comparison between sophisticated modeling methods and more standard ones.* For each of these approaches, the essential feature responsible for the quality of the prediction is highlighted. Two examples are chosen: (1) local differential heating of the ground and its effect on the dispersion of a heavy gas cloud; (2) complex topography and variability in the wind direction. The results of predictions are given in terms of the probability of given concentrations in the case of variations in wind angle. It is shown that such an approach gives a better indication of the ratio between mean and maximum concentrations. It is suggested that a fully probabilistic approach to the problem would be of greater practical value and that further efforts would be required for implementing such a method of prediction.

Several research group have been investigating the best way to evaluate air quality and gas models (e.g., Fox[182], 1984). Generally a combination of statistical analyses and scientific review is applied. Fractional bias and normalized mean square error criteria are applied to a table of observed and predicted parameters. The ratio of the predicted to observed variable is best used since the variable itself is a strong function of downwind distance. The ratio is likely to remain close to unity over that range. It is also of interest to calculate the 95% confidence intervals on these individual parameters. Another difficult statistical problem relates to the best way of combining performance measures when several different types of field experiments are being analyzed.

There has been several field and laboratory experiments carried out in order to study the transport and dispersion of plumes and puffs of dense gases or neutrally-buoyant gases. A variety of statistical techniques and other measures for assessing the performance of air quality models have been proposed. A review has been made by Ermak[183] (1989) of the validation techniques used within the U.S. meteorological and atmospheric science communities and a review of the specific comparison parameters used in recent heavy-gas dispersion model validation studies. This covers heavy gas dispersion model evaluation, limits to model accuracy, statistical methods for model evaluation and validation. A methodology has been proposed to predict the atmospheric dispersion from continuous releases of denser-than-air toxic gases.

The development and application of procedures for estimating the uncertainties of hazardous gas models is progressing. Formulas and software for assessing statistical performance measures have been undertaken and reported by Hanna[184] (1990) and Hanna[185] et al. (1992). So-called pre-processing and post-processing codes were written to perform the task of providing proper input conditions to the different models to be tested. Of the 25 or 30 field experiments that were initially considered, some were eliminated due to various problems and only nine datasets were finally retained. The data sets are pertaining to boiling liquid (LNG, LPG), 2-phase jet (ammonia), gas (krypton[85], freon & N_2), gas jet (krypton[85], SO_2), and diluted gas (freon). Instantaneous as well as continuous release were considered. A modeler's data archive was prepared by Hanna, that contains for all experiments, a comprehensive set of observed parameters sufficient to run the model, and a set of observed parameters for evaluating the model prediction behavior.

Methodology for Evaluating Heavy Gas Dispersion Models

Koopman[186] (1988), Ermak[187] (1989) have made a *review of recent work in the field and present a methodology for evaluating the effectiveness of mathematical models in predicting the atmospheric dispersion of heavier-than-air vapor releases.* The methodology is based upon ratio comparisons of the model-predicted value to the observed value in field-scale experiments involving continuous releases of denser-than-air gases. Plume characteristics used in the ratio comparisons include maximum concentration, centerline concentration, plume half-width, and plume height, all as a function of downwind distance from the source. Also included in the report is a review of the scientific efforts in the field of atmospheric dispersion model evaluation during the past two decades.

A report by Ermak[84] (1989) presents the results from selected field-scale, heavy gas dispersion test, which can be used to validate atmospheric dispersion models. A total of 26 benchmark tests were chosen from the Lawrence Livermore National Laboratory Burro (LNG), Coyote (LNG), Desert Tortoise (ammonia), and Eagle (nitrogen tetroxide) tests; the Shell Research Ltd. Maplin Sands tests (LNG and LPG); and the British Health and Safety Executive Thorney Island test (freon-air mixtures). Only continuous finite duration releases were selected for the test summaries as they were considered to better represent typical accidental releases from a storage vessel or transportation vehicle. The test summaries describe the manner of release, the ambient meteorological conditions, and the resulting vapor cloud characteristics including peak concentration, average centerline concentration, and average height and width of the cloud, all as functions of downwind distance.

Model Validation

An overview is presented by Smith[188] *(1985) of the problems of model development, validation, and interpretation from the viewpoint of the emergency response coordinator and contingency planner.* The consequence analysis process is outlined in order to compare priorities for information acquisition. The optimum characteristics of models used to support contingency planning and emergency response are listed. The value of measurement data is also discussed.

Results From a Statistical Examination of Wind Tunnel Modeling of the Thorney Island Trials

Davies and Inman[189] *(1987) have prepared and reported the results of a statistical examination of wind tunnel modeling of the Thorney Island trials.*

Wind tunnel simulation has emerged as an effective predictor of full-scale field experiments, both from the qualitative phenomenological standpoint and with certain qualifications on the quantitative side. A large number of wind tunnel simulations of Thorney Island heavy gas trials have been performed to establish overall trends with model scale and source conditions. Most of the analysis is based on the comparison of peak concentrations at model and at full scale. Within the confidence of determining trends from such a set of results support is given to the validity of relatively small scale modeling, particularly in the presence of sharp edged dispersing elements such as fences.

The statistical examination of wind tunnel modeling of the Thorney Island trials has revealed that the major quantitative qualification concerns the variability of the results, which is believed to be an essential part of the physics rather than to be due to measurement inaccuracy. The average trend from wind tunnel simulations for the Thorney Island trials was generally of conservative predictions of the limits of the cloud with a slight tendency to reducing the conservatism at smaller scales. In the presence of obstacles quite acceptable predictions were found down to the lowest Reynolds number utilized.

The fundamentals of modeling of dense gas dispersion at reduced scale has been discussed at some length in earlier work by Hall[190] (1979). The most important parameter in the modeling is the size of the release, the relative density ratio, and the velocity at a reference height. This leads to the following dimensionless parameters:

Reynolds number	:	$(U_{ref} L) / v$
Density difference ratio	:	$(\rho_{gas} - \rho_{air}) / \rho_{air}$, and
Froude number	:	$U_{ref} / (g L)^{0.5}$.

Using bulk Richardson number:

$$Ri_{bulk} = (\Delta\rho \, L g) / (\rho \, U_{ref}^2) \tag{5.1}$$

allows the wind tunnel model to operate at a higher velocity, when the difference density ratio is scaled accordingly.

In the wind tunnel model the Reynolds number is always smaller than at full scale. In case of dense gas dispersion, the gas has a strong negative buoyancy. This has a stabilizing influence on turbulence. When running at lower Reynolds number in a wind tunnel the flow can get locally laminar. This is not the case in the full scale trials. The presence of a fence or other obstacles introduces new turbulence into the flow and this opposes the stabilizing effect.

The scaling with bulk Richardson number is not strictly correct for differences in density larger than about 5% according to the Boussinesq approximation. This is the case when the gas cloud has spread out and the gas concentration has become low. But the error made by applying this approximation to the near field is indicated to be small.

To be able to compare full scale results with model scale results (wind tunnel), the latter results must be scaled to full scale. It can be concluded that it is possible to scale the Thorney Island using the bulk Richardson number. In such a case the gas is present for a longer period in the model experiment than in full scale (due to the difference in Reynolds numbers).

The experiments indicate that the maximum concentration level in the wind tunnel model is underestimated and that the dilution of gas tends to be faster than in the full scale tests.

5.6.3. COMPARISON OF FIELD EXPERIMENT DATA WITH MODEL PREDICTIONS

The early field test experiments were conducted by the U.S. Bureau of Mines (1970) in the USA with small quantities of LNG release (300 kg) on water. These tests and the subsequent scientific debate on the results have promoted more research on heavy gas spills.

Field experiments on dense gas dispersion are performed because there is a great need for data to confirm or contradict theoretical dispersion predictions. A survey by Havens[10] in 1978, showed the wide range of predictions made by a number of models which had at time been proposed to describe dense gas dispersion. The importance of heavy gas behavior was soon recognized, as well as the failure of the passive dispersion models, such as the Gaussian models to properly predict the observed cloud size or the concentration variation with distance.

Laboratory experiments can also be useful, but sometimes scaling-up becomes a problem, for instance for air entrainment. They are of two types: (a) experiments which try to isolate some process occurring in the dispersion and study it in details; (b) other laboratory experiments involve detailed simulation of dense gas spills in a wind tunnel or water flume. Such experiments need also to be carefully checked against field experiments. An extensive list of field experiments has been tabulated by Puttock and Blackmore[191] (1982) and a review was compiled by Raj[74] (1985) of which some are reported below:

- ESSO LNG spill experiments on water (1972).
- AGA (American Gas Association) LNG spill on land (1974).
- Shell Gadilla (1973).

These test have shown that the gas cloud was spreading laterally much more than could be predicted by models prevailing at that time, and dispersed with very little cloud expansion in the vertical direction.

- Freon experiment at the "Maasvlakte" (Van Ulden[14], 1974). The test were conducted with 1000 kg of cold Freon gas released quickly from essentially cylindrical geometry. The tests' results indicated that the cloud area expanded proportional to the time after release. Important dilution was taking place due to the vapor liberated by Freon boiling on water. Very little entrainment of ambient air was observed in the subsequent spreading phase.

- Experiments with Chlorine made in 1973 in the Netherlands.[192]

- Shell Maplin (1980).

- Freon experiment at "Porton" (Picknett[193], 1981). It consisted of 42 trials in each of which 40 m^3 of gas was released. The influence of initial relative density, wind speed, stability, ground roughness and ground slope were investigated. The concentration within the cloud was found to be relatively uniform during the gravity slumping phase. Significant dilution (by a factor 10) was taking place during the initial stage of dispersion.

- China Lake spill tests with LNG (on water and on land). These test (Koopman[194] et al., 1982); (Ermak[195] et al., 1981) can be construed to be semi-continuous release tests with spill rates varying from 11.3 m^3/min to 18 m^3/min. The quantities of LNG spilled varied from 9 m^3 to 39 m^3. The principal objective of these tests was to determine the down-wind distance and lateral spread extent to the lower flammability limit (LFL) as a function of different parameters. The source strength and wind speed have considerable influence on the LFL (from 180 m to 320 m for the above tests). The peak-to-mean concentration within the cloud were found generally to be in the range between 2 and 3.

- Thorney Island (McQuaid[196], 1984). It involves the release of 2000 m^3 clouds of gas.

- Laboratory experiments with Freon (Havens and Spicer[8], 1984; Havens and Spicer[197], 1985).

- Low wind cases of LNG, LPG experiments at "Maplin Sands" (Colenbrander and Puttock[198], 1984).

- Wind tunnel experiments carried out at the Warren Spring Laboratory, using the gas BCF (Hall[199], 1977).

In the majority of experiments the material spilled was liquefied natural gas (LNG), liquid propane or nitrogen, and freon. Ammonia although not denser than air, even at its boiling point, can form a dense cloud if the release produces an aerosol which then evaporates in the air.

There are two main types of release: instantaneous and continuous (also called steady-state release) with possible intermediate cases which are more difficult to handle mathematically. The concentrations from both extreme calculations are conservative estimates of the dispersion of an intermediate spill. *A number of models attempt to model the time dependence of a release explicitly* (Colenbrander[200], 1980; Havens[20], 1982).

The range of predictions of the models is wide. Cloud-centerline calculation has been performed with the ATMAS atmospheric transport code.[201] Laboratory simulation of negatively buoyant emissions into the earth's boundary layer is a valuable predictive tool to describe the motions of potentially hazardous chemicals. Wind-tunnel data can be correlated in a manner that yields an empirical prediction of vapor dispersion from full scale releases. Scaling criteria suggest that wind-tunnels can simulate a wide range of release situations. Further effort is needed to quantify fully the effects of non-adiabatic heat-transfer and humidity on cold plume model behavior. There are however limitations which must be recognized when interpreting the results. Dispersion in the atmospheric boundary layer can be simulated in meteorological wind tunnels with sufficient accuracy to permit realistic scaling of dense gas escape hazards.

The effect of negative buoyancy on plume behavior and resulting downwind concentrations will be greatest when crosswinds are light, and turbulence intensities are low. The entrainment of outside air into the plume, and the resulting diffusion, is a function of the interaction between plume and crosswind. A ground release of a dense gas is characterized by rapid slumping toward the surface. The initial potential energy of the dens gas is converted rapidly to kinetic energy; however this energy is also transmitted to the surrounding ambient fluid and dissipated by turbulence. The tendency of dense gas to remain near the ground enhances the importance of plume interaction with surface features (like houses, wears etc.). Slight changes of surface slope or the presence of buildings, fences, or dikes will affect plume behavior.

Tests/Simulation Results

Some results and findings of simulations and/or comparisons with field-scale experiments are briefly reported below:

Simulation of Large-scale Experimental LNG Spills Using the SLAB Model. The SLAB model has successfully simulated four large-scale experimental spills of liquefied natural gas (Morgan[202] et al., 1983). In particular, the calculated positions of the lower flammability limits (LFL) in the resultant vapor cloud agreed very well with the experimental measurements. The model is now being used to simulate other LNG spills. In addition, parameter studies are being conducted to determine the dependence of gas concentration, distance to the LFL, and the cloud dimensions on various quantities.

These quantities include source rate, wind speed, atmospheric stability, type of source gas, and source duration. Sensitivity studies are also being conducted to assess the effect of uncertainties in the submodels.

Burro Series LNG Spill Test Results. Ermak[195, 203] (1982) has made a *comparison of dense gas dispersion model simulations with Burro series LNG spill test results.* The ability to predict observed vapor dispersion over the flammable range of fuel-air mixtures is compared for a modified Gaussian plume model, a modified version of a 1-dimensional slab-average conservation equation model, and a fully 3-dimensional conservation equation model. The parts of these models needing improvement are identified.

Analysis of Catastrophic LNG Spill Vapor Dispersion. Havens[204] et al. (1983) have *analyzed catastrophic LNG spill vapor dispersion.* They compared two heavy gas dispersion models (Eidsvik and Olenbrander) for predictions of the downward atmospheric dispersion of liquefied CH_4 vapor following an instantaneous release of 25,000 m3 liquefied natural gas onto water in a marine vessel collision. For a wind speed of 2.24 m/s, with air temperature of 293 K and 68% relative humidity, and neutral atmospheric stability, the Eidsvik and Colenbrander models indicate max. distances to the lower flammability limit (time av. 5% concentration) of 4,900 and 5,800 m, respectively. The predictions show rough agreement with similar tests. made with the SIGMET model. All 3 models, which represent 3 different approaches to the dispersion prediction problem, indicate increased downwind distances to the lower flammability limit with increased wind speed for this scenario.

Simulation of LNG Vapor Dispersion over Variable Terrain. Modern models should incorporate the capability of modeling vapor dispersion over variable terrain. A three-dimensional, conservation equation model (FEM3) for simulating the dispersion of heavy gases has been described by Chen[205] (1982) and Chen[206] (1983) and used to simulate the vapor dispersion of two different LNG spill experiments (regarding the role of gravity-flow). Two numerical simulations of the LNG dispersion were carried out for each experiment. The first assumed a *flat terrain* and the second used a *numerical simulation of the actual terrain at the test site.* In general, good agreement between model predictions and field measurements, regarding maximum downwind distances to the LFL, time histories of temperature and concentration at several representative locations, and concentration contours on certain horizontal and crosswind surfaces was observed. The overall results obtained in the model calculations with the simulated actual topography were shown to correlate much better with the field data and, in particular, many important features of the vapor cloud observed under the light wind conditions of Burro 8 were reproduced in the variable terrain simulation. These include the vortex-induced high concentration regions resulting in the bifurcation of the LNG cloud and the deflection of the LNG cloud due to sloping terrain.

Test with Nitrogen Tetroxide. Large-scale spill tests of ammonia and nitrogen tetroxide were performed at the Nevada Test Site (NTS). The tests were extensively instrumented, resulting in large amounts of data which were used to quantitatively describe the observed phenomena. Preliminary results (Koopman[207], 1984) from both test series indicate that aerosols play a very important role in dense gas dispersion. Gaussian model calculations are found to be inadequate even at long distances downwind. New and unexpected phenomena were observed.

The evaporation rates and heavy gas dispersion aspects of N_2O_4 are reported by Havens[208] (1985). The vapor N_2O_4 is heavier-than-air and forms a nitric acid mist in vicinity of the spill area. As it disperses downwind, it dissociates to nitrogen dioxide. The field experiment is described and the accuracy of the experimental results are discussed. The predictions of several source strength and simple Gaussian dispersion models are compared to the results.

Analysis of Experimental Data from Field Trials. Hartwig[209] et al. (1984) have analyzed experimental data from field trials and the first results of this evaluation show that a heavy gas cloud is decoupled partly from the dynamics of the atmospheric boundary layer, even after the short gravity spreading phase. As expected, vertical diffusion coefficients are distinctly smaller as in the atmospheric boundary layer. Turbulence induced by the vapor cloud has a noticeable effect on diffusion. More data are needed and must be evaluated to support these findings.

Wind Tunnel Model Comparisons. Wind tunnel model comparisons with the Thorney Island dense gas release field trials and similar experiments at Porton Down were performed by Hall[210] (1984).

The SLAB model has been applied to determine the effects of varying a number of parameters on the dispersion of heavy gas in the atmosphere. In particular, one case was selected to simulate an actual spill conducted at China Lake[211] which compared with good agreement. The results obtained for the variations are explained in terms of the relevant physical phenomena. The SLAB model was also used to simulate three of the more recent Coyote series of LNG spills[212] and improved simulations of some of the Burro tests. The parameters studied include the principal physical parameters that determine the properties of the dispersing cloud: source rate, wind speed, atmospheric stability, type of source gas, and source duration, as well as the parameters important to certain physical submodels.

Burro Series of LNG Spill Experiments. The purpose of the Burro series of spill experiments, in 1980, and one of the purposes of the Coyote series, in 1981, was to *investigate the atmospheric dispersion of cold, dense LNG vapor resulting from an LNG spill onto water.* The SLAB model has been tested and compared to the experimental data by Morgan[213] (1983). Computer simulations of four of the Burro series large-scale liquefied natural gas (LNG) spill experiments at China Lake, California have been successful in predicting distances to the lower flammability limit (LFL). SLAB was also used in simulations of three of the Coyote series of experiments.

Various physical phenomena affecting LNG vapor dispersion were observed in LNG spill experiments (Morgan[214], 1984). Gravity flow of the cold dense vapor increased cloud width while density stratification and heat flow from the ground had substantial effects on the mixing rate with air. These phenomena led to a dependence of the maximum distance from the pounds to the LFL on source rate, wind speed, and atmospheric stability that was substantially different from the prediction of the Gaussian plume model. Studies employing the numerical SLAB model demonstrate the importance of including these phenomena in predictive models. Time-dependent features of the concentration field due to turbulence and rapid phase-transition explosions, which also affect pounds, were examined by applying a space-time interpolation scheme to the concentration data.

Phenomena affecting the maximum distance to the lower flammability limit (LFL), an important quantity which indicates the potential extent of an accidental combustion, have been investigated by Morgan[215] *(1984).* The LFL distance also depends on the spill parameters and meteorological conditions. Two additional phenomena, rapid-phase-transition (RPT) explosions and differential boil-off (producing increased ethane-to-methane ratio), that can lead to significant increases in the LFL distance were observed. Both the SLAB and FEM3 computer codes incorporate mathematical models of the physics that governs the dispersion phenomena. SLAB is a one-dimensional, crosswind-averaged, conservation-equation model that calculates cloud height and width, and then uses these values to determine the crosswind distribution of LNG vapor concentration. FEM3 is a fully three-dimensional, conservation-equation model that can include variable terrain. Both models are time-dependent. In spill simulations, both give results that are in agreement with the experimental data for downwind extent and duration of the flammable region and other cloud features. In addition, FEM3 can simulate the complicated three-dimensional structure of a cloud where heavy-gas dispersion and terrain effects predominate.

Evaluation of 3-D Hydrodynamic Computer Models for Prediction of LNG Vapor Dispersion in the Atmosphere. The FEM3, MARIAH II, SIGMET-N, and ZEPHYR Heavy Gas Dispersion Models have been transferred to the University of Arkansas and installed on a project-dedicated computer system for evaluation.(Havens[216] et al., 1984) *A series of test LNG release scenarios has been simulated to determine the models' predicted effects due to variation in dispersion scenario parameters such as gas release rate, wind speed, and atmospheric stability.* A series of isothermal, fixed volume, instantaneous releases of Freon-air mixtures has been performed, providing an extensive data base of concentration measurements as a function of radial and vertical position in the developing heavy gas cloud for evaluation of model predictions. A prototype gas sensor, which uses an aspirated flame ionization detector, suitable for concentration measurements in cold gas spills has been developed.

The above mentioned computer models were evaluated for prediction of atmospheric dispersion of LNG vapor clouds (Havens[217] et al., 1987). The *SIGMET-N model* is considered *unsuitable* for prediction of LNG vapor cloud dispersion due to the impracticality of controlling numerical diffusion by grid fine-zoning and the use of a turbulent mixing submodel which does not scale properly. The *ZEPHYR model* is also considered *unsuitable*, primarily because of the use of a Lagrangian particle method for solving the gas mass balance equation which is not applicable to such strongly density-stratified flows. The numerical diffusion properties and the turbulent mixing submodels of *MARIAH-II* and *FEM3* were evaluated by comparison of their predictions with analytical solutions, laboratory scale gas dispersion experiments comprising neutrally buoyant to strongly density-stratified two- and three-dimensional flows, and selected field experiments from the Thorney Island Heavy Gas Trials and the DOE Burro LNG tests. *A turbulent mixing model based on simplified second-order closure theory was evaluated, but modifications would be necessary before it could be recommended for LNG vapor cloud dispersion prediction.*

Air Entrainment Model. A model was developed by Jensen[218] et al. (1984) and a numerical treatment proposed for *air entrainment through the top of a heavy gas cloud.* The vertical growth of clouds having the same initial density excess but a different temperature difference between cloud and ground surface were considered. Denser-than-air often produce clouds which disperse in the atmosphere in a manner that is different than trace gases. These differences are due to density or gravity-induced effects such as turbulence damping from the stable density stratification, alteration of the ambient velocity field due to gravity flow, and the source momentum flux. Large scale tests involving releases of heavy gases have been conducted since the early 1970's. These tests have resulted in the discovery of previously unknown and important effects, the accumulation of data for model validation, as well as accident simulation and evaluation of accident mitigation equipment and techniques.

Remarks Concerning Heavy Gas Dispersion and Environmental Conditions. Gotaas[219] (1985) has investigated the dependence of heavy gas dispersion on environmental conditions, as revealed by the *Thorney Island trials data*, and how well this can be predicted by the *Eidsvik model* (Norwegian Institute for Air Research, NILU).

Time plots of average concentration values from the Thorney Island field experiments were used to draw cloud outlines. After the initial slumping, and the formation of a vortex ring, redistribution of mass took place. At later stages the highest concentrations were found to be well inside the cloud. Wind speed increment with height and surface drag sheared the cloud in the direction of the wind. They also created a high front and a low trailing edge. Some trial measurements suggest high gas concentrations below 0.4 m, which could be due to gas withheld in the grass at low wind speeds.

Field Test Validation of the Eidsvik Model

The field data verified the model assumptions and prediction that maximum distances to critical concentrations are independent of wind speed and air stability. Predicted distances were on the safe side and remarkably accurate. This suggest that a simple Box model is well suited for prediction of hazard distances. The too high decrease of concentrations with time could be offset by applying the wind speed at 10 m level and disregarding the wind profile.

Review of Field Experiments. Three of the most recent series of field experiments have been reviewed by Puttock[220] (1985):

- a study of the dispersion and combustion on release of dense flammable gases at the Maplin Sands in England
- the China Lake/Frenchman Flat experiments in California and Nevada studying the dispersion of liquefied natural gas, ammonia, and nitrogen tetroxide, and
- the Freon/nitrogen spill tests at an abandoned airfield on Thorney Island, England.

The significant findings from each series of experiments are described. The scaling rules used in dense-flow simulation are summarized and the limits to wind-tunnel and water-flume modeling are discussed. Future developments in field experimentation are outlined, focusing on the Thorney Island and Frenchman Flat facilities.

Modeling the Phase I Thorney Island Experiments. A simulation of *the Thorney Island Trials-I,* using a mathematical model developed for incorporation in the U.S. Coast Guard Hazard Assessment Computer System HACS[261], was compared with field experiments by Spicer[221] et al. (1985). The model used has been adapted from Shell HEGADAS model described by Colenbrander. A lumped parameter model of the initial formation of a heavy gas source cloud, which incorporates air entrainment at the gravity spreading front using a frontal entrainment velocity, was substituted for the source description recommended for HEGADAS. This model includes three parts: (a) The heavy gas source formation, simulated by a box model, (b) the downwind dispersion model, (c) the quasi-steady treatment of transient gas releases. The model requires as input the volume to be released and its dimensions, the initial density (gas concentration), wind velocity at specified height, ambient temperature, pressure and humidity, Pasquill stability category, and surface roughness.

Validation of the Dense Gas Model Incorporated in HACS
The model predicts the maximum (peak) concentrations vs. distance observed in the simulated trials sufficiently accurately to justify the model's use for prediction of heavy gas peak concentrations and decay with distance expected for releases with wind speeds of 1.7 to 7.0 m at 10 m height and atmospheric stabilities ranging from slightly unstable to stable.

Assessment of FEM3. Chan[222] et al. (1985) have reported the *assessment of FEM3 - A three-dimensional numerical model for the dispersion of heavy gases over complex terrain, with field test data.* During the past few years, FEM3 has been assessed, using data from the Burro and Coyote series of LNG spill experiments conducted by LLNL (Lawrence Livermore National Laboratory) at China Lake, California. In general, the model has been found to perform very well and it greatly complements the field experiments in enhancing the understanding of the phenomena associated with LNG vapor dispersion, including gravity spreading, heating from the ground surface, and terrain effects. The FEM3 model has further been applied to simulate the dispersion of *nitrogen dioxide (NO_2)* for one of the LLNL conducted nitrogen tetroxide (N_2O_4) spill tests and also to simulate the dispersion of propane gas for four of the refrigerated liquid propane spills conducted by SHELL Research Limited at Maplin Sands. The main purpose of the NO_2 simulation is to demonstrate the heavy gas effects in this test and the latter simulations are for assessing the performance of the current model for simulating the dispersion of *propane gas*.

Evaluation of 3-D Hydrodynamic Computer Models. The MARIAH-II, SIGMET-N, ZEPHYR, and FEM3 3-D *hydrodynamic computer models for atmospheric dispersion of LNG vapor* were evaluated (Havens and Schreurs[223], 1985). The models show important differences in the prediction of LNG vapor dispersion; the differences are attributed primarily to numerical diffusion and modeling of turbulent air entrainment. The importance of controlling numerical diffusion to prevent masking the physical diffusion is demonstrated. The turbulent air entrainment submodel used in MARIAH-II, SIGMET-N, and ZEPHYR is considered unsuitable for simulation of the general LNG dispersion scenario. An improved turbulent entrainment submodel is being developed, with initial testing against laboratory mixing experiment and wind tunnel data. Validation of models against controlled experiment data will precede comparison of simulations with a wide range of heavy gas field test data.

Review Large Scale Spills (Cryogenic or Pressurized). A review of recent tests concerning large scale spills of volatile cryogenic or pressurized ambient temperature liquids and some of the important results are presented by Koopman[224] *(1986).* The status of computer modeling in this field is also reviewed with an emphasis on the problems associated with Gaussian models, and recent progress in intermediate and 30 conservation equation models. Two of these models, FEM3 and SLAB are compared to data so that the observed phenomena can be understood and so that predictions can be made regarding the consequences of accidental spills of sizes of concern to industry. FEM3 simulates the dispersion of the released gas by solving the time-dependent, three-dimensional conservation equations of mass, momentum, energy, and species. This model is used as a research tool and is designed to simulate as closely as possible the physics, and thermodynamics governing the dispersion process. Steady-State SLAB is based on the crosswind-averaged form of the conservation equations, using similarity profiles to determine the crosswind dependence. It also conserves mass, momentum, energy and species, but is inherently one-dimensional, and is computationally simple and fast to use. (ERA citation 12:005227)

Review of the Status of Heavy Gas Dispersion Modeling. Schnatz[103] et al.(1986) have prepared a review, with 29 references, on the status of heavy gas dispersion modeling, including comparative calculations and experimental verifications.

Recent Developments of FEM3 and SLAB Atmospheric Dispersion Models. Lawrence Livermore National Laboratory (LLNL), under the sponsorship of the U.S. Department of Energy and other agencies, has been conducting research in the area of atmospheric dispersion of heavier-than-air gases over the past eight years. This research has successfully produced a number of field scale test series and *two state-of-the-art dense gas dispersion models called FEM3 (a fully three-dimensional model) and SLAB (a quasi three-dimensional model).* Over the past few years, the predictions from both models have been compared by Chan[225] (1986) and Ermak[226] (1986) with the data obtained from a variety of field scale experiments. The entrainment submodel in SLAB were further evaluated and an improved turbulence submodel for FEM3 has been presented, which was assessed by using the data obtained from two laboratory-scale dense gas dispersion experiments conducted by McQuaid.

FEM3 is a three-dimensional numerical model for simulating the atmospheric dispersion of heavy gases over complex terrain. During the past few years, FEM3 has been assessed, using data from the Burro and Coyote series of LNG spill experiments conducted by LLNL at China Lake, California. In general, the model has been found to perform very well and it greatly complements the field experiments in enhancing the understanding of the phenomena associated with LNG vapor dispersion, including gravity spreading, heating from the ground surface, and terrain effects. The FEM3 model has further been applied to simulate the dispersion of *nitrogen dioxide* (NO_2) for one of the LLNL conducted *nitrogen tetroxide* (N_2O_4) spill tests and also to simulate the dispersion of *propane gas* for four of the refrigerated liquid propane spills conducted by SHELL Research Limited at Maplin Sands. The main purpose of the NO_2 simulation is to demonstrate the heavy gas effects in this test and the latter simulations are for assessing the performance of the current model for simulating the dispersion of propane gas.

Assessment of Wind Tunnels Experiments. A study was undertaken by Davies[227] (1986) in order *to assess the ability of wind tunnels to predict vapor dispersion from accidental releases.* The Thorney Island Heavy Gas Dispersion Trials (performed at scales between 1:40 and 1:250) are quite often used for assessing models. Most of the analysis has been based upon the use of peak concentration statistics. An overall picture of good model - full scale comparisons has emerged from a pattern of significant variability between individual realizations at both model and full-scale. Within the confidence of determining trends from such a set of results, support is given to the validity of relatively small scale modeling, particularly in the presence of sharp edged dispersing elements such as vapor fences.

Guideline LNG Fluid Modeling. A Guideline for of Liquefied Natural Gas cloud dispersion (vol. I and II) has been prepared by Meroney[228] (1986). The capabilities and limitations of fluid modeling for dense gas cloud behavior are summarized and standards to be followed during the preparation of risk analysis studies were recommended.

Dense Gas Plume Interaction with Surface Features and Buildings. There exist many predictive models for downwind dispersion of dense gas of an isolated source release. The variation of such predictions is significant in assessing credibility of potential hazards, and this uncertainty may increase due to the presence of buildings and other obstacles.

A three-dimensional numerical program has been developed by Jacobsen[57] (1987) for simulation of heavy gas dispersion. The *K-epsilon model with standard constants* has been used for calculation of the diffusion coefficients where the effect of density gradients on the mixing process is accounted for by applying correction factors given as functions of local Richardson number. The results of a numerical simulation of the Thorney Island Trial 008 are in good agreement with experimental data.

The Heavy Gas Mixing Process in Still Air, at Thorney Island, and in the Laboratory. A *dynamic integral model* is described by Van Ulden[229] (1987), that includes a time dependent radial momentum budget and a turbulent kinetic energy budget. These budgets are used to predict radial gravity spreading and cloud generated turbulent entrainment. The measured area-averaged concentrations from the Thorney Island Trials 12 and 34 and from the laboratory experiments by Havens and Spicer were analyzed. In this comparison it appears that the model accurately describes radial gravity spreading. Evidence is provided that measured concentrations depend strongly and systematically on the measuring height. This implies two things: first, the height of the center of mass of the cloud was not great in comparison with the measuring heights; and second, the "true" surface concentrations are likely to be significantly higher than the concentrations measured at Thorney Island and in the bulk of the laboratory experiments. From the measured data a preliminary normalized concentration profile was deduced. When this profile is used in Van Ulden's model a fair and consistent simulation of the measured concentrations is obtained, both for the two Thorney Island trials and for the laboratory experiments.

Predictive Capabilities of Various Classes of Dispersion Models. The major objectives of the work of Cornwell[230] et al. (1987) were to provide *a state of the art information* on the dispersion of large-scale instantaneous releases of heavier-than-air gases and to quantify the predictive capabilities of the various classes of dispersion models, using the Thorney Island data. Five analytical methods for comparing the test data to model calculations were developed. The methods compared various physical characteristics of the trial gas cloud to the predicted physical characteristics of the model cloud. Four models that represent state-of-the-art techniques were selected for comparison with data from the Thorney Island trials. The models were divided into 3 categories: box, extended box, and 3-dimensional fluid dynamic models. From the box model group, the Cox and Carpenter[15] (1979) model and the Eidsvik[231] (1978) model were selected. The Colenbrander[19] (1980) model, HEGADAS II, is an extension of the box model concept, considers concentrations and velocity profiles, and uses the K-theory eddy diffusivity approach. From the fluid dynamic group of models that use K-theory for turbulent closure, the MARIAH II model was selected for this study.

Critique of the Thorney Island Dataset. Brighton[232] (1987) has given a user's critique of the Thorney Island Heavy Gas Dispersion Trials, whose primary purpose was to obtain reliable data at large scale to test the validity of the mathematical and physical models. The trials were also intended for improving the understanding of the physical mechanisms in heavy gas dispersion and for testing the fundamental hypothesis in mathematical models. The most widely used models are the *integral or Box models,* which represent the dispersing cloud by a single volume of gas of uniform concentration. Such models do not predict concentrations at individual positions. The basic data gathered during field experiments cannot be used directly for validating such models. Rather complex data-processing must be carried out to determine quantities such as cloud area, mean concentration and translational speed. Thus there is considerable scope for diverging interpretation of the raw data (due to different definitions of these quantities and due to sampling errors). The other type of mathematical model is the *three-dimensional time dependent numerical solver of the Navier-Stokes equations with a suitable turbulence closure scheme.* In principle, it is much easier to compare models of this category with the experiments since they can calculate the time-history of concentration at individual points. Physical wind tunnel modeling also has this feature.

Interpretation of the Thorney Island Phase 1 Trials with the Box Model Cigale2. CEA /IPSN have been investigating the atmospheric dispersion of heavy gases. Three different ways of approach are under development: box modeling, three dimensional modeling and small-scale simulation in water channel. *An interpretation of the Thorney Island phase 1 trials with the box model DENZ and the consequent development of the improved box model CIGALE2 has been undertaken* and the final report presented by Crabol[233] et al. (1987). It has previously appeared that the main disagreement between the DENZ code and the experiments comes from an erroneous modeling of the motion of the cloud leading to much faster cloud travel (i.e., shorter time to reach a given distance) than is observed. The assumed reason for this is the omission by the code of a significant inertia effect of the cloud accelerating from its initial position to reach a constant advection velocity after a certain time. The inertia of the cloud obviously does not affect this advection velocity, but the time to reach this velocity. In the DENZ code, the cloud is supposed to travel with the wind velocity at half height from the instant release. The DENZ code model the cloud area correctly, but tends to overestimate the height of the cloud at small elapsed time, and also to overestimate the concentration at a level of 2.4 m above ground.

The changes incorporated into the new code CIGALE2 tend to improve such predictions, i.e., a significant improvement of the realism of the CIGALE2 code concerning the mean position of the cloud, the cloud surface, cloud height, and mean concentration at 0.4 and 2.4 m height. The idea is also to model the cloud acceleration by assuming that it is entirely due to the momentum of the air entrained into the cloud. The interpretation of the Thorney Island phase 1 trials have shown that particularly the mean deviation between the code and the experiment was reduced to 10% concerning the position of the cloud vs. time and less than a factor 2 for the ground concentration vs. distance. The validity of the improved code was confirmed using the Porton trials data.

Analysis and Simulation of Thorney Island Trial 34. The results of Thorney Island trials 34, which *approximated an instantaneous release of heavy gas in zero wind conditions*, were summarized by Havens[223] et al. (1987). The gravity spreading and cloud dilution are predicted by laboratory scale instantaneous releases in calm air. The laboratory scale model experiment and Trial 34 were simulated with MARIAH-II mathematical code which has been modified to incorporate a simplified second-order turbulence closure. Other models such as SIGMET-N, ZEPHYR and FEM3 were also used, and the result of simulations and comparisons were reported elsewhere. For this simulation the turbulence submodel originally used in MARIAH-II has been replaced with a local turbulence model derived from a second-order formulation incorporating simplifying approximations. A 2-D cylindrical coordinate version of MARIAH-II was used for the calculations.

Field Test Validation of the MARIAH-II Model
Cloud time-of-arrival and peak concentration vs. distance-from-release measurements from Thorney Island Trial 34 have been compared with similar measurements from laboratory-scale releases in air. MARIAH-II predictions of the gravity spreading and dilution of the laboratory-scale and Trial 34 experiments are in good agreement when scaled by the characteristic length and time. Predictions of the cloud volume-averaged concentration vs. time are in good agreement with the laboratory-scale observations. MARIAH-II predictions of the cloud time of arrival are later than were observed in the field trial, but are in a reasonable agreement if allowance is made for cloud advection by the wind. The predictions of peak concentrations vs. time are higher than observed in both laboratory-scale and Trial 34 experiments.

Field Test Validation of the DEGADIS Model. The DEGADIS model has been used to simulate a large collection of field experimental dense gas releases. DEGADIS is an adaptation of the Shell HEGADAS model by Colenbrander and Colenbrander and Puttock; DEGADIS incorporates some techniques used by van Ulden.

Field test validation of DEGADIS were reported by Spicer[234] et al. (1987). The model-predicted downwind gas concentration decay has been shown consistent with the field scale experimental data available. Application of DEGADIS has been primarily directed to the *prediction of concentrations in the lower flammability limit range (1-5%).* Application of DEGADIS to the prediction of concentration levels of interest for toxic gases (0 and <100 ppm) is less complete. However using the Gaussian plume model generally underpredicts the concentration due to an overprediction of the vertical mixing present, while DEGADIS predictions were consistent with the observed concentrations.

FEM3 Model Simulations of Selected Thorney Island Phase I Trials. The FEM3 model has been actively used *for simulating both continuous and instantaneous heavy gas releases.* Assessment of the model was done by Chen[235] et al.(1987) using some of the Thorney Island Phase I trials data, ranging from low wind speed and stable atmospheric conditions to strong wind and neutral stability, with initial density varying from 1.6 to 4.2.

FEM3 is a three-dimensional computer model that was designed to simulate the atmospheric dispersion of large heavier-than-air gas releases. Since it is fully three-dimensional, FEM3 can also treat complex flow and dispersion scenarios over variable terrain and around obstructions such as buildings. FEM3 model include a simple submodel for treating *aerosol effects in the pressurized ammonia spills*, a phase-change submodel to treat *humidity in the ambient atmosphere*, and can be used as a *tool for emergency-response planning for potential accidental releases of liquid chlorine.*

Field Test Validation of the FEM3 Model
Due to the inappropriateness of the present K-theory turbulence submodel for simulating the early phase of an instantaneous release, FEM3 consistently underpredicts the peak values of concentration, sometimes by as much as a factor of 2. The shapes of the predicted peak concentration curves, however, agree favorably with field data. The FEM3 code appears to predict well most of the salient features present in large, instantaneous releases; specifically gravity slumping and spreading, formation of the doughnut-shaped cloud, and cloud bifurcation. The predicted major cloud characteristics, including cloud area increase rates, critical distances and cloud arrival times, also agree well with field measurement. All the simulations were performed on a CRAY-1 computer.

Model for Instantaneous Releases of Denser-Than-Air Clouds. Carpenter[236] et al. (1987) have developed a simple mathematical *"box-type" model for instantaneous releases of denser-than-air clouds*. Data from the Phase I Thorney Island trials were compared with observations of other experiments. The data were then used to calibrate the model for this type of release. The model was developed jointly by Bradley[237] (1983), British Gas, and Cremer and Warner. It incorporates a momentum relationship to determine the cloud advection speed. It assumes an idealized geometry (cylindrical form) for the dispersing cloud, and attempts to follow the bulk cloud advection and dilution in the given atmospheric concentration conditions. The approximation used are valid provided the cloud height is not substantially different from 10 m. The edge mixing and momentum parameters (not present in the equations of Bradley) take well defined values but the model is relatively insensitive to the values of the top surface parameters for the Thorney Island Phase I trials. However, a comparison with additional data, previously calculated from records of the trials, shows that this model produces reasonable results. The inclusion of a momentum relationship produces a significant improvement to an earlier model. The chosen method of calibration used the time of arrival of gas and the peak observed concentration at the fixed sensor locations

It is concluded that other sources of data should be examined to see if the model parameter values deduced here may be used for other release configurations. The model is used to assess the hazards associated with the storage and transmission of flammable gases (LPG, LNG) and to predict the dispersion of denser-than-air clouds.

Enhanced Box-Type Model. Deaves[238] (1987) has demonstrated how the data which were obtained from the Thorney Island trials were used to enhance a box-type model and how comparison between Phase I and Phase II results can suggest ways in which more complex mathematics modeling can be applied. Particular emphasis is placed on the manner in which each type of model is likely to be used during consequence analysis studies.

Heat Transfer Effects in Wind Tunnel. The *possibility of using cold gases in a wind tunnel to correctly model heat transfer to the plume and subsequent plume density changes* was reported by Britter[239] (1987). A literature survey is provided of the previous modeling of dense-gas dispersion including heat transfer effects. The scaling laws for the physical modeling of the dispersion of dense gases in the absence of heat transfer effects and those incorporating heat transfer are developed. As a result of the scaling laws developed Britter found that the modeling of flows in which free convective heat transfer is important at full-scale is not possible. The application of the scaling laws to possible release scenarios is undertaken. When forced convection heat transfer is dominant at full-scale physical modeling is feasible. However, the resulting model will have heat transfer dominated by free convection and, therefore, be unacceptable.

Comparison of an Accidental Release Modeling Technique with Recent Spill Test Data. A Comparison of an accidental release modeling technique with recent spill test data has been reported by Chikhliwala[240] (1987). The use of mathematical models to evaluate and analyze the consequences of a hazardous spill, has become a crucial part of Emergency Response and Planning. Simple models for the evaporation and subsequent dispersion of the cloud serve a useful role for screening purposes. However, to model realistic scenarios one has to design a comprehensive system capable of handling various situations. The authors focus on one aspect of the complete spectrum of scenarios: they attempt to study the *instantaneous release of a heavier-than-air gas*. It has been argued that this scenario does not truly reflect a real-life situation and is most unlikely to occur. However, some precise trials have been conducted where large quantities of heavy gas were instantaneously released and these provide an excellent data base for evaluating and testing the assumptions made by a model. *They describe a simple box-model that can take into account the effects of gravity slumping, ground heating and has a very simple entrainment formulation.* They use this model to compare with real-life trials conducted at Thorney Island and Porton Downs. The exercise of simulating these trials illustrate the complex phenomena that actually occur during a release, and highlight the limitations of predictions due to inherent model assumptions.

Numerical Simulation of the Mitigating Effects of an LNG Vapor Fence. The U.S. Department of Transportation (DOT) and the Gas Research Institute (GRI) initiated a program to evaluate methods for predicting LNG dispersion distances for realistic facility configurations. *FEM3A, a fully three-dimensional numerical model for simulating the atmospheric dispersion of heavy gases involving complex geometry, has been used to investigate the mitigating effects of a vapor fence for LNG storage areas* (Chan[241], 1990). The simulation of the vapor dispersion of four large-scale LNG vapor barrier field experiments was compared with the relevant field data. The model was able to reproduce the major results of the experiments within a factor of two under most circumstances. An intercomparison among the results from numerical simulations (with and without the vapor fence) and field data (with vapor fence) was made. The numerical results indicate that, with the present fence configuration, the maximum concentration on the cloud centerline was reduced by a factor of two or more within 250 m behind the fence, and the downwind distance to the 2.5% concentration was reduced from 365 m to 230 m. However, a vapor fence could also cause the vapor cloud to linger considerably longer in the source area, thus increasing the potential for ignition and combustion within the vapor fence and the area nearby over time.

For Falcon-1 (Nevada Field Experiment, 1987), with additional heat flux over the source area to model the superheating effects, results consistent with field observations were obtained (Chan[242], 1992). In particular, a vapor cloud overfilling the fenced enclosure was reproduced, in contrast with vapor clouds essentially contained within the fence at all time observed in a pre-spill wind tunnel simulation. The simple approach currently taken to model turbulence and heat transfer in the source area has performed reasonably well; however, more sophisticated modeling of the source may be necessary for more accurate predictions at all locations.

Results from simulations of the Falcon-4 experiment indicate that an LNG vapor fence can significantly reduce the downwind distance and hazardous area, thus increasing the potential for ignition and combustion within the vapor fence and the area nearby.

Model Comparison. A comparison has been reported by Britter[243] (1991) between *experimental data* obtained from spills of heavy gas or volatile liquids and *predictions of five different models of heavy vapor cloud dispersion*:

MARIAH (Deygon-Ra), ZEPHIR (Energy Resources Co., Inc.), HEGADAS-II (Shell), Eidsvik's "top hat" model (Norwegian Institute for Air Research, and the Germeles and Drake "top hat" model (Cabot Corp.). *The Eidsvik's model has been recommended as one of the most advanced of the "top hat" class of models.* The HEGADAS-II cannot predict the calm wind cases, the Germeles and Drake model seriously overpredicts air entrainment or cloud dimensions. It has been recognized that models developed for the dispersion of neutrally buoyant clouds are inadequate to describe the dispersion of heavy vapor clouds formed from accidental spills of volatile liquids. The above mentioned models are representative of the K-theory (eddy diffusivity) type (MARIAH and ZEPHIR) and of the "top hat" or uniform concentration type models (Eidsvik, and Germeles and Drake). The Germeles and Drake are of the first generation models which incorporate air entrainment in the vertical via constant coefficients.

It has been found that each model is limited in its range of applicability. A favorable comparison of model prediction against cloud radius alone, is not an indication that the model will correctly predict concentrations in the cloud. Models in general and top-hat models in particular can predict cloud radius vs. time without reference to air entrainment terms (air entrainment being directly related to cloud height in top-hat models). Concentration predictions are very sensitive to wind profile assumptions, just as cloud dimensions are. The K-theory models are more versatile than top-hat models.

Evaluation of Fourteen Hazardous Gas Models with Ammonia and Hydrogen Fluoride Field Data. The evaluation of fourteen hazardous gas models has been reported by Hanna[244] et al. (1991). These models were compared and tested using data from the Desert Tortoise *ammonia* and Goldfish *hydrogen fluoride field experiments,* which involved *horizontal releases of aerosol jets.* Seven experiments are available for analysis, with data at three downwind monitoring arcs, at distances ranging from 100 m to about 3000 m.

The models include eight publicly available models, (ADAM, AFTOX, ALOHA, Britter and McQuaid, DEGADIS, HEGADAS, OB/DG, and SLAB) and six proprietary models (CHARM, EAHAP, PHAST, SAFETI, TRACE, and WHAZAN). In addition, the methods of initializing the ALOHA, CHARM, DEGADIS, and HEGADAS models were modified to account for initial dilution in an aerosol jet and these revised predictions were included in the evaluation. About one-half of the models yield relatively good performance in their predictions of maximum concentrations on monitoring arcs with relative mean biases of ± 30% or less and root mean square error (rmse) values that are about 40% to 60% of the mean. It is interesting that the simple Britter and McQuaid model performs just as well as some of the more sophisticated models, indicating that the simple model has captured the essence of the plume thermodynamics. Because this data set is not large, no significant differences can be shown (at the 95% confidence level) among the better models. This analysis will be expanded in the future to include other field datasets (e.g., Thorney Island and Maplin Sands).

Modeling of Heavy Gas Cloud Transport in Sloping Terrain. A model has been presented by Kukkonen[154] et al. (1992) for the transport of *denser-than-air gas clouds released instantaneously on a slope.* The numerical model was introduced into a modified version of the heavy gas dispersion model DENZ. The structure of the model is discussed and detailed model equations for the special case in which the wind is directly uphill or downhill are derived. The model was designed as hazard analysis tool, and its computer implementation can be used as a subprogram in heavy gas dispersion models. Model predictions were compared with results of the Thorney Island phase I field experiments. Although these trials were conducted on flat terrain, the comparison is useful for understanding the cloud transport processes. Numerical calculations of heavy gas cloud dispersion on a slope were also analyzed.

The predictions of the present model are clearly better compatible with the data, compared to the original DENZ predictions. However the predicted cloud speeds are somewhat larger than experimental results for all trials except No 5. *The influence of the slope of the terrain is noticeable in the near range. At sufficiently large distances, the cloud speed tend to the same value irrespectively of the slope and wind direction (uphill or downhill or flat terrain). The currently available experimental data is insufficient for validating the present model.*

ALWAS, A Multi-Compartment Model. The Air Land Water Analysis System (ALWAS) is a *multi-media environmental model for describing the atmospheric dispersion of toxicants, the surface runoff of deposited toxicants, and the subsequent fate of these materials in surface water bodies.* ALWAS depicts the spatial and temporal distribution of contaminant concentrations in a watershed and the air above it. Linked in ALWAS are three submodels that provide for independent and partially coupled usage modes - Dispersion and Deposition of Toxics (DiDOT); a modification, called NPSDEP, of the Nonpoint Source Model; and Exposure Analysis Modeling

System (EXAMS). ALWAS is appropriate for evaluating multi-media water quality problems for watersheds as large as 10,000 sq. km. The full range of ALWAS capabilities may be exercised for hydrophobic organic chemicals of relatively low vapor pressure. By careful selection and linkage of the submodels, ALWAS may be applied to investigate the environmental behavior of a broad range of toxicants including highly soluble or volatile organics and heavy metals. The report incorporates detailed software documentation and a users manual.

5.7. Reports and Studies on Accidental Release of Toxic Vapor Clouds

5.7.1. CASE I: SCENARIOS FOR H_2S RELEASE

A statistical model was developed by Monserco Ltd.[245] *in 1981, that calculates the probability of short-term H_2S-concentrations exceeding the lethal level given the downwind range from the release, the short-term averaging time of interest, the long-term average concentration, and meteorological and terrain conditions.* An extensive and up-to-date review of H_2S toxicity was conducted, with emphasis on *acute* and *sub-acute poisoning*. The literature on animal studies and cases of human exposure were used to derive a lethal dose relationship (concentration-exposure time) appropriate for the general population. Results were obtained for passive releases of H_2S under a range of hypothetical conditions. Interpretation of these results is given in terms of the overall probability of lethal exposure during a 30-minute H_2S gas release. The probabilistic consequence assessment of hydrogen sulfide releases from a heavy water plant were reported.[246] As scenario, an accidental release of hydrogen sulfide to the atmosphere following a pipe or pressure envelope failure, or some other process upset, at a heavy water plant was considered.

5.7.2. CASE II: SCENARIO FOR ANHYDROUS AMMONIA AND NITROGEN DIOXIDE RELEASES

Abbott[247] *(1992) has evaluated potential atmospheric and human health impacts that may result from accidental releases of anhydrous ammonia and nitrogen dioxide at the Idaho Chemical Processing Plant (ICPP) NO_x Abatement Facility.* Excess process gas releases are evaluated using a traditional Gaussian puff model.

The SLAB results are also compared to those using the neutral-buoyancy puff model. A SLAB sensitivity analysis is presented which examines various combinations of ambient temperatures and wind speeds in order to determine worst-case downwind air concentrations.

Case Study
Dense two-phase aerosol releases from an 18,000 gallon liquefied ammonia storage tank and a 6,000 gallon tanker truck accident were evaluated using the refined vapor dispersion model, SLAB.

The results from the storage tank releases indicated that potentially serious ammonia concentrations (greater than 1000 ppm) could result at downwind distances ranging from 150 meters (relief valve malfunction) to approximately 3 kilometers (catastrophic tank failure). The tank failure scenario produced concentrations that could be rapidly fatal (greater than 5000 ppm) out to 1.3 kilometers. Under worst-case meteorological dispersion conditions, recognized exposure limits (IDLH, TLV-STEL) were exceeded for very large distances (greater than 15 kilometers).

5.7.3. CASE III: A DISPERSION MODEL FOR HYDROGEN FLUORIDE AND FLUORINE

Dispersion models for *hydrogen fluoride* and *fluorine* have been developed and integrated into the *Air Force Dispersion Assessment Model ADAM system* (Raj[248], 1990). The thermodynamic aspects of polymerization reaction and dissociation of the chemical vapors, when mixed with air, have been modeled and considered in dispersion calculations. The dispersion results have been compared with test results from the Goldfish Series of field tests. The agreement is good between predicted and measured parameters such as cloud temperature, cloud width, and downwind concentration. The mixing of fluorine with ambient air has been modeled. Dispersion results for fluorine are presented; however, due to the absence of any field data, no verification of predicted results are possible.

5.8. Risk Data Bases and Documentation Centers

The reader will find below some addresses of documentation centers and of databases related to industrial risk.

5.8.1. THE COMMUNITY DOCUMENTATION CENTER ON INDUSTRIAL RISK (CDCIR)

The Commission of European Communities (CEC) has recognized the importance of information as a means for risk management and accident prevention. The *Directive 82/501/EEC* was adopted in 1982. It is known as the *Major Accident Hazards Directive* or the *Seveso Directive*. For systematic diffusion of information concerning the particle implementation of the directive in the twelve EEC member states, relevant documents were centralized in a new documentation center, *the Community Documentation Center on Industrial Risk (CDCIR).*[249, 250] Besides different kinds of reports on safety, emergency plans, fire and explosion, corrosion, fertilizers and gas processing, there is also an accident documentation available for chlorine, petroleum refining, etc. Major accident case histories and the lessons learnt from these are also collected.

5.8.2. MAJOR ACCIDENT REPORTING SYSTEM (MARS)

The Major Accident Reporting System by the Seveso directive (MARS), (see CDCIR above), has been described by Drogaris[251] (1991).

5.8.3. DANISH PRODUCT REGISTER DATA BASE (PROBAS)

The *Danish Product Register Data Base (PROBAS)*[252] is a governmental database for information and evaluations concerning chemical substances, materials and products, which was started in 1980. The product register is located at the Danish National Institute of Occupational Health (DNIOH), which is part of the Danish Labour Inspection. The database contains information on about 53,000 chemical products (January 1991). The technical configuration of PROBAS is a Local Area VAX-cluster. The data base management system is ADABAS using inverted lists for data access with applications written in NATURAL.

5.8.4. DECARA, A MANAGEMENT TOOL FOR SEVERE CHEMICAL ACCIDENTS

A computer code for consequences analysis in chemical installations, DECARA, has been developed.[253] DECARA provides an integrated risk analysis including source-term strength evaluation, estimation of the hazardous cloud dispersion, and quantification of health impacts (QRA†).

The code comprises the following modules: plant damage state module; outflow module with instantaneous release or continuous release at constant rate (liquid outflow model based on the Bernoulli equation, or vapor outflow TNO-model), or a continuous release with a time dependent rate; evaporation module (spreading pool or evaporation); weather module; topography module assuming flat terrain; dispersion module (box-model); dose module (as a function of time); dose-response module (calculates the fatality probability for an individual); risk-assessment module. This last module integrates the partial results of each module into a quantitative risk index. The liquid outflow module is based on the Bernoulli equation which estimates the initial release rate. This model is used for liquid outflows for both pipe breaks and tank breaks. The vapor outflow model is based on the model resented in the Yellow book by TNO[254] (1980).

† QRA: Quantified Risk Assessment

> ***Features of the DECARA Code***
> • *Multiple accidents, each with a certain probability occurrence can be handled.*
> • *Dispersion of heavier as well as lighter-than-air gases released instantaneously or continuously, can be simulated. Time varying release rates are possible.*
> • *Uncertainty analysis can be performed including both parameter and modeling uncertainty. The code calculates the unconditional fatality probability at any point around the site of release. Isorisk curves or maximum individual risk versus distance from the source can be generated.*
> • *The code is written in FORTRAN (Lahey compiler) and requires extended memory to run on an IBM-PC.*
> • *The computer package is portable and available for running on personal computers. Other computer code such as DENZ, CRUNCH, WHAZAN, ZZB, EFFECTS are commercially available, but their QRA possibilities are limited.*

A Gaussian distribution within the plume is assumed for buoyant cloud, and a simple box model is applied to heavy gas clouds: it takes into account the formation of the cloud and initial entrainment of air, the slumping phase, the transition to a passive dispersion phase, and the passive dispersion period. The number of possible scenarios and their corresponding frequencies are determined by the user. References regarding the models cited are given in the paper.

> *DECARA has been tested using ammonia, and compared to the results of the WHAZAN code. There is a very good agreement between the two models. The simulation with ammonia has shown a significant dependence on weather conditions. Air temperature, weather stability category, wind speed directly affect the evaporation rate and the dispersion of ammonia. Furthermore, the wind speed direction affects the expected concentration at each point surrounding the site of installation. All these parameters exhibit a stochastic variability, not only in the value they take, but also on the possible combination of these values.*

5.8.5. RISKMOD, A CONSEQUENCE ASSESSMENT MODEL FOR THE TRANSPORT OF DANGEROUS GOODS

The computer model RISKMOD was developed to provide assistance in the evaluation of policies for the movement of dangerous goods, by estimating objective risk. (Stewart et al.[255], 1990) The model represents individual vehicle shipments of dangerous goods on the truck and rail networks for which the associated risks are estimated in a series of steps. The first step involves an accident rate prediction, followed by a spill rate prediction, given an accident. In subsequent steps, the damages are evaluated of an accident alone, and of an accident followed by a spill of goods. The final output tables provide link by link risk estimates and a summary of the total risk for the entire route.

The code has now been improved. Specifically, the risk associated with the mechanical aspects of an accident are included separately from the risk due to the release of the dangerous cargo. In addition, more detailed truck release data have been included to better reflect the range of consequences which follow a release event. Both modifications assist in providing a more accurate and representative account of the risks associated with transporting dangerous goods.

> *A novel feature of the RISKMOD model is the valuation of consequences of transport accidents, which provides a common denominator in terms of expressing risk, so that all risks may be summed on a transport route.*

5.8.6. HAZARD ASSESSMENT COMPUTER SYSTEM "HACS"

There are a number of hazard assessment systems with varying degree of applicability regarding (chemical conditions, etc.), complexity and cost to operate.[256] Some are government owned and a number of systems are commercially available. Several reviews are available in the literature AIPE[257] (1989), Hanna and Drivas[258] (1987)

An integrated computer system for hazard analysis has been developed for the U.S. Coast Guard (Raj[259] et al., 1990). The system HACS (Hazard Assessment Computer System) consists of state-of-the art compendium of mathematical models describing different sources for chemical releases (pressurized storage tanks, bulk storage tanks, barge, rail car, and road transports). The HACS code (Potts[260], 1981) can also assess hazardous behavior in the environment (two-phase jet flow, instantaneously released vapor cloud, dispersion of heavy gases with liquid aerosols, reactive chemical dispersion in air, jet fires, pool fire, thermal radiation, explosion effects, water dispersion in rivers and streams, etc. *Attach to the system are a physical and chemical property data base (including temperature dependent properties) for over 1200 chemicals.* The code runs on a VAX mainframe based system.

The Original HACS system is applicable only to the case of chemical releases on water. Another modified version of the code runs on both IBM-PC compatible and Micro VAX platforms. These systems are called *SAFEMODE* (Safety Assessment For Effective Management Of Dangerous Events) and *MicroHACS.* While the overall goal of this two new products are very similar to the original HACS system of the U.S. Coast Guard, they differ significantly in features, models, and the range of applicability:

- General information
- Release environment
- Chemical information
- Accident conditions
- Release on/into Land/Water
- Chemical container information
- Model selection
- Tank specifications

The user selects various options through a series of menu driven panels. The main topics are:

- Emergency response
- Contingency planning

- Chemical property value calculations
- Default data modification

The types of hazards evaluated by the system are:

- Fire radiation model
- Explosion models
- Models for the dispersion of vapors in the atmosphere
- Water dispersion models

The dispersion of heavy gas and clouds containing aerosols is based on the heavy gas box model modified by the superposed finite source Gaussian model. This submodel is called *ADAM.*[261]

The computerized systems (HACS, SAFEMODE, etc.) can provide useful, effective and rapid means of estimating the potential hazard areas without too much effort. As with all models the results of these new models have to be viewed with caution and interpreted carefully. They give good estimates of potential hazards which may be used for planing or response guides.

Summary (Chapter 5)

This chapter deals with a review of various atmospheric dispersion models involving gases for the major release mechanisms that may result in the quasi-instantaneous formation of a cloud depending on the mode of storage. Various types of models are included (e.g., box models, K-theory models, K-theory models, 3-D time dependent conservation models. A number of practical conclusions to the risk engineering specialist on the use of heavy gas dispersion models are included. Some procedures for estimating the uncertainties of hazardous gas models are considered jointly with the comparison of field experiment data with model predictions. Presentation of some reports and studies on accidental release of toxic vapor clouds, available data bases, computer software and documentation centers are included in this chapter.

References (Chapter 5)

[1] Hanna, S.R., Review of atmospheric diffusion models for regulatory applications, Technical Note No. 177, Rep. No. WMO 581. Secretariat of the World Meteorological Organization, Geneva - Switzerland, (1982).

[2] Turner, D.B., Atmospheric dispersion modeling: A critical review, J. Air Poll. Control Assoc. 29 (1979) 502-519.

[3] Simpson, J.E., Gravity currents in the laboratory, Atmosphere and Oceanic Ann. Rev. Fluid mech., 14 (1982) 213-234.

[4] Van Ulden, A.P., The spreading and mixing of dense gas clouds in still air, Thesis of the Technische Universiteit Delft, The Netherlands, (1988).

[5] Picknett, R.G.; Carpenter, R.J., Field Experiments on the Behavior of Dense Clouds, Chemical Defense Establishment, Porton Downs, UK, September 1978, Rep. No. Ptn. IL 1154/78/1.

[6] Fannelop, T.K., The Dynamics of Heavy Gas Clouds, Division of Aero and Gas Dynamics, University of Trondheim, Norway, May 1980, Report IFAG B-124.

[7] Fannelop, T.K.; Jacobsen, O., Gravitational spreading of heavy gas clouds instantaneously released, Z. Angew. Math. Phys., 35 (1984) 559-584.

[8] Havens, J.A.; Spicer, T.O., Gravity spreading and air entrainment by heavy gases instantaneously released in a calm atmosphere, in Ooms and Tennekes (eds.), Atmospheric Dispersion of Heavy Gases and Small Particles, Springer Verlag, (1984), pp. 179-189.

[9] Kaiser, G.D.; Walker, B.C., Releases of anhydrous ammonia from pressurized containers - The importance of denser-than-air mixtures, Atmospheric Environment, 12 (1978) 2289.

[10] Havens, J.A., An assessment of the predictability of LNG vapour dispersion, Journal of Hazardous Materials, 3, (1980) 257.

[11] Havens, J.A., An Assessment of predictability of LNG vapor dispersion from catastrophic spills onto water, J. Hazard. Mater., 3 (1980) 267-278.

[12] Webber, D.M., The Physics of Heavy Gas Cloud Dispersal. Safety and Reliability Directorate, UKAEA, Culchetch, GB, Rep. No. SRD R243, (1983).

[13] Mc Quaid, J., The box model of heavy gas dispersion: a useful and practical tool, J. Occ. Acc., 6 (1984) 253-261.

[14] Van Ulden, A.P., On the spreading of heavy gas released near the ground. Proc. Int. Loss Prevention Symp., Ed.: C.H. Buschman, Elsevier, Amsterdam, (1974), pp. 221-226.

[15] Cox, R.A.; Carpenter, R.J., Further development of a dense vapour cloud dispersion model for hazard analysis, Proc. of the Symposium on Heavy Gas Dispersion, Frankfurt, (1979).

[16] Raj, P.K., On the near Field dispersion of Heavy Vapor/Aerosol Plume Generated by the Release of Pressurized Liquid Ammonia, Interim Report to U.S. Naval Weapons Center, China Lake, California, November 1980.

[17] Raj, P.K.; Aranamuden, K., Theoretical Models Supporting the Design of Ammonia Spill Experiments, Report to the Fertilizer Institute, Washington, DC, (1980).

[18] Jagger, S.F., Development of CRUNCH: a dispersion model for continuous release of a denser-than-air vapour in the atmosphere, UKEA Rep. No. SRD R229, (1983).

[19] Colenbrander, G.W., A mathematical model for the transient behaviour of dense vapour clouds, 3rd International Symposium, Loss Prevention and Safety Promotion in the Process Industries, Basel, Switzerland, (1980).

[20] Havens, J.A., A review of mathematical models for prediction of heavy gas atmospheric dispersion, I. Chem. E. Symposium Series No. 71, 14 - 16 April 1982.

[21] Wheatley, C.J.; Webber, D.M., Aspects of the Dispersion of Denser-than-Air Vapours Relevant to Gas Cloud Explosions, SRD Report to the European Atomic Energy Community, (1984).

[22] Raj, P.K., Atmospheric Dispersion of Chemical Agents: A Model Including Thermodynamic and Heavy Gas Effects. Report to the U.S. Army by Technology & Management Systems, Burlington, Mass., (1983).

[23] Havens, J.A., A review of mathematical models for prediction of heavy gas atmospheric dispersion, International Chemical Symposium, Serial No.71, Manchester, England, (1982).

[24] Carslaw, H. S.; Jaeger, J.C., Conduction of Heat in Solids, 1st edition, Clarendon Press, Oxford, (1947).

[25] Rottman, J.W.; Simpson, J.E.; Hunt, J.C.R.; Britter, R.E., Unsteady gravity current flows over obstacles - Some observation and analysis related to the phase II trials, Proc. of Symp. on Heavy Gas Dispersion Trials at Thorney Island, Univ. of Sheffield 3-5 April 1984, J. Hazard Mater., (1985).

[26] Havens, J.A., A Description and Assessment of the SIGMET LNG Vapor Dispersion Model, U.S. Coast guard, Report CG-M-3-79, February 1979.
[27] Havens, J.A., A description and computational assessment of the SIGMET LNG vapor dispersion model, J. Hazard. Mater, 6(1) (1982) 197-212.
[28] Jagger, S.F., The dispersion of dense, toxic, and flammable clouds in the atmosphere, I. Mech. E. Seminar, September 1982.
[29] Robins, A.G., Introduction to the Numerical Modelling of Dispersion in the Atmospheric Boundary Layer, in "von Karman Inst. For Fluid Dynamics", Introduction to Numerical Solution to Industrial Flows, Vol. 1 Leatherhead (England), (1986).
[30] Blackmore, D.R.; Herman, M.N.; Woodward, J.L., Heavy gas dispersion models, J. Hazard. Mater., 6(1-2), (1982) 107-128.
[31] Fannelop, T.K.; Waldman, G.D., The dynamics of oil slicks - or "creeping crude", 9th Aerospace Science Meeting A.I.A.A. , New York, January 1971. See also The dynamics of oil slick, A.I.A.A. Journal, 10 (1972) 506-510.
[32] Van Ulden, A.P., The unsteady gravity spread of a dense gas cloud in a calm environment, in Proceedings of 10th. I.T.M. on Air Poll. Mod. and its Appl., NATO-CCMS, Rome, (1979).
[33] Byggstoyl, S.; and Saetran, L.R., An Integral Model for gravity Spreading of Heavy Gas Clouds, Atm. Env. 17(9), (1983) 1615-1620.
[34] Van Ulden, A.P., A new bulk model for dense gas dispersion: two-dimensional spread in still air, in Atm. Dispersion of Heavy Gases and Small particles, in G. Ooms and H. Tennekes, (eds.), Springer Verlag, pp. 419-440, (1984).
[35] Havens, J.A.; Spicer-T.O., Development of an Atmospheric Dispersion Model for Heavier-Than-Air Gas Mixtures. Volume 1. Final rept. Sep 80-May 85. Arkansas Univ., Fayetteville. Dept. of Chemical Engineering, Rep. No. USCGD2285.
[36] Jagger, S.F.; Kaiser, G.D., The accidental release of dens flammable and toxic gases from pressurized containment - transition from pressure driven to gravity driven phase, Proc. 11th Int. Tech. Meeting on Air Pollution Modelling and its Apllications, Amsterdam, NATO, (1981).
[37] Thaning, L.; Winter, S.; Nyren, K., Uppkomst och Utbredning av Explosiva och Giftiga Gasmoln: Inventering av Kunskapslaege och Forskningsbehov (Formation and Spreading of Flammable and Toxic Gas Clouds: A Survey of the Current Knowledge and Need for Research). Rep. No. FOAE40036, (1988).
[38] Kaiser, G.D., Examples of the successful application of a simple model for the atmospheric dispersion of dense, cold vapours to the accidental release of anhydrous ammonia from pressurized containers, UKAEA Rep. No. SRD R150, (1979).
[39] Griffiths, R.F.; Kaiser, G.D., The accidental release of anhydrous ammonia to the atmosphere - A systematic study of factors influencing cloud density and dispersion, UKAEA Rep. No. SRD R154, (1979).
[40] Kaiser, G.D., The accidental release of anhydrous ammonia to the atmosphere - Evidence for the occurence of denser-than-air-mixtures, Loss Prevention Bull., 38 (1981) 1-22.
[41] Havens, J.A., A review of mathematical models for prediction of heavy gas atmospheric dispersion, I. Chem. E. Symposium Series No. 71, Manchester, 14-16 April 1982.
[42] Van Ulden, A.P., On the spreading of heavy gas released near the ground, First International Loss Symposium, The Hague, Netherlands, (1974).
[43] Germeles, A.E.; Drake, E.M., Gravity spreading and atmospheric dispersion of LNG vapor clouds, Fourth International Symposium on Transport of Hazardous Cargo by Sea and Inland Waterways, Jacksonvillle, Florida, (1975).
[44] Eidsvik, K.J., A Model for Heavy gas dispersion in the atmosphere, Atmospheric Environment, 14 (1980) 769-777.
[45] Cox, R.A.; Carpenter, R.J., Further development of a dense vapor cloud dispersion model for hazard analysis, in S. Hartwig, D. Reidel (eds.), Heavy Gas and Risk Assessment, Dordrecht, Holland, (1979).
[46] Te Riele, P.H.M., Atmospheric dispersion of heavy gases emitted at or near ground level, 2nd International Symposium on Loss Prevention and Safety Promotion in the Process Industries, Heidelberg, (1977).
[47] Flothmann, D.; Nikodem, H.J., A heavy - gas dispersion model with continuous transition from gravity spreading to tracer diffusion, in S. Hartwig. D. Reidel (eds.), Heavy Gas Risk Assessment, Dordrecht, Holland, pp. 89-102 (1980).
[48] Ooms, G.; Mathieu, A.P.; Zelis, F, Plume paths of heavy gases, First Loss Prevention Symposium, The Hague, Netherlands, (1974).

[49] Havens, J.A., A Description and Assessment of the SIGMET LNG Vapor Dispersion Model, U.S. Coast Guard, Rep. No. CG-M-3-79, February 1979.
[50] Energy Resource Company, La Jolla, California.
[51] Deygon-Ra, La Jolla, California.
[52] Te Riele, P.H.M., Atmospheric dispersion of heavy gases emitted at or near ground level, Second International Symposium on Loss Prevention and safety Promotion in the Process Industries, Heidelberg, Germany, (1977).
[53] Zeman, O., The Dynamics and Modeling of Heavier-than-Air, Cold Gas Releases, Lawrance Livermore Laboratory (University of California), Rep. No. UCRL-15224, April 17, (1980).
[54] Rosenzweig, J.J., A Theoretical Model for the Dispersion of Negatively Buoyant Vapour Clouds, Ph.D. Dissertation, Massachusetts Institute of Technology, (1980).
[55] Havens, J.A., An assessment of predictability of LNG vapor dispersion from catastrophic spills onto water, J. Hazard. Mater., 3 (1980) 267-278.
[56] Deaves, D.M., 3-Dimensional model predictions for the upwind building trial of Thorney Island Phase II, J. Hazard. Mater., 11 (1985) 341-346.
[57] Jacobsen, O., 3-D numerical simulation of heavy gas dispersion, J. Hazard. Mater., 16 (1987) 215-230.
[58] Havens, J.A., An assessment of predictability of LNG vapor dispersion from catastrophic spills into water, Paper C2, Proc. 5th Int. Symp. on the Transport of dangerous goods by Sea and Inland Waterways, Hamburg, F.R.G., (1978).
[59] Schnatz, G.; Flothmann, D., A K-model and its modification for the dispersion of heavy gases, in Sylvius Hartwig, (ed.), Heavy Gas Risk Assessment, Proc. Symp., (1979), Reidel, Dordrecht, Netherlands, pp. 125-39 (1980).
[60] Deaves, D.M., Application of a turbulence flow model to heavy gas dispersion in complex situations, in Sylvius Hartwig (ed.), Heavy Gas Risk Assessessment II, Proc. Symp., 2nd Meeting 1982, Reidel, Dordrecht, Netherlands, pp. 91-102, (1983).
[61] Deaves, D.M., Application of advanced turbulence models in determining the structure and dispersion of heavy gas clouds, in Ooms, G. and H. Tennekes (eds.), Atmospheric Dispersion Heavy Gases Small Part., Symp., 1983, pp. 93-103. Springer Verlag, Berlin, Federal Republic Germany, (1984).
[62] Knox, J.N., The modeling of dispersion of heavy gases, NATO Challenges Mod. Soc., 5 (Air Pollution Model. Its Appl.), pp. 285-94, (1984).
[63] Tasker, M.N., A review of the basic concepts of dense gas dispersion with special regard to modelling of heat transfer, Sci. Tech. Aerosp. Rep. 22(16) (1984), Abstr. No. N84-25952, (1984).
[64] Tasker, M., Preliminary Wind Tunnel Experiments to Investigate the Effect of Heat Transfer on the Dispersion of Cold Dense Gases, Oxford Univ. (England). Dept. of Engineering Science, (1984).
[65] Farmer, C.L., A survey of turbulence models with particular reference to dense gas dispersion, Rep. No. SRD R221, Safety and Reliability Directorate, UKAEA, Culcheth, (1982).
[66] Raj, P.K., Heavy Gas Dispersion - A State-of-the-Art Review of the experimental Results and Models, Heavy Gas Dispersal Lecture Series, 1982-83, Von Karman Institute, Belgium, (1982).
[67] Woodward, J.L.; Havens, J.A., et al., A comparison with experimental data of several models for dispersion of heavy vapor clouds, J. of Hazardous Materials, 6 (1982) 161-180.
[68] Wheatley, C.J.; Webber, D.M., Aspects of the Dispersion of Heavier-than-air Vapours that are of Relevance to Gas Cloud Explosions, Rep. No. 007 SRUK, Commission of the European Communities (DG XII), Brussels, (1984).
[69] Wheatley, C.J.; Webber, D.M., Aspects of the dispersion of denser-than-air vapours relevant to gas cloud explosions, Rep. No. SR/ 007/UK/H Final Report, Commission of the European Communities (DG XII), Brussels, (1984).
[70] Gaffen, D.J.; Benocci-C.; Olivari-D., Application of a Lagrangian Dispersion Model to Environmental Problems, Von Karman Institut for Fluid Dynamics, Rhode-Saint-Genese (Belgium). Rep. No. VKITM38, (1985).
[71] Kunkel, B.A., Development of an Atmospheric Diffusion Model for Toxic Chemical Releases, Environmental Research Papers, Oct 84-Sept. (1985), Air Force Geophysics Lab., Hanscom AFB, MA; Rep. No. AFGLTR850338, AFGLERP939.
[72] Frayne, R., Heavy gas dispersion: Applied safety, in R.V. Portelli (ed.), Proceedings of the Heavy Gas (LNG/LPG) Workshop, pp. 108-116, (1985). Rep.No. CONF-8501127--, CE--03673, CSC--CE303673.

[73] McQuaid, J., Overview of current state of knowledge on heavy gas dispersion and outstanding problems/-issues, in R.V. Portelli (ed.), Proceedings of the Heavy Gas (LNG/LPG) Workshop, (1985), pp. 5-28. Rep. No. CONF-8501127--, CE--03673, CSC--CE303673.
[74] Raj, P.K., Summary of heavy gas spills modeling research, in R.V. Portelli (ed.), Proceedings of the Heavy Gas (LNG/LPG) Workshop, (1985), pp. 51-75. Rep. No. CONF-8501127--, CE--03673, CSC--CE303673.
[75] Redondo, J.M., Effects of ground proximity on dense gas entrainment, J. Hazard. Mater., 16 (1987) 381-393.
[76] Webber, D.M.; Wheatley, C.J., The effect of initial potential energy on the dilution of a heavy gas cloud, J. Hazard. Mater., 16 (1987) 357-380.
[77] Dunst, M.; Fischer-Bruns, I.; Luksch, U., Box Model Describing the Dispersion of Heavy Gases after Accidents and Its Verification by Experiments, Meteorology Inst., Hamburg Univ. (Germany, F.R.), Rep. No. NP8770246, (1987).
[78] Vanulden, A.P., Spreading and Mixing of Dense Gas Clouds in Still Air, Doctoral thesis, Royal Netherlands Meteorological Inst., (1987). Rep. No. KNMIWR8712.
[79] Fischer-Burns, I., Untersuchungen Zur Ausbreitung Spontan Freigesetzter Schwerer Gase, Meteorology Institut, Hamburg University (Germany, F.R.), Rep. No. WA81, ETN8790373, (1987).
[80] Schreurs, P.; Mewis, J., Numerical aspects of Lagrangian particle model for atmospheric dispersion of heavy gases, J. Hazard. Mater., 17 (1987) 61-80.
[81] Tasker, M.N., Effect of Heat Transfer on the Dispersion of Cold Dense Gases, Oxford University, England, Thesis, (1987). Rep. No. OUEL168887, ETN8790974.
[82] Mathematical models of the spreading of liquefied hydrocarbons and of the atmospheric dispersion of heavy gases, Sherbrooke, P.Q., Univ. of Sherbrooke, Thesis (July 1987). Availability: MF National Library of Canada, 395 Wellington St., Ottawa, ON, CAN K1A 0N4.
[83] Ausbreitungsrechnungen im Rahmen des Vollzugs der Stoerfall-Verordnung. Colloquium on Dispersion Evaluations in the Framework of the Implementation of the Nuclear Accident Ordinance, Muenchen (Germany, F.R.), 23-24 Nov. 1987, Umweltbundesamt - Texte, no. 1/89.
[84] Ermak, D.L.; Merry, M.H., Methodology for Evaluating Heavy Gas Dispersion Models: Final Report, November 1985 - February 1988, Lawrence Livermore National Lab., CA; Rep. No. AFESCESLTR8837.
[85] Rodean, H.C., Toward more realistic material models for release and dispersion of heavy gases, Rep. No. UCRL--53902, (1989).
[86] Ermak, D.L., Atmospheric dispersion models for dense gas releases, International System Safety Society (SSS), 10th conference, Dallas, TX, USA, 18-22 July (1991), Lawrence Livermore National Lab., CA; Rep. No. UCRLJC107536, CONF91071143.
[87] Touma, J.S.; Guinnup, D.; Spicer, T., Guidance on the Application of Refined Dispersion Models for Air Toxics Releases, Final report, Rep. No. EPA450491007, (1991).
[88] Ermak, D.L., Concept for treating dense-gas dispersion under realistic conditions of terrain and variable winds, Lawrence Livermore National Lab., CA. Rep. No. UCRLJC104039, CONF90062101, (1990).
[89] Ermak, D.L.; Lange-R., Treatment of denser-than-air releases in an advection-diffusion model: Thermodynamic effects, 84th Annual meeting and exhibition of the Air and Waste Management Association (AWMA), Vancouver, Canada, 16-21 June (1991), Lawrence Livermore National Lab., CA; Rep. No. UCRL-JC--106798, CONF-910659--12, (1991).
[90] Matthias, C.S., Dispersion of dense cylindrical cloud in calm air, J. Hazard. Mater., 24 (1990) 39-65.
[91] Bidokhti, A.A., A numerical model of heavy gas dispersion, Proceedings of the 8. Brazilian Meeting on Reactor Physics and Thermal Hydraulics, (1991), pp. 87-90. Rep. No. CONF-910983--, INIS-BR--2846.
[92] Deaves, D.M.; Hall, R.C., The effects of sloping terrain on dense gas dispersion, J. Loss Prev. Process Ind. 3 (1990) 142-145.
[93] Nikmo, J.; Kukkonen, J., Modelling of heavy gas cloud advection in complex terrain, Finnish Meteorological Institute, Helsinki, (preprint 1991).
[94] Webber, D.M.; Jones, S.J.; Martin, D., A model of the motion of a heavy gas cloud released on a uniform slope, J. Hazard. Mater., 33 (1993) 101-122.
[95] Pape, R.; Nussey, C., A basic approach for the analysis of risk for toxic risks, The Assessment and Control of Major Hazards, I. Chem. Eng. Symposium, 22-24 April 1985.
[96] Webber, D.; Brighton, P., A mathematical model of a spreading vaporising liquid pool, 3rd Symposium on Heavy Gases and Risk Assessment, Bonn, November 1984.

[97] Shaw, P.; Briscoe, F., Evaporation from spill of hazardous liquids on land and water, UKAEA Rep. No. SRD R100, (1978).
[98] Chay, H.R.; Reid, R.C., Spreading boiling model for instantaneous spills of liquified petroleum gas (LPG) on water, J. of Haz. Materials, 7 (1982) 19-35.
[99] Hall, S.F., A simple homogeneous equilibrium critical discharge model applied to multi-component, two-phase systems - The computer programs CRITS and CRITTER, UKAEA Rep. No. SRD R127, (1978).
[100] Cox, R.A.; Roe, D.R., A model of the dispersion of dense vapour clouds. 2nd International Symposion on Loss Prevention and Safety Promotion in the Process Industries, Heidelberg, (1977), p. 359-.
[101] Niggli, S., DispTool - Part I: User Manual and Part II: Theory Manual, (1992). Available from Swiss Reinsurance Company, Risk Management Service Center, Mythenquai 50/60, Zürich, P.O. Box, CH-8022, Switzerland.
[102] Schiegl, W.E.; and Schorling, M., TA Luft, Vorschriften und Erläuterungen zum Immissionschutz (Ecomed Verlagsgesellschaft mbH, 8910 Landsberg/Lech, BRD, (1986).
[103] Schnatz, G.; Rohbock, E., Dispersion of Heavy Gases - Experiments and Models, VDI-Bericht, 558 (1986) 143-66.
[104] Fryer, L.S.; Kayser, G.D., DENZ A computer program for the calculation of the dispersion of dense toxic or explosive gases in the atmosphere, UKAEA Rep. No. SRD R 152, July 1979.
[105] McQuaid, J., Objectives and Design of the Phase I Heavy Gas Dispersion trials, J. Hazard. Mater., 22 (1985) 13.
[106] Brighton, P.W.M.; Prince, A.J.; Webber, D.M., Determination of cloud area and path from visual and concentration records, J. Hazard. Mater., 11 (1985) 155-178.
[107] Fryer, L.S.; Kayser, G.D., DENZ A computer program for the calculation of the dispersion of dense toxic or explosive gases in the atmosphere, UKAEA Rep. No. SRD R 152, July 1979.
[108] Jagger, S.F., The application of the computer code DENZ, U.K. Atomic Energy Authority, Safety & Reliability Directorate, SRD, Report No. SRD R 277, (1985).
[109] Crabol, B.; L'Homme-V.; Roux-A., Interpretation of the Thorney Island Phase 1 Trials with the Box Model Cigale2. Symposium on Analysis and Interpretation of Results of the Thorney Island Trials, Sheffield, UK, 23 Sept. 1986, CEA Centre d'Etudes Nucleaires de Fontenay-aux-Roses (France), Dept. d'Analyse de Sûreté, (1986).
[110] Alp, E., COBRA: An LNG (liquefied natural gas) model. Proceedings of the Heavy Gas (LNG/LPG) Workshop, (1985), pp. 76-92. Rep. No. CONF-8501127--, CE--03673, CSC--CE303673.
[111] Alp, E.; Matthias,-C.S., COBRA. A heavy gas/liquid spill and dispersion modelling system. Journal of Loss Prevention in the Process Industries, United Kingdom, 4(3) (1991) 139-150.
[112] Witlox, H.W.M.; Puttock, J.S.; Colenbrander, G.W., HEGADAS: Heavy Gas Dispersion Program, User's Guide, Shell Research Ltd., Chester (England), (1988). Rep. No. EPASWDK89027A.
[113] Marsden, A.; Guinnup, D., HEGADAS: Heavy Gas Dispersion Model (for Microcomputers), Model-Simulation. Rep. No. EPASWDK89027, (1989).
[114] Singh, M.P.; Mohan, M.; Panwar, T.S.; Chopra, H.V., Estimation of vulnerable zones due to accidental release of toxic materials resulting in dense gas clouds, in Risk-Analysis, United-States, (1991).
[115] Singh, M.P., Vulnerability analysis of airborne accidental release of toxic chemicals, in B.W. Gay, Jr. (ed.), EPA/AWMA International Symposium on Measurement of Toxic and Related Air Pollutants. Environmental Protection Agency, Washington, DC, USA, (1991); See also Rep. No. CONF-8907103--(1989), LA--11728-C.
[116] Havens, J.A., A Description and computational assessment of the "SIGMET LNG Vapor Dispersion Model", J. of Haz. Materials, 6 (1982) 181-195.
[117] Havens, J.A., A description and assessment of the SIGMET LNG vapor dispersion model, U.S. Coast Guard, Rep. No. CG-M-3-79, Feb. 1979.
[118] Smith, T.B., Howard, S.M., Methodology for treating diffusivity, Final Report to Systems, Science and Software, LaJolla, CA, by Meteorology Research, Inc., Altadena, CA (MRI 72 FR-1030).
[119] Havens, J., A dispersion model for elevated dense gas jet chemical releases, Volume II. User's guide, U.S. Environ. Prot. Agency, Office Air Quality Planning Standards, Rep. No. EPA, EPA-450/4-88-006b, (1988). See also Volume 1, PB88-202387.
[120] Guinnup, D., Dispersion Model for Elevated Dense Gas Jet Chemical Releases (Ooms/DEGADIS) (for Microcomputers), Environmental Protection Agency, Research Triangle Park, NC. Office of Air Quality Planning and Standards, (1988). Rep. No. EPASWDK88048 (Software).

[121] Havens, J.A.; Spicer-T.O., Development of an Atmospheric Dispersion Model for Heavier-Than-Air Gas Mixtures, Volume 1. Final rept. Sep 80-May 85 (AD-A171 522), Arkansas Univ., Fayetteville, Dept. of Chemical Engineering. Rep. No. USCGD2285.

[122] Havens, J.A.; Spicer, T.O., Development of an Atmospheric Dispersion Model for Heavier-Than-Air Gas Mixtures, Volume 2 (AD-A171 523), Laboratory Calm Air Heavy Gas Dispersion Experiments, Dept. of Chemical Engineering, Arkansas Univ., Fayetteville. Final report, Sept. 80-May 85.

[123] Havens, J.A.; Spicer, T.O., Development of an Atmospheric Dispersion Model for Heavier-Than-Air Gas Mixtures, Volume 3, DEGADIS User's Manual, Dept. of Chemical Engineering, Arkansas Univ., Fayetteville, Final report, Sept. 80-May 85 (AD-A171 524).

[124] Spicer, T.; Havens, J. A., User's guide for the DEGADIS 2.1 dense gas dispersion model, U.S. Environ. Prot. Agency, Off. Air Qual. Plann. Stand., EPA, Rep. No. EPA-450/4-89-019, (1989).

[125] Hofmann, J.E., Dense gas dispersion modeling on the IBM-PC, in Proc. - APCA Annu. Meet., 81st(8), Paper 88/146.7, (1988).

[126] Wheatley, C.J.; A user's guide to TRAUMA - A computer code for assessing the consequences of accidental two-phase releases of NH3 into moist air, UKAEA Rep. No. SRD/HSE R394, London, (1987).

[127] LNG Vapor Dispersion Prediction with the DEGADIS Dense Gas Dispersion Model (for Micro-computers). Model-Simulation. Gas Research Inst., Chicago, IL. Rep. No. GRISWDK91002(1991).

[128] Havens, J.A.; Spicer, T.O., LNG Vapor Dispersion Prediction with the DEGADIS Dense Gas Dispersion Model, Topical Report, April 1988 - July 1990, Dept. of Chemical Engineering, Arkansas Univ., Fayetteville. Rep. No. GRI890242, GRISWDK91002A.

[129] Spicer, T.O.; Havens, J.A., Development of Vapor Dispersion Models for Nonneutrally Buoyant Gas Mixtures Analysis of USAF/N2O4 Test Data, Dept. of Chemical Engineering, Arkansas Univ., Fayetteville, Final report, 1 Feb. - 31 Jul. 85, (1986). Rep. No. AFESCESLTR8624.

[130] Ermak, D.L., Denser-Than-Air Dispersion Modeling in the Atmosphere, JANNAF Safety and Environmental Protection Meeting, Livermore, CA, USA, 8th March 1983, Rep. No. UCRL88782, CONF8303213.

[131] Morgan, D.L.; Morris-L.K.; Ermak-D.L., SLAB: A Time-Dependent Computer Model for the Dispersion of Heavy Gases Released in the Atmosphere. Lawrence Livermore National Lab., CA, (1983). Rep. No. UCRL53383.

[132] Chan, S.T., FEM3: A Finite-Element Model for the Simulation of Heavy-Gas Dispersion and Incompressible Flow, User's Manual. Lawrence Livermore National Lab., CA, (1983). Rep. No. UCRL53397.

[133] Chan, S.T., FEM3; Heavy Gas Dispersion Incompressible Flow, Lawrence Livermore National Lab., CA. (1985). Rep. No: ESTSC--000198CY00100; NESC--9903, (Software).

[134] Rodean, H.C.; Chan, S.T., Generalized phase-change submodels in the FEM3 model for gas dispersion, Rep. No. UCID-20817, Energy Res. Abstr. 11(18), Abstr. No. 41571, (1986).

[135] Ermak, D.L.; Chan-S.T., FEM3 Dispersion Calculations: Evaluation of Turbulence Submodel. Lawrence Livermore National Lab., CA, (1986). Rep. No. UCID20933.

[136] Gresho P.M.; Chan, S.T.; Upson, C.D.; Lee R.L., A modified finite element method for solving the time-dependent, incompressible Navier-Stokes equations: Part 1, Theory; Part 2, Applications, (1984).

[137] Chan, S.T., FEM3A: A Finite Element Model for the Simulation of Gas Transport and Dispersion: User's Manual. Lawrence Livermore National Lab., CA, (1988). Rep. No. UCRL21043.

[138] Chan, S.T.; Gresho-P.M., Ensuring mass conversion in a heavy-gas dispersion model using the generalized anelastic equations. Lawrence Livermore National Lab., CA, (1991). American Institute of Astronautics and Aeronautics/American Society of Mechanical Engineers (AIAA/ASME) National Fluid Dynamics Congress, Los Angeles, CA, 15-18 Jun 1992. Rep. No. UCRL-JC--107535; CONF-920605--3.

[139] ARCHIE, Handbook of Chemical Hazard Analysis Procedures, Federal Emergency Management Agency (FEMA), U.S. Department of Transportation, U.S. Environmental Protection Agency, (1987).

[140] Raj, P.K.; Morris, J.A., Source Characterization and Heavy Gas Dispersion Models for Reactive Chemicals, Vol. I, (1987). Rep. AD-A200121, AFGL-TR-88-0003 (I) From: Gov. Rep. Announce. Index (U.S.) (1989), 89(5), Abstr. No. 911,263 (1987). See also volume 2, AD-A192 209.

[141] Ille, G.; Springer, C., The Evaporation and Dispersion of Hydrazine Propellant From Ground Spills, CEEDO-TR-78-30, AD A059407, (1978).

[142] Meroney, R.N., Transient characteristics of dense gas dispersion. Part I: a depth-averaged numerical model, J. Hazard. Mater., 9(2), (1984) 139-57.

[143] Meroney, R.N., Transient characteristics of dense gas dispersion. Part II: numerical experiments on dense cloud physics, J. Hazard. Mater., 9(2), (1984) 159-70.
[144] Wheatley, C.J.; Webber, D.M., Aspects of the dispersion of denser-than-air vapours relevant to gas cloud explosions, Commission of the European Communities, Brussels. Rep. No. EUR 9592, (1984).
[145] Davies, D.M., Development and application of heavy gas dispersion models of varying complexity, J. Hazard. Mater., 16 (1987) 427-438.
[146] Schatzmann, M., Experimental determination of the effect of building construction on the spread of heavy gases. Final report. Rep.No. BMFT 13RG8603, (1990).
[147] Schatzmann, M.; Marotzke, K., Dispersion modeling within risk assessment studies, NATO Challenges Mod. Soc., 15 (Air Pollut. Model. Its Appl. 8), (1991), pp. 639-42.
[148] Ermak, D.L., Dense-gas dispersion advection-diffusion model. JANNAF Safety and Environmental Subcommittee meeting, Monterey, CA (United States), 10-14 Aug. 1992, Lawrence Livermore National Lab., CA. Rep. No. UCRLJC109697, CONF92081122.
[149] Gudiksen, P.H.; Edwards, L.L.; Ermak, D.L.; Leone, J.M., LLNL atmospheric dispersion model developments in support of emergency response, Topical Meeting on Emergency Preparedness and Response (3rd), Chicago, IL (USA), 16-19 Apr. (1991). Lawrence Livermore National Lab., CA. Rep. No. UCRLJC106282, CONF9104342.
[150] Ermak, D.L., User's manuai for SLAB: An atmospheric dispersion model for denser-than-air-releases. Lawrence Livermore National Lab., CA, (1990). Rep. No. UCRL-MA--105607.
[151] Puttock, J.S., Comparison of Thorney Island data with predictions of HEGABOX / HEGADAS, J. Hazard. Mater., 16 (1987) 439-455.
[152] Riou, Y., Comparison between MERCURE-GL code calculations, wind tunnel measurements and Thorney Island field trials, J. Hazard. Mater., 16 (1987) 247-265.
[153] Nikmo, J.; Kukkonen, J., Modelling heavy gas cloud advection in complex terrain, in M. Kulmala, and J. Kukkonen (eds.), Workshop on the Atmospheric Chemistry and Physics, (1990), pp. 36-41. Rep. No. CONF-9008169--, LTKK-EN--D19.
[154] Kukkonen, J.; Nikmo, J., Modeling heavy gas cloud transport in sloping terrain, J. Hazard. Mater., 31(2), (1992) 155-76.
[155] Webber, D.M.; Kukkonen, J.S., Modelling two-phase jets for hazard analysis, J. Hazard. Mater., 23 (1990) 167-182.
[156] Kukkonen, J., Modelling source terms for the atmospheric dispersion of hazardous substances, in Commentationes Physico Mathematicae Societas Scientiarum Fennica, Finland, 115 (1990)1-111.
[157] Grand, D.; Villand, M.; Silveira, A., Numerical simulation of the dispersion of pollutants with the three-dimensional code Trio, in Advanced Modelling and Computer Codes for Calculating local-Scale and Meso-Scale Atmospheric Dispersion of Radionuclides and their Applications, Organisation for Economic Co-Operation and Development. Paris (France), (1991).
[158] Sedefian, L.; Benett, E., A comparison of turbulence classification schemes, Atmos. Environ., 14 (1980) 741-750.
[159] Chaudhry, F.H.; Meroney, R.N., A laboratory study of diffusion in a stably stratified flow, Atm. Env., 74 (1973) 443.
[160] Martin, J.E., The correlation of wind tunnel and field measurements of gas diffusion using Kr-85 as tracer, Ph.D. Thesis, MMPP 272, University of Michigan, (1965).
[161] Ysiumov, N.; Jondali, T.; Davenport, A.G., Model studies and the prediction of full-scale levels of stack gas concentration, J. of APCA, 26(10), (1976) 956.
[162] Kothari, K.M.; Meroney, R.N.; Bouwmeester, R.J., An algorithm to estimate field concentrations in the wake of power plant complexes under nonsteady meteorological conditions from wind-tunnel experiments, J. Applied Meteorology, 20(8), (1981) 934-943.
[163] Yingst, J.C.; Swanson, R.N.; Mooney, M.L., et al., Review of five wind-tunnel modeling results in complex terrain, 5th Symp. on Turbulence, Diffusion, and Air Pollution, Atlanta, Georgia, (1981), pp. 148.
[164] Weil, J.C.; Cermak, J.E.; Petersen, R.L., Plume dispersion about the windward side of a hill at short range: Wind tunnel versus measurement, 5th Symp. on Turbulence, Diffusion, and Air Pollution, Atlanta, Georgia, (1981), pp. 159.
[165] Andreiev, G.; Neff, D.E.; Meroney, R.N., Heat Transfer Effects during Cold Dense Gas Dispersion. Final report, Aug. 82 - Sept. 1983, Colorado State University, Fort Collins, Dept. of Civil Engineering; Rep. No. CER8384GADENRNM3, GRI830022.

[166] Hall, D.J.; Hollis, E.J.; Ishag, H., Wind Tunnel Model of the Portion Dense Gas Spill Field Trials, Warren Spring Lab., Stevenage, England, (1982). Rep. No. LR394AP, ISBN0856242764.

[167] Hall, D.J.; Waters, R.A., Wind tunnel model comparisons with the Thorney Island dense gas release fiels trials, J. Hazard. Mater., 11 (1985) 209-235.

[168] Van Heugten, W.H.H.; Duijm, N.J., Some findings based on wind tunnel simulation and model calculations of Thorney Island trials No. 008, J. Hazard. Mater., 11 (1985) 409-416.

[169] Baynes, C.J., Calculation of near-field concentrations of hydrogen sulphide, (1985). Rep. No. INFO--0163.

[170] Krogstad, P.A.; Pettersen, R.M., Windtunnel modelling of a release of a heavy gas near a building, Atmos. Environ., 20(5), (1986) 867-78.

[171] Britter, R.E., Fluid modeling of dense gas dispersion over a ramp, J. Hazard. Mater., 18 (1988) 37-67.

[172] Hall, D.J.; Waters, R.A., Investigation of Two Features of Continuously Released Heavy Gas Plumes. Warren Spring Lab., Stevenage, England, (1989). Rep. No. LR707PAM, ISBN0856245658.

[173] Neff, D.E., Physical modeling of heavy plume dispersion. Ph.D. Thesis (1989). Colorado State Univ., Fort Collins, CO (USA). Availability: University Microfilms, PO Box 1764, Ann Arbor, MI 48106, Order No. 90-00,477.

[174] Murphy, M.C.; Heidorn, K.C.; Irwin, P.A., Scale model studies and development of prediction procedures for heavy gas dispersion in complex terrain, in Proceedings of the Technology Transfer Conference, Toronto, ON. Session A: Air quality research, (1988), pp. 109-147. Rep. No. CONF-8811291--, MICROLOG--89-02641.

[175] Murphy, M.C.; Heidorn, K.C.; Irwin, P.A.; Davies, A.E., Scale model studies of heavy gas dispersion, in Proceedings of the Air and Waste Management Assoc., 82nd Annual Meeting and Exhibition (Abstracts), Technical Rep. No. 89-55.1., (1989), pp. 57. Rep. No. CONF-890692--.

[176] Murphy, M.C.; Heidorn, K.C.; Irwin, P.A.; Davies, A.E., Heavy gas dispersion in terrain with obstacles. Proceedings of the sixth technical seminar on chemical spills, (1989), pp. 19-43. Rep. No. CONF-8906284--, MICROLOG--89-04698, CONF-8906284--.

[177] Krogstad, P.A.; Jacobsen, O., Dispersion of heavy gases, in D.N. Cheremisinoff, Encyclopedia of Environmental Control Technology, New Jersey Inst. of Tech., Newark, NJ, USA, Gulf Publishing Company, (1989), pp. 631-678.

[178] Gudivaka, V.; Kumar, A., An evaluation of four box models for instantaneous dense-gas releases, J. Hazard. Mater., 25 (1990) 237-255.

[179] Shin, S.H.; Meroney, R.N.; Neff, D.E., LNG Vapor Barrier and Obstacle Evaluation: Wind-Tunnel Simulation of 1987 Falcon Spill Series, Final Report, July 1987 - February 1991, Dept. of Civil Engineering, Colorado State Univ., Fort Collins.

[180] Heidorn, K.C.; Murphy, M.C.; Irwin, P.A., Effects of obstacles on the spread of a heavy gas - Wind tunnel simulations, J. Hazard. Mater., 30 (1992) 151-194.

[181] Riethmuller-M.L., Critical Confrontation of Standard and More Sophisticated Methods for Modelling the Dispersion in Air of Heavy Gas Clouds: Evaluation and Illustration of the Intrinsic Limitations of both Categories, Commission of the European Communities, Luxembourg, Final report, (1983). Rep. No. EUR8423.

[182] Fox, D.G., Uncertainty in air quality modelling, Bull Am. Meteorol. Soc., 65 (1984) 27-36.

[183] Ermak, D.L., Field validation of dispersion models for dense-gas releases, NATO Challenges Mod. Soc., 13 (Air Pollut. Model. Its Appl. 7), (1989), pp. 383-92.

[184] Hanna, R.S.; Chang, J.C., Uncertainties in Hazardous Gas Model Predictions, Int. Conf. and Workshop on Modelling and Mitigating the Consequences of Accidental Releases of Hazardous Materials, New-Orleans, Louisiana, May 20-24, (1991).

[185] Hanna, S.R.; Strimaitis, D.G.; Chang, J.C., Evaluation of 14 hazardous gas models with ammonia and hydrogen fluoride field data. To appear in J. Hazard. Mat. (1992- ...)

[186] Koopman, R.P.; Ermak-D.L.; Chan-S.T., Review of Recent Work in Atmospheric Dispersion of Large Spills, Lawrence Livermore National Lab., CA, (1988). Rep. No. UCRL97377, CONF88051035.

[187] Ermak, D.L.; Merry-M., Methodology for Evaluating Heavy Gas Dispersion Models. Lawrence Livermore National Lab., CA, Final report Nov 85-Feb 1988. Rep. No. AFESCESLTR8837 (1989).

[188] Smith, D.G., Emergency response/contingency planning considerations, in R.V. Portelli (ed.), Proceedings of the Heavy Gas (LNG/LPG) Workshop, (1985), pp. 216-223. Rep.No. CONF-8501127-, CE--03673, CSC--CE03673.

[189] Davies, M.E.; Inman, P.N., A statistical examination of wind tunnel modelling of the Thorney Island trials, J. Hazard. Mater., 16 (1987) 149-172.

[190] Hall, D.J., Further experiments on a model of an escape of heavy gas, Warren Spring Laboratory, Stevenage, Herts, (1979). Rep. No. LR 312 (AP).
[191] Puttock, J.S.; Blackmore, D.R.; Coelnbrander, G.W., Field Experiments on Dense Gas Dispersion, in R.E. Britter and R.F. Griffifths (eds.), Dense Gas Dispersion, Chemical Engineering Monographs 16, Elsevier, (1982).
[192] "Experiments with Chlorine": Report published by the Directorate General of Labour of the Ministery of Social Affairs, Voorburg, The Netherlands, (1975).
[193] Picknett, R.G., Dispersion of dense gas puffs released in the atmosphere at ground level, Atm. Env., 15 (1981) 509-525.
[194] Koopman, R.P.; Cederwall, R.T.; Ermak, D.L., et al., Analysis of Burro series 40-m3 LNG spill experiments, J. Hazard. Mater., 6 (1982) 43-83.
[195] Ermak, D.L.; Chan, S.T.; Morgan, D.L.; Morris, L.K., A comparison of dense gas dispersion model simulations with Burro series LNG spill test results, J. Hazard. Mater., 6(1-2), 129-60 (1982).
[196] McQuaid, J., Observation on the current status of field experimentation on heavy gas dispersion, in Ooms and Tennekes (eds.), Atmospheric Dispersion of Heavy Gases and Small Particles, Springer Verlag, (1984), pp. 241-267.
[197] Havens, J.A.; Spicer, T.O., Development of an atmospheric gas model for heavier-than-air gas mixtures. U.S. Dept. of Transport, U.S. Coast Guard, (1985). Rep. No.CG-D-23-85.
[198] Colenbrander, G.W.; Puttock, J.S., Maplin Sands experiments 1980: Interpretation and modelling of liquified gas spills into the sea, in Ooms and Tennekes (eds.), Atmospheric Dispersion of Heavy Gases and Small Particles, Springer Verlag, (1984), pp. 277-295.
[199] Hall, D.J., Further experiments on a model of an escape of heavy gas, Warren Spring Laboratory, (1977). Rep. CR 1314 (AP).
[200] Colenbrander, G.W., A mathematical model for the transition behavior of dense vapor clouds, in 3rd Proc. Int. Symp. Loss Prevention and Safety pronmotion in the Process Industries, Basel, (1980).
[201] Ermak, D.L; Nyholm, R.A.; Lange, R., ATMAS: A three-dimensional atmospheric transport model to treat multiple area sources, Lawrence Livermore Nat. Lab., Livermore, CA, (1978). Rep. No. UCRL-52603.
[202] Morgan, D.L.; Kansa, E.J.; Morris, L.K., Simulations and Studies of Heavy-Gas Dispersion Using the SLAB Model. American Meteorological Society symposium on turbulence and diffusion, Boston, MA, USA, 22 March 1983. Rep. No. UCRL88009, CONF83030711, (1983).
[203] Ermak, D.L., LNG Vapor-Dispersion Research at LLNL: Remaining Questions. Gas Research Institute workshop on vapor dispersion, Cambridge, MA, USA, 23 Mar 1982. Rep. No. UCRL87668, CONF8203781.
[204] Havens, J.A.; Spicer, T.O., Further analysis of catastrophic LNG spill vapor dispersion, in S. Hartwig (ed.), Heavy Gas Risk Assessessment, II., Proc. Symp., 2nd, Meeting (1982), Reidel, Dordrecht, Netherlands, (1983), pp.181-210.
[205] Chan, S.T.; Rodean, H.C.; Ermak, D.L., Numerical simulations of atmospheric releases of heavy gases over variable terrain. 13th International Technical Meeting on Air Pollution Modeling and its Application, Toulon, France, 14 Sept. (1982). Rep. No. UCRL87256, CONF8209362.
[206] Chan, S.T.; Ermak, D.L., Recent results in simulating LNG (Liquefied Natural Gas) vapor dispersion over variable terrain, Revision 1, International Union of Theoritical and Applied Mechanics, Symposium on Atmospheric Dispertion of Heavy Gases and Small Particles, Delft, Netherlands, 28 Aug. (1983). Rep. No. UCRL88495REV1, CONF8308024-Rev1.
[207] Koopman, R.P.; McRae, T.G.; Goldwire, H.C.; Ermak, D.L.; Kansa, E.J., Results of recent large-scale NH_3 and N_2O_4 dispersion experiments. Symposium on Heavy Gases and Risk Assessment, Bonn, F.R. Germany, 12 Nov. (1984).
[208] McRae, T.G., Analysis and Model/Data Comparisons of Large-Scale Releases of Nitrogen Tetroxide. (1985). Lawrence Livermore National Lab., CA. Rep. No. UCID20388.
[209] Hartwig, S.; Schnatz, G.; Heudorfer, W., Improved understanding of heavy gas dispersion and its modeling, Atmos. Dispersion Heavy Gases Small Particles, Symp., Meeting Date 1983, 139-55. Edited by: Ooms, Gijsbert; Tennekes, Hendrik. Springer: Berlin, Fed. Rep. Ger., (1984).
[210] Hall, D.J.; Waters, R.A., Wind Tunnel Model Comparisons with the Thorney Island Dense Gas Release Field Trials, Air Pollution Abstracts, Warren Spring Lab., Stevenage, England, (1984). Rep. No. LR489APM, ISBN0856243434.
[211] Morgan, D.L.; Kansa, E.J.; Morris, L.K., Simulations and parameter-variation studies of heavy-gas dispersion using the SLAB Model, International Union of Theoritical and Applied Mechanics

Symposium on Atmospheric Disperation of Heavy Gases and Small Particles, Delft, Netherlands, 28 August, (1983). Lawrence Livermore National Lab., CA; Rep. No. UCRL90150, CONF8308021Sum.
[212] Morgan, D.L.; Kansa, E.J.; Morris, L.K., Simulations and parameter variation studies of heavy-gas dispersion using the SLAB model. International Union of Theoritical and Applied Mechanics Symposium on Atmospheric Disperation of Heavy Gases and Small Particles, Delft, Netherlands, 28 August, (1983). Rep. No. UCRL88516Rev1, CONF8308021Rev1.
[213] Morgan, D.L.; Kansa, E.J.; Morris, L.K., Simulations and parameter variation studies of heavy gas dispersion using the SLAB model, Condensed. International Union of Theoritical and Applied Mechanics Symposium on Atmospheric Disperation of Heavy Gases and Small Particles, Delft, Netherlands, 28 August (1983), Lawrence Livermore National Lab., CA. Rep. No. UCRL90150, CONF8308021Sum.
[214] Morgan, D.L., Dispersion Phenomenology of LNG Vapor in the Burro and Coyote LNG Spill Experiments, Lawrence Livermore National Lab., CA, (1984). Rep. No. UCRL91741, CONF8412O119.
[215] Morgan, D.L.; Morris, L.K.; Chan, S.T.; Ermak, D.L.; McRae, T.G., Phenomenology and Modeling of Liquefied Natural Gas Vapor Dispersion, Lawrence Livermore National Lab., CA, (1984). Rep. No. UCRL53581.
[216] Havens, J.A.; Schreurs-P.J., Evaluation of 3-D Hydrodynamic Computer Models for Prediction of LNG Vapor Dispersion in the Atmosphere, Annual Report PB85-118503, Mar 83 - Feb 84, Dept. of Chemical Engineering Arkansas Univ., Fayetteville, (1984).
[217] Havens, J.A.; Spicer, T.O.; Schreurs-P.J., Evaluation of 3-D Hydrodynamic Computer Models for Prediction of LNG (Liquefied Natural Gas) Vapor Dispersion in the Atmosphere, Dept. of Chemical Engineering, Arkansas Univ., Fayetteville, Final Report March 1983 - April (1987).
[218] Jensen, N.O.; Mikkelsen, T., Entrainment through the top of a heavy gas cloud, numerical treatment, NATO Challenges Mod. Soc., 5 (Air Pollut. Model. Its Appl.), (1984), pp. 343-51.
[219] Gotaas, Y., Heavy gas dispersion and environmental conditions as revealed by the Thorney Island experiments, J. Hazard. Mater., 11 (1985) 399-408.
[220] Puttock, J.S.; Colenbrander, G.W., Dense gas dispersion: Experimental research, in R.V. Portelli (ed.), Proceedings of the heavy gas (LNG/LPG) workshop, (1985), pp. 32-50. Rep. No. CONF-8501127--, CE--03673, CSC--CE303673.
[221] Spicer, T.O.; Havens, J.A., Modeling the phase I Thorney Island experiments, J. Hazard. Mater., 11 (1985) 237-260.
[222] Chan, S. T.; Ermak, D.L., Further Assessment of FEM3: A Numerical Model for the Dispersion of Heavy Gases over Complex Terrain, JANNAF Safety and Environmental Protection Subcommittee meeting, Monterey, CA, USA, 4 Nov. (1985). Rep. No. UCRL92497, CONF85111103.
[223] Havens, J.A.; Schreurs-P.J., Evaluation of 3-D Hydrodynamic Computer Models for Prediction of LNG Vapor Dispersion in the Atmosphere, Annual Rep. March 1984 - February 1985, Arkansas University, Fayetteville.
[224] Koopman, R.P., Atmospheric Dispersion of Large Scale Spills, AIChE Winter Annual Meeting, Miami, FL, USA, 2 Nov. (1986). Lawrence Livermore National Lab., CA. Rep. No. UCRL95091, CONF8611463.
[225] Chan, S.T.; Ermak, D.L., Further Assessment of FEM3: A Numerical Model for the Dispersion of Heavy Gases over Complex Terrain, Lawrence Livermore National Lab., CA, (1985).
[226] Ermak, D.L.; Chan, S.T., Institute of Mathematics and its Applications, Conference on stably stratified flows and dense gas dispersion, Chester, UK, April 9, (1986).
[227] Davies, M.E.; Inman-P.N., Wind Tunnel Modelling of the Thorney Island Heavy Gas Dispersion Trials, British Maritime Technology Ltd., Teddington, England, Applied Fluid Mechanics Div., Final Rep. January 1985 - July 1986. Rep. No. BMT25003, GRI860264.
[228] Meroney, R.N., Guideline for Fluid Modeling of Liquefied Natural Gas Cloud Dispersion, Volume 2, Technical Support Document, Final report (July 1985). See also Volume I: Instruction Guide for the fluid model prediction of liquefied natural gas (LNG) storage and transportation hazards, (1986). Rep. No. CER8485RNM50B, GRI8501022.
[229] Van Ulden, A.P., The heavy gas mixing process in still air at Thorney Island and in the laboratory, J. Hazard. Mater., 16 (1987) 411-425.
[230] Cornwell, J.B.; Pfenning, D.B., Comparison of Thorney Island data with heavy gas dispersion models, J. Hazard. Mater., 16 (1987) 315-37.

[231] Eidsvik, K.J., Dispersion of Heavy gas Cloud in the Atmosphere, Rep. NILU OR 32/78, Norvegian Institute of Air Research, 1978. (see also Eidsvik, K.J., A model for heavy gas dispersion in the atmosphere, Atmos. Environ., 14 (1978) 769-777).

[232] Brighton, P.W.M., A user's critique of the Thorney Island dataset, J. Hazard. Mater., 16 (1987) 457-500.

[233] Crabol, B.; Roux, A., Interpretation of the Thorney Island Phase 1 Trials with the Box Model Cigale2, J. Hazard. Mater., 16 (1987).

[234] Spicer, T.O.; Havens, J.A., Field test validation of the DEGADIS model, J. Hazard. Mater., 16 (1987) 231-245.

[235] Chan, S.T.; Ermak, D.L.; Morris, L.K., FEM3 model simulations of selected Thorney Island phase I trials, J. Hazard. Mater., 16 (1987) 267-292.

[236] Carpenter, R.J.; Cleaver, R.P.; Waite, P.J.; English, M.A., The calibration of a simple model for dense gas dispersion using the Thorney Island Phase I trials data, J. Hazard. Mater., 16, 293-313 (1987).

[237] Bradley, C.I.; Carpenter, R.J.; Waite, P.J.; Ramsey, C.G.; English, M.A., Recent development of a simple box-type model for dens vapour cloud dispersion, in D. Reidel (ed.), Proc. Symposium on Heavy Gases and Risk Analysis, May 25-26, 1982, Battelle Institute, Frankfurt, (1983).

[238] Deaves, D.M., Development and application of heavy gas dispersion models of varying complexity, J. Hazard. Mater., 16 (1987) 427-38.

[239] Britter, R.E., Assessment of the Use of Cold Gas in a Windtunnel to Investigate the Influence of Thermal Effects on the Dispersion of LNG Vapour Clouds, Cambridge Univ., England, (1987). Rep. No. CUEDAAEROTR14.

[240] Chikhliwala, E.D.; Oliver,-M.; Shinde, A.N., Proceedings Annual Meeting, Air Pollution Control Association, USA, (1987), pp. 1-26. Rep. No. CONF-870695--.

[241] Chan, S.T., Numerical simulation of the mitigating effects of an LNG vapor fence, Joint Army Navy NASA Air Force Safety and Environmental Protection Conference, 18-22 June 1990, Livermore, CA (USA), (1990).

[242] Chan, S.T., Numerical simulations of LNG vapor dispersion from a fenced storage area, J. Hazard. Mater., 30 (1992) 195-224.

[243] Britter, R.E., The ground level extent of a negatively buoyant plume in a turbulent bondary layer, Atm. Env. 14 (1980) 779.

[244] Hanna, S.R.; Strimaitis, D.G.; Chang, J.C., Evaluation of fourteen hazardous gas models with ammonia and hydrogen field data, J. Hazard. Mater., 26 (1991) 127-158.

[245] Influence of Short-Term Concentration Peaks on Exposure Risks in the Vicinity of an Episodic Release of Hydrogen Sulphide, Monserco Ltd., Montreal, Quebec, (1981).

[246] Probabilistic Consequence Assessment of Hydrogen Sulphide Releases from a Heavy Water Plant, Scope Determination, Atomic Energy Control Board, Ottawa, Ontario, (1983). Rep. No. INFO01021.

[247] Abbott, M.L., Toxic vapor cloud impacts from accidental releases of anhydrous ammonia and nitrogen dioxide at the ICPP NO_x Abatement Facility, EG and G Idaho, Inc., Idaho Falls, (1992).

[248] Raj, P.K., Hydrogen Fluoride and Fluorine Dispersion Models Integration Into the Air Force Dispersion Assessment Model, Technology and Management Systems, Inc., Burlington, MA; Volume 1, Final report, 1 March 89 - 30 Nov. (1990). Rep. No. GLTR900321VOL1.

[249] Rasmussen, K., European Community Documentation Centre on Industrial Risk, Toxicol. Environ. Chem., 25 (1990) 213-219.

[250] Rasmussen, K.; Gow, H.B.F., The importance of information on industrial risk: A new documentation center, J. Hazardous Materials 30 (1992) 355-359.

[251] Drogaris, G., Major Accident Reporting System: Lessons learned from Accidents, Notified EUR 13385 EN, Commission of the European Communities, Luxembourg, (1991).

[252] Flyvholm, M.-A.; Andersen, P.; Beck, I.D.; Brandorff, N.P., PROBAS: The Danish Product Register Data Base - A national register of chemical substances and products.

[253] Papazoglou, I.A.; Christou, M.; Nivolianitou, Z.; Aneziris, O., On the Management of Severe Chemical Accidents, DECARA: A computer code for consequences analysis in chemical installations, Case study: Ammonia plant, J. Hazard. Mater., 31 (1992) 135-153.

[254] TNO 1980, 1992. Methods for the calculation of physical effects resulting from releases of hazardous materials (liquids and Gases), TNO - The Netherlands Organization of Applied Scientific Research, Directorate-General of Labour of the Ministry of Social Affairs and Employment, Postbox 90804, 2509 LV, The Hague, Rep. No. CPR 14E (1992), Second Edition, (ISSN 0921-9633/2.10.014/9203).

[255] Stewart, A.M.; Van Aerde, M.; Shortreed, J.H., Enhancements and updates to the RISKMOD risk analysis model, J. Hazard. Mater., 25 (1990) 107-119.
[256] Raj, P.K.; Kalelkar, A.S., Assessment Models in Support of the Hazard Assessment Handbook (CG-446-3, Technical support to the U.S. Coast Guard, Washington DC, 20590, (Jan. 1974).
[257] AIPE, Environmental Software Product Review, J. of Am. Institute of Plant Managers - Facility Management, Operations & Engineering, 16(4), (1989), pp. 20-23.
[258] Hanna, S.; Drivas, P., Guidelines for Use of Vapor Cloud Dispersion Models, Report to the center for Chemical Process Safety (CCPS), (1987). AIChE, 345 E, 47th Street, New York 100017.
[259] Raj, K.; Morris, J.A., Computerized Spill Hazard Evaluation Models, J. Hazard. Mater., 25 (1990) 77-92.
[260] Potts, R., Hazard Assessment Computer System - HACS, User Operation Manual, Vol. 2, Report to the U.S. Coast Guard, Washington DC, 20590, (Sept. 1981).
[261] Raj, P.K.; Morris, J.A., Source Characterization and Heavy Gas Dispersion Models for Reactive Chemicals, Vol. 1, Rep. to the U.S. Air Force, Air Force System Command, Hanscom AFB, MA 01731, (Dec. 1987).

CHAPTER 6

TOWARDS INTEGRATED RISK ASSESSMENT AND SAFETY MANAGEMENT AT REGIONAL LEVEL

6.1. The Potential Use of Comparative Risk Assessment

Methodological key issues (Gheorghe[1], 1994) to ensure a credible comparison to be made among different technologies or technological systems in a given region can be summarized as follows:

a) The delineation of consistent and comparable boundaries for different technologies or chains of technologies. The definition of principles and delineation of "reference technological chain" for risk comparison purposes is a priority issue in this regard.

b) Methods and dose-effect relationship for the risk assessment of the health effects of different technological systems (e.g., chemical plants, electricity generation systems, biotechnological installations) including, in particular, the deviation of consistent indicators of health risks for comparative assessment.

c) Methods and relationships for the assessment of impacts of different technologies (chemical works, electricity generation plants, biotechnological installations, etc.) on the natural environment and associated ecosystems, and particularly the derivation and application of indicators of environmental risks for comparative assessment.

d) How to deal, in the comparative risk assessment process, with the time and technological dependencies of risk estimation, given variations in technologies and in the state of technological development (over time). Can the principles of "risk discounting" be used to compare existing risks by postulating and accounting for possible future technological development. How can future technological development be accounted for in the comparative assessment process.

e) How to deal in the comparative risk assessment process with the issue of "uncertainties". Such uncertainties exist in particular in the type, nature and extent of various impacts and with the variable nature of some impacts over time. Uncertainties also exist in the estimation (both qualitatively and quantitatively) of the various risks, both in terms of consequences and probabilities.

f) Comparative risk assessment for severe accidents in different technological systems. The main issues here relate to the tools and criteria for comparison; the methods for presenting the results and the comprehensiveness of the comparative results, particularly in relation to environmental impacts from severe accidents.

g) The integration of the different elements of risks, including the question as to whether one overall integrated indicator of risk is possible or feasible; the applicability of comparative risk assessment to regional safety management studies; the weighting of the various dimensions of different risks in the comparative risk assessment process.

h) Methods and tools for the presentation of the results of comparative risk assessment.

6.1.1. SETTING BOUNDARIES FOR COMPARATIVE RISK ASSESSMENT OF DIFFERENT TECHNOLOGIES FOR REGIONAL SAFETY MANAGEMENT

The definition of consistent boundaries for the different technological chains or technologies which might operate in a given area is important for credible comparative risk assessment results. When dealing with comparative risk assessment of various technologies which might coexist in a given region, a consistent boundary definition has to be chosen.[2] When comparing technological or activity risks in a given area there is obviously a need to agree on some boundaries.

The principles for setting system boundaries are as follows:

- The setting of boundaries has to be adapted to the purpose of the assessment
- The boundaries should be set taking into account:
  - time (horizon)
  - space (workers and public exposure, environment)
  - phases of the systems under investigation
  - impacts to be considered
  - material balance consideration in terms of risk
  - per net production (e.g., energy) output
  - acceptable threshold criteria.
- At a regional level, the setting of boundaries will greatly benefit from structured public scoping/participation process, so that all concerned could be involved in boundary setting. This process would greatly enhance the credibility and acceptance of the study process.

6.1.2. ASSESSMENT OF ENVIRONMENTAL RISKS (IMPACTS)

Usually in a region where complex risk assessment studies are made, one has to take in consideration that a large number of compounds are released into the atmosphere and water and disposed on land with associated environmental impacts (risks). These impacts may be immediate or long term. Implicit "environmental impacts" are often a direct or indirect "inference" of "health impacts". TABLE 6.1and TABLE 6.2 show a generalized environmental transfer model outlining the various essential components of health and environmental risk estimations in large industrial complexes /regions.

TABLE 6.1. Categories of risks usually adopted to assess and compare the health impacts of different pollutant sources

Health Risk			
Source	**People at Risk**	**Exposure**	**Effects**
Routine or accidents	Workers and the public	Short, or medium and long term	Fatal and non-fatal Immediate/delayed respectively Long-term/delayed

TABLE 6.2. Categories of risks usually adopted to assess and compare the environmental impacts of different pollutant sources

Environmental Risk		
Source	**Effects**	
	Duration	**Extent**
Routine or Accidents	Short or medium and long term	Local, regional and global

A linear pollutant pathway model[3] indicating the amount reaching the receptor (target at risk) as a function of the amount emitted, altered by dilution and removal and enhanced by environmental accumulation factors is shown in *Figure 6.1*.

As shown, distance (space) and rate of movement (time) are critical parameters to an environmental impact assessment study. The complex interdependencies between time, space and feedback mechanisms (the degree of resilience of a given environment to external factors) are not fully known; very often it is difficult to normalize them on a common scale for comparison.

In comparative risk assessment of different technologies one has to "compare risks", which may be different in a subjective way from impacts, effects, emissions, etc. Two major limitations have to be considered when dealing with the assessment of environmental impacts of different technologies or technological chains, namely:

– the effects are not always susceptible to quantification
– there is no general agreement on what should be quantified.

Methods of relevance for making comparisons of environmental impacts in integrated regional risk assessment and safety management studies are:

– ranked matrix environmental assessments
– emission values and ambient quality indices
– critical loads and critical levels.

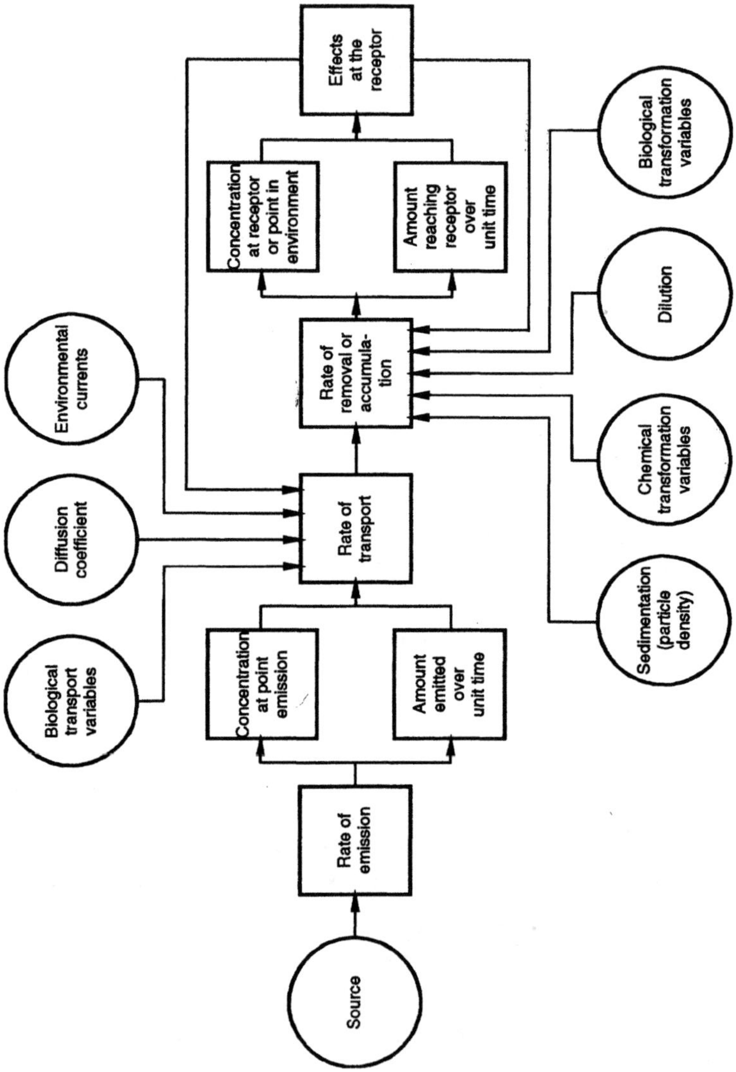

Figure 6.1. Pollutant pathway model used for assessing environmental impacts from emissions

- *A* ***matrix approach*** *is a ranking method to quantification of environmental impacts. It attempts to relate dose to response.*
- *The* ***emission values method*** *is a dose oriented approach.*
- ***Critical load model*** *whilst being dose based, is in fact essentially of a response type, the response being the outset of damage systems.*
- *The* ***Critical Loads concept*** *provides an adequate scientific basis to set limits to environmental impacts above which damage would occur.*

Critical Levels. It expresses the concentration of pollutants in the atmosphere above which direct adverse effects on receptors such as the ecosystem, may occur, according to present knowledge.

Critical Loads. It is a quantitative estimate of an exposure to one or more pollutants below which significant harmful effects on specified sensitive elements of the environment which do not occur, according to present knowledge.

The Critical Loads approach is a procedure for developing optimized abatement strategies by which leveled emission reductions are obtained on the basis of scientifically derived critical values.

Additional concepts associated with the critical loads and levels are:

- target levels and loads
- actual loads and levels.

The critical loads approach for regional environmental impact assessment requires the following types of information:

- inventories of current emissions and projections of future emissions rates
- estimates of the potential emissions reductions
- long range transport models
- maps of critical loads and target loads
- integrated assessment modeling.

In general it is considered that, for example, large scale electricity production systems which might exist at the regional level have an exponentially increased value for the environmental impacts due to the "effect of scale". There are economic and technological limitations to the elimination of all emissions or waste products and their associated impacts. Some emissions could be "imported" from long distances (e.g., transboundary effects of heavy metals, or acid rain); this has to be stated in the risk assessment studies for regional areas in view of proper results regarding comparative risk assessment of various technologies.

Indicators of environmental risks: Qualitative and quantitative indicators could be used for environmental impact indicators of different technologies. Such indicators are: site specific, society specific, value judgment dependent. Synergistic effects could be taken into consideration when dealing with comparative environmental impact assessment.

Environmental performance measurements already accepted by international organizations and practice and which can be used in comparative analysis are:

- environmental performance indicators (e.g., river quality, air quality, soil quality etc.)
- environmental goals (critical loads, sustainability index)
- environmental emissions (SO_x, NO_x, CO_2, Trade in Forest Products, etc.).

6.1.3. COMPARATIVE HEALTH RISK ASSESSMENT

The assessment of health risk is usually based on:

i) data from epidemiological studies for particular situations
ii) estimations based on dose-effect relationships for different levels of exposure.

Mathematical models enable the determination of ambient concentrations of pollutants in different media, and the prediction of health effects through dose-effect relationships. The state of knowledge concerning exposure paths, dose-effect relationships and the health risks for different technologies varies amongst different systems.

Both approaches to estimate health effects from various technological systems which can coexist in a given region, only rarely deliver quantitative risk coefficients with high confidence intervals. Dose-effect relationships for many chemical pollutants are not available.

There is a particular lack of data concerning the delayed (chronic) health effects of many chemical pollutants. The health effects at the low spectrum of continuous exposure need particular attention in this regard. Another issue of research need relates to the indirect health effects of pollutants where relatively little information is available.

A short list of parameters, being used as indicators of health risks for comparative assessment of various technological systems, includes:

- number of fatalities (acute)
- number of fatalities (delayed)
- incidence of injury and disease
- projected incidence of genetic effects
- years of lost life (YOLL)
- relative utility loss (RUL).

The following comments are offered in relation to the use of the above indicators:

■ Although compound indices such as YOLL and RUL offer the prospect of improved reliability in assessing and comparing health detriments, simple indicators such as numbers of fatalities and serious injuries should continue to be used in parallel, at least for the time being.

■ Some classes of detriment should not be aggregated (e.g., prompt and delayed fatalities).

■ Health risks to different groups and of different types should be distinguished, although it may in addition be permissible to aggregate in some cases.

Time Dependence in Comparative Risk Analysis
In comparative risk assessment (CRA), appropriate time frames must be defined for different stages of technological systems (e.g., impacts of thousands of years). In this respect one can consider micro-time (e.g., fraction of seconds) and macro-time (e.g., decades, centuries) as possible concepts to be used in comparative risk assessment.

6.1.4. UNCERTAINTIES IN COMPARATIVE RISK ASSESSMENT

When dealing with comparative risk assessment for activities which take place in large (regional) areas, uncertainties occur in the prediction of scenarios, models and data.

Uncertainties in scenarios:

- erroneous probabilities
- factors not considered
- factors screened out.

Uncertainties in models:

- imperfect conceptual model
- imperfect mathematical model
- imperfect computer model.

Uncertainty in data:

- general vs. site specific data
- measurement errors
- data reduction.

Expert judgment is inherent in the evaluation of uncertainties. Uncertainties must be delineated and exposed whenever appropriate and attempts made to deal with them. Uncertainties take many forms and it is essential that a coherent and clearly visible approach is adopted, both in the computational process and in the interpretation of results.

Within the complex study process of integrated regional risk assessment and safety management, the uncertainties arise from a number of sources:

Data. All data are subject to sampling errors. Statistical uncertainty requires the reporting of range or confidence levels.

Models. Estimates of risks through scientifically based models (lack of direct observations) brings additional uncertainty in a comparative risk assessment. When models are the only tool to be used for risk assessment, uncertainty analysis of the results should be exposed and taken into consideration.

Knowledge. Lack of knowledge or uncertainty in the foundation of basic physical phenomena can lead to erroneous results and faulty predictions. At the present time, this uncertainty in results due to knowledge is solved by using simulation procedures or inter-comparative computer exercises.

There are some techniques already in use which can help to evaluate uncertainties; among them are sensitivity analysis, inferential statistics (Bayesian methods), complex statistical and probabilistic analyses, Monte Carlo simulation. In case of lack of knowledge, bounding calculation and conservative assumptions can be used. As a rule of thumb, one can consider that risk assessment should be realistic and, if this is not possible, then the results should be considered as conservative.

There are different mathematical approaches to dealing with uncertainty: probability theory, Markow models, possibility theory (fuzzy sets), event trees, influence diagrams, heuristic models, etc.

The fact that uncertainties exist does not invalidate the use of comparative risk assessment as a useful tool for overall assessment of the impacts of various technologies in a given area, as long as such uncertainties are clearly understood, exposed and acknowledged in the presentation of the results of the study.

6.1.5. UNCERTAINTY ESTIMATION IN ENVIRONMENTAL REGIONAL RISK ASSESSMENT

In risk estimate, uncertainty pervades each element of environmental risk assessments, as an examination of estimations of the risks to humans from the release of a harmful substance into the environment.

The mathematics in a quantitative risk estimate may enter the following elements:

*Risk = Probability * Quantity * (1/Dispersion) * Population* Uptake* (Dose vs. Response)*

It is clear that an increase in any of these six elements leads to higher estimates of risk levels. There is always some uncertainty about each of these elements in the risk estimation process. In the current discussion, the terms "extreme certainty" and "extreme uncertainty" denote opposite ends of the spectrum of how well a particular phenomenon is understood. Qualitative modifiers "high", "fair", and "moderate" depict gradations of this understanding . These gradations, though arbitrary, are appropriate to the range of uncertainties found in environmental risk assessment.

1) *Uncertainties on the probability of a release of a harmful substance:*

   - May be unnecessary for many environmental risk assessment
   - Estimating the probability or frequency of such accidental releases is the first step in many environmental risk assessment studies
   - There is sizable uncertainty about estimates of the probability of accidental releases of harmful substances (N.B., historical data offer some guide to this

probability for frequently occurring accidental releases, but are less information on some kinds of accidents involving certain substances)
- Mathematical models produce fairly uncertain estimates of the probability of accidental releases of radioactive substances, it would appear that most estimates of the probability of chemical releases would be no better than fairly uncertain.

2) *Uncertainty on the quantity of a harmful substance released:*

- Such estimates improves with the quantity of the information gathered on the harmful substance released into the environment.

3) *Dispersion of a harmful substance and uncertainties on the resulting concentrations of that substance in the environment:*

- In practical situations estimates of concentrations at particular points in the dispersion pathway range from fairly to highly uncertain
- When measurements of concentrations are available, risk analyst can omit estimation of the probability of release of a given substance
- Actual measurements of concentrations of particular substances in the environment reveal that concentrations are highly variable ambient concentration of some pollutants are moderately invariable to moderately variable from day to day in a particular season. These concentrations may vary by a factor of 5 to 10 between warm and cold seasons and among different years. Human exposure to some chemicals in the workplace may be extremely variable among individual work situations.

4) *Uncertainty on the population exposed to release of a harmful substance:*

- In the absence of site-specific data, any estimate of the population potentially exposed to the release of a harmful substance from such sites would be extremely uncertain.

5) *Uncertainties on the uptake of harmful substances by humans and other organisms:*

- Pharmaco-kinetic models reflect how the physiology of humans differs from that of test animals with respect to uptake, metabolism, and excretion of particular chemicals. Pharmaco-kinetic models may modify risk estimates in significant ways, but their structure and the data they contain make these modifications moderately to extremely uncertain.

6) *Uncertainty on the relationship between dose of a harmful substance and adverse toxicological response:*

- The toxicity of a particular dose of a substance not only varies across species, but among individuals of the same species.
- Most acute toxicity estimates are moderately uncertain.
- Cancer potency estimates are fairly to extremely uncertain.

7) *Measurement error:*

- Measurement error increases the uncertainty associated with each of the above elements in an environmental and health risk assessment.
- Among all the elements of risk assessment, measurement error generally introduces the least uncertainty.

8) *Overall uncertainty:*

- Uncertainty in each of the elements of a quantitative risk assessment combines to produce an overall uncertainty in the estimate of a particular risk.
- Only a few studies have carefully integrated the uncertainty associated with individual elements of a risk estimation into an overall estimate of a risk and its overall uncertainty.
- Estimates of human cancer risks due to exposure to chemicals have been shown to be moderately to extremely uncertain.

9) *Credibility of Risk Analysis; Uncertainty on Policy Implications:*

- Uncertainties in environmental and health risk assessment are present in all elements of the assessment process.

There are several approaches to deal with uncertainty:

i) Risk analysts should present the particular level of uncertainty in the risks they are estimating.
ii) Risk managers should reduce uncertainty when the reductions would be cost-effective and would affect the choice of risk management option.
iii) When conservative assumptions must be applied, this must be done in the context of the particular uncertainties of the risk being addressed.

Reducing the uncertainty in any one or more of the elements of a risk assessment will reduce the overall uncertainty of a risk estimate; not all such reductions are equally cost-effective. Risk managers can also use the economic concept of the efficient frontier to determine which reductions in uncertainty are likely to reduce overall uncertainty the most and at the lowest cost:

- with regards to making conservative assumptions in the context of uncertainty, risk mangers should consider when and how to apply these assumptions
- some critics assert that conservative assumptions have resulted in grossly inflated risk estimates and, thus, in regulations that are too stringent; others assert that these assumptions are needed in order to compensate for various uncertainties.

Some guidelines to deal with uncertainty

■ Action 1. Explicitly acknowledge and approximate the uncertainties in each element of the risk estimation process:

■ Action 2. Carry the uncertainty through the calculation process and calculate not only the median or mean risk, but its overall uncertainty as well:

■ Action 3. Develop and apply decision criteria that explicitly address acceptable risk levels for individuals and the overall population.

■ Action 4. Offer to the decision makers a description of the nature and the magnitude of risks, including the uncertainty of these risks.

TABLE 6.3 presents some verbal description pertaining to the range of uncertainty.

TABLE 6.3. Range of uncertainty and variability in environmental risk assessments

Range of uncertainty / Variability	**GSD Descriptive Interpretative term phrases**	
More than 10.00	Extremely uncertain	Extremely variable
10.00	Highly uncertain	Highly variable
5.00	Fairly uncertain	Fairly variable
3.00	Moderately uncertain	Moderately variable
2.00	Moderately certain	Moderately invariable
1.50	Fairly certain	Fairly invariable
1.25	Highly certain	Highly invariable
1.05	Extremely certain	Extremely invariable
1.00	There is absolute certainty	

6.1.6. INTEGRATED ENVIRONMENTAL AND HEALTH RISK ASSESSMENT IN LARGE INDUSTRIAL AREAS

Uncertainty in Risk Analysis and Results Representation in Data Bases

There is a need to represent the uncertainty bounds of such data and knowledge in comprehensive studies and in specialized data bases. The purpose of this section is:

i) to help in identifying (direct) uncertainty measures of data and knowledge specific to health and environmental risk assessment studies in large industrial regions and
ii) their representation in documented and specialized data bases.

No measured or calculated value related to environmental and health impact (risk) assessment has any meaning unless it is accompanied by a declaration of quality, in form of an uncertain statement. The uncertainty declaration(s) can be regarded as the quality assurance tool for that process.

Uncertainties associated with model predictions in risk assessment

Uncertainties in models may result from transformation of the given model into a mathematical model. One has to distinguish between parameter, model and conceptual uncertainties, and treat them separately. Often results from modeling activities are stored in a data base. Models for assessment of environmental, health and safety impacts are important elements in integrated and comparative risk assessment for different technologies which might coexist in a given region.

A formal procedure of parameter uncertainty analysis is given next:

1) List all the parameters that are potentially important contributors to uncertainty in the final model prediction.

2) For each parameter listed, specify the maximum conceivable range of possibly applicable alternative values:

   - Specify the degree of belief (in percentage) that the appropriate parameter value is not larger than specific values selected from the range established above and select a probability distribution that best fit the quoted degrees of belief.
   - Account for dependencies among model parameters by introducing suitable restrictions, by quoting appropriate degree of belief, or by specifying suitable measure of the degree of association.

3) Set up a subjective *probability density function (pdf)* for the combined range of parameter values. This will subsequently be referred to as a joint pdf. Propagate this joint pdf through the model to generate a subjective probability distribution of predicted values:

   - Derive quantitative statements about the effect of parameter uncertainty on the model prediction:
   - Rank the parameters with respect to their contribution to the uncertainty in model prediction:
   - Present and interpret the results of the analysis.

   Expert judgment elicitation is preferred when one of the following situations exists:

   - no other means are available for quantifying an important issue.
   - the information available is characterized by high variability.
   - some expert question the applicability of the available data.
   - the existing results from code calculations need top be supplemented.
   - analysts need to determine the current state of knowledge.

Uncertainty declarations in the databases
There are reasons for including uncertainty indicators in the databases which store information relevant to integrated risk assessment or comparative risk assessment of various technological systems. Uncertainty declarations are quality declarations and a good description of uncertainties is decisive for good practice and scientific acceptance.
Uncertainty criteria may be used as acceptance criteria in quality assurance for choosing data to be accepted and included in the data base. A good scientific practice is a strategy for accepting material in the data base.

Validation. The advanced assumptions (theories) should be subject to a validation procedure.
Criticizable. The assumption in risk assessment should be formulated in a well defined manner so that it can be criticized.
Acceptability. The assumptions for risk estimation should be discussed when necessary in a forum (e.g., steering committee).

6.1.7. INTEGRATION OF THE DIFFERENT ELEMENTS OF RISK AND THE ROLE OF COMPARATIVE RISK ASSESSMENT FOR SAFETY MANAGEMENT

Lack of a single integrated risk indicator; the elements of risks from various technological systems can be broadly categorized as indicated in TABLE 6.3. In reviewing risk assessment and management results for their use in area safety planning, one should take in consideration the followings:

■ The various elements and dimensions of environmental and health risks cannot be integrated into one overall indicator of total risk. No comparison on the basis of a single indicator is possible. The comparative risk assessment process must specify on which basis (indicator) the comparison is being made.

■ It is necessary to expose all the dimensions and elements of health and environmental risks in the comparative process. Differences between regions and societies make the development of one overall indicator meaningless. There is no "global" risk value. The results should not be transferred from one study to another without appropriate investigations of differences between regions and countries.

■ One approach to develop an overall indicator is the estimation of "external costs" of impacts in terms of monetary value.

■ The integrated approach to health and environmental risk means all risks should be identified, assessed and considered in the comparative risk and management process.

More Comments on Continuous Emissions

The most serious risks (impacts/effects) from continuous emissions from different industrial activities (e.g., power generation, chemical plants, heating, etc.) may be expected to be those related to human health and of environmental effects. The following impacts were identified by at large for making priority risk analysis due to continuous emissions:

- effects of atmospheric pollution on:
    - human health
    - materials
    - crops
    - forests
    - freshwater fisheries
    - unmanaged ecosystems
- effects of altered land use on unmanaged ecosystems
- assessment of global warming potential of greenhouse gas emissions
- effects of discharges to water and abstraction on health and water supplies
- effects of discharges to water and abstraction on aquatic ecosystems
- effects of noise on human health
- effects of noise on amenity.

The above list is not exhaustive and the process of prioritization is subjective.

A damage function approach can be used to assess and value the impacts resulting from activities which are taking place in a large industrial area. This type of approach allows calculation of marginal damages; many impacts are very often side and technological dependent. A pathway is chosen to show the progression of impact from the industrial activity and corresponding emission through effects on plants in natural ecosystems to economic valuation.

In order to quantify different risks from a specific industrial site it is necessary to identify the distribution of receptors that may be affected by operation of the plant in some detail. The type of data necessary to make this calculations are of the following form:

- Distribution of people in different risk categories
- Distribution of various crop species
- Distribution of various forest species
- Distribution of various species in different risk categories.

It is important that health and environmental damages are assessed over as much of the area as possible. One has to be aware that the artificial truncation of the reference environment for risk assessment will lead to underestimation of impacts.

Impact estimation can be performed on a marginal basis using at least three types of procedures

- Statistical relationships (proportion of the population who will develop respiratory symptoms due to low level pollution in the analyzed region)
- Dose-response functions (linking the rate of corrosion of steel to pollutant levels, known as critical level concept)
- Mathematical models containing a series of cause-effect relationship (e.g., *magic model* used to predict the rate and extent of ecosystem acidification and chemical changes for different pollutant deposition rates).

At the present time, international team of experts were convened (e.g., in EEC) to discuss, agree and recommend the most applicable dose response functions and models for risk assessment. Large degree of uncertainty is associated to the estimation of acidification impacts on forestry and the calculation of global warming damages. Valuation can be performed most easily when the commodity is directly traded (agricultural produce, timber, building repairs, etc.). When this is not possible, it is necessary to assess values indirectly in terms of *Willingness to Pay (WTP)* for improved environmental quality, or *Willingness to Accept (WTA)* increased environmental damage.

External Costs. They are defined to be the costs which fall on one group of people due to the social or economic activities of another group, and where the latter group does not take these costs into account. In the case of electricity generation, this is equivalent to costs which are not reflected in the market price of the electricity generated. There are often difficulties in defining to what extent damage to health or the environment is an external cost.

TABLE 6.4. Indicators representing damage categories

Public health:	
- PM_{10} mortality	- PM_{10} respiratory hospital admission
- SO_2 mortality	- SO_2 cough episodes for children
- PM_{10} bronchitis in children	- SO_2adult chest discomfort days
- PM_{10} symptoms days	- SO_x-eye irritation
- PM_{10} asthma attack	- NO_x-eye irritation
- PM_{10} restricted activity day	- Transport accidents
- PM_{10} emergency room visits	- Construction accidents
Occupational health:	
- Mining (respiratory diseases; mortality and morbidity)	- Transport (death, major/minor injuries)
- Mining (death, major/minor injuries)	- On road accidents (death, major/minor injuries).
Agriculture:	**Aquatic impacts:**
- SO_2 -direct effects on wheat, barley, rye, oats, peas and beans, or other crops	- Mining (groundwater and surface water)
- Acidic deposition -soil liming	- Acid deposition -fisheries
- Effects of ozone	**Forests:**
	Biodiversity:
Materials:	
- Stone refacing	- Galvanized steel replacem0ent
- Mortar repointing	- Aluminum replacement
- Concrete replacement	- Cultural impacts
- Repainting	- Building damage from subsidence
- Steel replacement	
Noise:	**Global warming impacts:**
- Plant operation	- Aquatic life
- Road traffic	- Air quality (ozone)
- Rail traffic	- Glacier regression
	- Agricultural/Forest growth
	- Desertification

In order to be comprehensive, consistent and transparent it is necessary to consider the environmental impacts and their monetarized values over all space and time (the life cycle analysis). Some risks due to technological systems (e.g., electricity generation, production of different chemicals, etc.) due extend far beyond the operating lifetime, for example forest decline and fisheries or the global warming effect.

All part of the analysis are subject to uncertainty, some of which is very considerable. In many cases it is not even possible to describe the uncertainty in terms of distribution of values. The valuation of uncertain events (risk evaluation) poses considerable challenges. Discounting of the monetary values of environmental and health impacts is controversial.

Impact Categories
A list with guiding priority impact categories is given next:

- atmospheric pollution/human health
- accidents (continuous emissions) occupational and public
- atmospheric pollution/materials
- atmospheric pollution/crops
- atmospheric pollution/forests
- atmospheric pollution/freshwater fisheries
- land use/natural ecosystems
- global warming
- mining/ground and surface waters
- emissions to water/drinking water
- emissions to water/ecosystem
- noise/human health
- noise/amenity.

Damage Categories Indicators
Appropriate indicators for damage categories related to the above impact categories' list are given in TABLE 6.4. They have to be considered as indicative, which can guide the analyst in his effort to quantify consequences in the risk assessment stage.

6.2. Some Methodological Issues in CRA for Severe Accidents

The assessment and hence the comparative risk assessment (CRA) of different technologies or technological systems as regard major accidents, focused almost entirely on estimating acute fatalities to people from historical records or using techniques of Probabilistic Safety Assessment (PSA) as a predictive tool.

Methodologies for the estimations of the late health effects to people and of the environmental impacts from major accidents are limited, or in need of significant development and applications (with the exception of nuclear technologies).

A definition for severe accidents should encompass all the elements of health, environmental risks and damage to plant, equipment, buildings and should be expressed in terms of both the potential as well as the actual damage and risk.

6.2.1. METHODS AND TOOLS OF COMPARATIVE RISK ASSESSMENT FOR SEVERE ACCIDENTS

The comparison cannot be made on the basis of the consequence of such accidents in isolation. The likelihood (or probability) of occurrence should also be taken into account. Hence, estimation of the frequency of such accidents is relevant. Such estimations necessitate reliable information on the past records of such accidents and their effects and/or the application of probabilistic methods that predict the likelihood of their future occurrence.

It is difficult to assess and compare the frequency and the health and environmental damages caused by severe accidents because such data are not systematically collected by a single national agency.

There are no data concerning in particular the delayed effects on health from severe accidents for non-nuclear energy systems. All health effects in such cases are reported in terms of immediate fatalities, with immediate injuries reported in a few cases. This makes complete comparison difficult, since the total impact may be underestimated.

The ultimate long term environmental effects, particularly from severe accidents, are difficult to establish; it may be difficult to establish whether the effect is irreversible or whether a recoverable effect is possible.

Presentation of Results of Comparative Risk Assessment for Severe Accidents
In regional risk assessment and safety management studies, in presenting the comparative risk assessment results for severe accidents, the following principles should be adopted:

a) Risks from severe accidents should be presented and compared separately from the risks resulting from routine operations.
b) Data based on historical (actual) occurrences should not be compared directly with data on probabilistic predictions of likely future events.
c) Comparative results should not be presented in terms of aggregate risk in isolation.
d) All dimensions and elements of environmental and health risks should be exposed and presented.

The presentation of results in the form of F-N curves (frequency distribution) is a useful presentation technique. However, the values usually presented without additional information may not by themselves be sufficient since other dimensions of risks may not be exposed. Another method of presenting the results in tabular form should ensure the completeness of all relevant dimensions of risk rather than relying on isolated information (e.g., on fatalities).

The above principles are to be considered for accidental situations, which are mainly investigated in the second volume of this monograph.

6.3. Basic Aspects in Decision Aiding Techniques

It is already acknowledged that the process of initiating, promoting risk analysis and implementing safety management studies for regional areas involves complex decisions as well as the participation of a large number of actors (e.g., public, experts, systems analysts, environmentalists, safety engineers, administrators, politicians etc.). The process of *integrating various aspects of risk* such as environment, health, hazardous installations, safety culture, management, involves decision aiding techniques which are close to the management science field known as decision analysis.[4]

There are, in general, positive and negative aspects associated with decisions. Indeed, many decisions are made intuitively by experts and do not use structured processes or techniques. For many decision problems related to risk assessment and safety management in large and complex industrial areas, the solutions and then advantages and disadvantages may not be immediately apparent because of the complexity of the issues involved. There is a need for systematic processes to be followed that help structure thinking and analysis, and allow different viewpoints to be taken into consideration. Structuring helps avoid inappropriate *ad hoc* decisions and allows the process of reaching a decision to be more open and the decision itself to be more readily defensible (decisions which are taken today very often have long term effects). In the end, the use of various decision aiding techniques and the overall process and technology of decision analysis allows the integration of various risk at regional and area level. The integration of various risks into the decision making process is the appropriate mechanism which allows displaying various risks and choosing the most appropriate safety improvement strategy.

There are many inputs, influences and constraints that a decision maker will consider when deciding which actions to initiate regarding risk reduction or safety management to a particular plant or for the region under analysis. Decision aiding techniques (DAT) are tools for decision makers; they are decision aiding techniques, but not decision making techniques. In the content of regional safety management such techniques could i) assist structuring the complexity of issues derived from hazard identification and prioritization, ii) assist in improving safety of individual installations, iii) siting of new installations to comply with numerous criteria and performance risk indicators, etc. There are complex decisions that need to be made in the field of regional safety management which may involve conflicting positions. A large number of tools are available to assist in solving and structuring decisions of such complexity.

The main purpose of this section is:

- to emphasize and strengthen the use of DAT for integrated regional risk assessment and safety management
- to provide guidance as to where such processes may or may not be appropriate (to avoid the application of sophisticated processes to trivial problems).
- to recognize that DATs are tools, not panaceas.

The main stages of DAT in relation to the integrative process of various risks which could coexist in a given region, are:

- <u>Step 1</u>: Definition and description of the problem.
- <u>Step 2</u>: Consideration and definition of appropriate quality assurance requirements.
- <u>Step 3</u>: Formalizing the descriptive model of the problem.
- <u>Step 4</u>: Obtaining the necessary information for modeling.
- <u>Step 5</u>: Analyze in order to determine the set of alternatives and criteria.
- <u>Step 6</u>: Ensure the selection of the proper method to make the decision regarding the proper integration of various risks in the region and their minimization

■ Step 7: Establish a clear record of the process and any decisions taken as a result of the integration process of various types of risk and appropriate safety measurements.

To make a decision means to select a way of action out of a set of possible alternatives. The decision process is complex and sometimes iterative; the set of alternatives or criteria may vary from an interaction to iteration.

Decision making involves three major elements:

a) alternatives among which the "best" one will be chosen
b) criteria for judgment
c) methods for selecting one alternative from the whole set.

Decision aiding process (DAP) is needed because it helps to generate a degree of shared understanding among interest parties who are concerned with issues on risk assessment and safety management in regions with complex and large industrial installations. There are a few issues to be highlighted: in this domain one has to involve complexity, uncertainty or even fuzziness, multiple criteria and group interests with the objectives in conflict. DAP can provide models that integrate these features into an adequate methodology. DAP provide a framework within which informed discussions of key issues (e.g., environmental or health risks) can be conducted, and facilitate generation of consensus and integration of such issues. The relative costs and benefits to the user of applying a decision aiding process should be compared to the costs and benefits of not using one of these decision aiding processes. The cost of not using a decision aiding process can be high, especially in terms of inefficient use of human resources and in the end the incapacity of integrating various types of risks and selective safety improvement measures in the overall decision process.

In the risk management phase for studies concerning complex industrial areas, all possible options (e.g., technological, organizational, managerial etc.), have to be listed exhaustively as well as all relevant choice factor from which the appropriate comparison of the set of options will be made. After having identified the sets of options and factors, each option has to be assessed and weighted according to each factor. These assessments involve generally uncertainties, especially when probabilistic events are concerned for which occurrence probabilities and possible associated consequences cannot be easily evaluated.

When the best option has been selected, one has to check the robustness of this choice in relation with the variety of risk impact indicators, safety improvement measures and a sound cost effectiveness strategy. Various uncertainties or value judgments have to be tested in a sensitivity analysis phase. At the end of this selecting process, results have to be presented as simple and as clear as possible.

The following steps have to be considering when involving DAP for integrated regional risk assessment and safety management:

- Choice of options and relevant technical, environmental, health and economical factors.
- Assessment of options according to each relevant factor and the relative importance of each factor in the political or technical decision environment.

- Choice of the most appropriate and relevant decision aiding techniques.
- Estimation of weighting factors.
- Use of sensitivity analysis.
- Presentation and interpretation of results.

The wide practice of the use of decision analysis shows that no decision aiding technique is appropriate for all circumstances. A variety of techniques which could be used in risk management in complex industrial areas are presented in this chapter. Examples of the types of problems in the field of risk assessment and safety management where DAT can be helpful are outlined below:

■ Siting of new industries, land fills, or other hazardous installations, taking into account factors such as alternative land uses, environmental effects, and population density.

■ Development of strategies for emergency planning (e.g., sheltering, evacuation, relocation) taking into account the social implications of moving people in all types of weather, as well as financial and logistical considerations.

■ Deriving levels of pollution contamination in water and ground at which emergency countermeasures may be introduced following an accident.

■ Determining requirements for additional safety equipment.

■ Determining the appropriate degree of redundancy of safety devices and plant equipment, taking into account possible safety effects (e.g., domino effect) on other plant equipment.

■ Optimizing the protection and other aspects of maintenance and repair operations, choosing between major alternatives such as replacement or repair of large components.

■ Selection of optimum plant equipment and procedures for normal and abnormal circumstances.

Decisions related to risk management in large (regional) areas involve negotiations between people with different viewpoints or value judgments, such as representatives of the cantons, health officials, the environment departments, representatives of the workers and the management of a given facility, and the public. One objective of the negotiations is to determine the "choice factors" which are representative of the different viewpoints.

The range of possible decisions in any investigated area is generally limited by certain constraints which may be imposed, for example, by a regulatory authority at the federal or cantonal level, by operating policies, or by generally accepted good practice.

Other constraints may be less tangible, such as acceptance by the public or by the industry practice, limited perceived threats to environmental quality and health risks, and political conditions. New techniques are now available (decision conferencing) which enable the integration of various aspects into a comprehensive decision policy.

6.3.1. OVERVIEW OF DECISION AIDING TECHNIQUES

Before carrying out any decision aiding process for integrated risk assessment and safety management at regional level, it is essential to identify all of the available options and relevant factors that would influence the outcome. "Brainstorming" can present a constructive approach. The use of multi-disciplinary scientific and engineering expertise groups should be considered for dealing with the more complex assessment problems, particularly those involving the need to choose between availability of alternative or competing action pathways.

Generally, one option is devoted to a given objective function (e.g., reduction of the probability of occurrence of an undesired event, reduction of various releases, etc.). One can thus structure a brainstorming session as follows:

- What are the involved functions?
- Is each function necessary?
- It is possible to ensure a given function by different ways?
- For each way, is it possible to ensure the function differently?

The selection of the relevant factors can be aided by the application of analytical hierarchy process or value trees which enable the complexity of a given problem to be broken down into smaller constituent parts. As an example, TABLE 6.1 and TABLE 6.2 illustrate the criteria hierarchy for the overall health risks' As shown, distance (space) and rate of movement (time) are critical parameters to an environmental impact assessment study. The complex interdependencies between time, space and feedback mechanisms (the degree of resilience of a given environment to external factors) are not fully known; very often it is difficult to normalize them on a common scale for comparison.

In comparative risk assessment of different technologies one has to "compare risks", which may be different in a subjective way from impacts, effects, emissions, etc. Two major limitations have to be considered when dealing with the assessment of environmental impacts of different technologies or technological chains, namely:

- the effects are not always susceptible to quantification
- there is no general agreement on what should be quantified.

Methods of relevance for making comparisons of environmental impacts in integrated regional risk assessment and safety management studies are:

- ranked matrix environmental assessments
- emission values and ambient quality indices
- critical loads and critical levels.

objective (e.g., routine, accidental, public, occupational etc.).

When different groups are directly involved in the decision making process, it is important to take into account the different objectives of each group. In this context it is necessary to recognize the fundamental objectives and the boundary conditions, together with any particular preferences with respect to these criteria. For instance, for the objective "minimize hazardous potential" the associated criterion is *"hazardous potential"* and the preference is for *minimization.*

When a particular option cannot be uniquely prescribed by a single figure within acceptable confidence limits, it may be possible to use alternative values or distribution of values, or distribution functions, or even possibility distributions or fuzzy sets evaluations.

6.3.2. DECISION AIDING TECHNIQUES IN USE FOR INTEGRATED RISK ASSESSMENT AND SAFETY MANAGEMENT FOR LARGE AND COMPLEX INDUSTRIAL AREAS

A number of decision-aiding techniques are currently available. For the purpose of this book, the application of Cost Effectiveness Analysis (CEA), Cost Benefit Analysis (CBA), Multi-Attribute Utility Technique (MAUT), and Multi-Criteria Outranking Technique (MCOT) are considered in further details.

Cost Effectiveness Analysis (CEA)
CEA technique can be applied when two related factors are considered, one of which being a cost. The relative interest of each option is assessed through a "cost effectiveness ratio" illustrating the amount of effort (cost) devoted to a decreasing of one unit of risk. This technique is relatively easy to apply and can be recommended when only two relevant factors are envisaged because in this case the sensitivity analysis is very easy to make.

Cost Benefit Analysis (CBA)
CBA. originates from the economic theory of welfare; it compares the benefits and harm associated with different options The principal characteristic of this technique is that all relevant factors have to be expressed in monetary terms. When all these factors are expressed in the same unit it is very easy to aggregate all these different costs in a total cost. The best option is then the option presenting the minimum total cost.

Multi-attribute Utility Technique (MAUT)
The essence of this technique is to define a scoring scheme (or a multi-attribute utility function), measured on a scale between 0 and 1, with the property that if the score (or utility) is the same for two options there is no preference for one or the other. If however the utility for *option i* exceeds that for *option j*, then *option i* is preferred to *option j*. Such utility functions are established for each relevant factor (or attribute) and then aggregate in a total utility function representing the global interest for each option. The MAUT is an interesting decision analysis model for high level strategy problems involving different factors (quantitatively or qualitatively). Recent experiments show that this technique can be implemented in practice when dealing with different actors involved in a complex decision making environment. Fuzzy sets theory can also be adopted to this technique in order to accommodate for imprecision in evaluating of the factors.

Multi-Criteria Outranking Technique (MCOT)
-The aggregation techniques described in the previous sections combine all the evaluations according to the different relevant factors into a single figure of merit expressing the global interest of each option. Instead of expressing the performances of each option in terms of a single overall figure of merit, outranking techniques compare each *option i* to every *option j* in order to evaluate whether *option i* outranks (or is preferable to) *option j*.

6.3.3. SENSITIVITY ANALYSIS

Any decision aiding procedure involves the use of data, models, techniques, value judgments which contain uncertainties of various types such as:

- Uncertainties associated with an imperfect knowledge of the performances of the options under different circumstances (scenarios) or of the parameters and data used in the assessments.
- Uncertainties associated with the use of models.
- Uncertainties due to imperfect knowledge (e.g., about the future of each option).
- Intrinsic uncertainties resulting from the statistical treatment of the variable (this is very important when assessing the expected outcomes from low probability events).
- Value judgments.

The sensitivity analysis highlights the aspects which have the greatest influence on the results, where it may be desirable to attempt to reduce the uncertainties, if possible.

Types of sensitivity analysis involve the following process:

- Modification of a given parameter.
- Assign probability distributions to the important variables (probability encoding).

6.3.4. PRESENTATION OF THE RESULTS FROM APPLYING A DECISION AIDING TECHNIQUE

Aspects to consider when presenting the results of integrated decision for risk assessment and safety management at regional level are as follows:

- List of the assumptions, hypothesis and initial conditions considered within the case study
- Comment on the weighting factors
- Comment and reference on the various models and techniques used as well as their relevance and integration within the overall study
- Indication of the main uncertainties, quality of data
- Presentation of the sensitivity analysis results.

From these different information the final decision will be made by the decision maker or the group involved in preparing recommendations and policy strategy at area level.

6.3.5. ADVANCED TECHNIQUES FOR DAP

Artificial intelligence models have been lately implemented for computer aided design for risk assessment and management, as well as for expert systems development and implementation. A few comments is worth considering:

i) Expert systems have been developed in assisting the probabilistic safety analysis for complex nuclear and other industrial facilities and processes; they are relevant in integrating various aspects concerning risk in a given region.

ii) Decision support systems by means of expert systems, operation research, multi-criteria decision models with objectives in conflict are already in use in making decisions concerning risk reduction and cost optimization for marginal safety measures implementation.

iii) Connectionist expert systems (neural networks)[5] which exhibit characteristics and capabilities not provided by any other technology; this a complementary technique designed to assist solving ill-designed risk assessment and management problems (e.g., risk to special targets at risk - people with asthma and bronchitis - exposed to low level pollution and a complex meteorological environment and possible changes as well as their integration).

iv) Integrated knowledge based decision support systems by using systems analysis techniques and procedures and by integrating in an efficient way expert systems and neural network technology into an advanced methodology of systems analysis and information technology.

v) Influence diagrams which enables capturing numerous elements which could lead to an unsafe operational environment.

Decision Conferencing
Decision Conferencing is an efficient tool for integrated regional health and environmental risk assessment and safety management. *Decision Conferencing* is a relatively new decision aiding tool, which has been developed from the practice and need of integrating various aspects of risk, which has grown up in a number of organizations or projects, when complex situations and difficult issues arise. In a recent report ("International Chernobyl Project - input from the Commission of the European Communities to the Evaluation of the Relocation Policy adopted by the former Soviet Union[6]) it is highlighted the followings:

"Rather than discuss the matter amid the hurly-burly of their day to day activities, the decision makers take time to go away from their regular working place in order to concentrate fully and solely on the issues before them. Sometimes straightforward discussion can lead to a clear decision and a view of

a way forward, but the complexity of the issues and the uncertainties involved may be too great for simple discussion to resolve. In such cases, it has become the practice in some organizations for the team responsible for the decision to meet together for two days or more away from their normal working environment to discuss and explorer the issues".

Decision conferencing uses the services of a moderator, who is skilled in the process of group decision. The moderator is seldom a person with experience in the context of the issues at hand or have expertise in the discussion. The moderator has a very definite role, namely to smooth the team's work, to help the process and make the team productive and more creative. Within a decision conference event, the content of the discussion comes from the team themselves. A decision conference is a two-day event in which all the 'owners' of the problem (e.g., integrating health and environmental risks in a given region and making appropriate decisions for safety improvements) gather together to agree upon a strategy. The decision makers, or advisors (the problem owners) are supported by a moderator and a decision analyst. Two important comments are made:

i) - the moderator has the role to lead the conference (he acts as a disinterested chairperson guiding the discussion forward constructively, building and interpreting decision models): this would help the decision makers or their advisors to appreciate various facets of the problem before them

ii) - the analyst deals with the details of the model building, by using information technology (models and computers).

The technology of decision conferencing brings together knowledge, skills and techniques arising from the following fields:

■ *Decision analysis* which provides a structure and language (e.g., multi-attribute value theory) in which decision makers can think and talk about their problem.

■ *Information technology* assists in "real time" the decision making process; the results of decision analysis can be presented to the decision makers in simple, easily comprehended ways.

■ *Group dynamics* an awareness of which helps the moderator and analyst ensure that the group of decision makers or advisors interact constructively.

■ *Within a decision conference* the team can discover the importance of differences in opinion between the team members and the sensitivity of their conclusions to these differences and to those judgments of which they are most unsure. Controversial discussions are unimportant when discussion is focusing on the critical issues without being deflected. The moderator is expert in group dynamics and rational decision theory, as well as a skilled communicator.

In practice, each decision conference is different. It evolves according to the needs of the team and not according to some fixed agenda. The topic of integrated regional risk assessment and safety management and its complexity is an ideal topic for decision conferencing. No model should be taken as definitive: revised models are build and much sensitivity analysis undertaken until no further insights are obtained and agreed conclusions are reached. Decision conferences are creative events; they create an environment where participants create, evaluate, modify and re-evaluate options, building a strategy which they all support.

In accordance with the practice in this field: "Surprisingly, the analysis in decision conferences needs much less hard data than one would, at first, think. Strategies have to be priced: that is clear. But the cost estimation need only be rough. It is a broad brush picture that the event seeks to create. Detail can be added at a later date".

The lack of a single risk indicator to deal with risk assessment, health and environmental impacts at the regional level, should be compensated by the use of this specialized model - *decision conferencing* - able to integrate objective and subjective decision aspects when dealing with complex issues and costly options and decisions.

Summary (Chapter 6)

Integrating various types of risks, and their sources, existing in a region is not an easy or trivial task. Various types of techniques, methods and tools could be employed in helping the integration process. One potential / relevant instrument for the integration process of various types of risks in a region is that of Comparative Risk Assessment. The chapter investigates the process of setting boundaries for comparative risk assessment of different technologies, the assessment of environmental risks / impacts, comparative health risk assessment, uncertainties in comparative risk assessment. Methods and tools of comparative risk assessment for severe accidents are also discussed.

Recently, Decision Aiding Techniques play a significant role in the integration process of various types of risks and their management at the regional level. This chapter reviews such techniques and their relevance to the risk integration process. Special attention is given to decision analysis and decision conferencing.

References (Chapter 6)

[1] Adrian Gheorghe - "Comparative risk assessment of the health and environmental impacts of various energy systems", in Int. J. of Environment and Pollution, 4(3-4) (1994) 329-349.

[2] Sam Haddad, Adrian Gheorghe - " Issues in comparative risk assessemnt of different energy sources", in Int. J. of Global Energy Issues, Vol. 4(3) (1992).

[3] United Nations Environment Programme, The Environmental Impacts of Production and Use of Energy. Part IV. The comparative assessment of the environmental impacts of energy sources. Phase I. Comparative data on the emissions, residuals and health hazards of energy sources, Rep. ERS-14-85, UNEP, Nairobi (1985).

[4] Adrian Gheorghe, Decision Processes in Dynamic Probabilistic Systems, Kluwer Academic Publishers, Dordrecht, (1990).

[5] Adrian Gheorghe, Connectionist expert systems for analysis of health problems associated with industrial activity and electricity generation systems, Int. J. Environment and Pollution, 4 xxx 107-124.

[6] International Chernobyl Project - Input from the Commission of the European Communities to the Evaluation of the Relocation Policy Adopted by the Former Soviet Union", Report EUR 14543 EN, Commission of the European Communities, 1992.

SUBJECT INDEX

C

E

F

H

I

J

K

L

M

N

O

S

T

INTEGRATED REGIONAL RISK ASSESSMENT

Vol. I: Continuous and Non-Point Source Emissions
■ Air, Water, Soil

CONTENTS (Vol. I)

Environmental Science and Technology Library

---

1. A. Caetano, M.N. De Pinho, E. Drioli and H. Muntau (eds.), *Membrane Technology: Applications to Industrial Wastewater Treatment.* 1995
   ISBN 0-7923-3209-1
2. Z. Zlatev: *Computer Treatment of Large Air Pollution Models.* 1995
   ISBN 0-7923-3328-4
3. J. Lemons and D.A. Brown (eds.): *Sustainable Development: Science, Ethics, and Public Policy.* 1995 ISBN 0-7923-3500-7
4. A.V. Gheorghe and M. Nicolet-Monnier: *Integrated Regional Risk Assessment.* Volume I: Continuous and Non-Point Source Emissions: Air, Water, Soil. 1995 ISBN 0-7923-3717-4
   Volume II: Consequence Assessment of Accidental Releases. 1995
   ISBN 0-7923-3718-2
   Set: ISBN 0-7923-3719-0

KLUWER ACADEMIC PUBLISHERS – DORDRECHT / BOSTON / LONDON

9 780792 337188

GPSR Compliance

The European Union's (EU) General Product Safety Regulation (GPSR) is a set of rules that requires consumer products to be safe and our obligations to ensure this.

If you have any concerns about our products, you can contact us on ProductSafety@springernature.com

In case Publisher is established outside the EU, the EU authorized representative is:

Springer Nature Customer Service Center GmbH
Europaplatz 3
69115 Heidelberg, Germany

Batch number: 10607204

Printed by Printforce, the Netherlands